MECHANICS OF Composite Materials

SECOND EDITION

Mechanical Engineering Series

Frank Kreith - Series Editor

Published Titles

Distributed Generation: The Power Paradigm for the New Millennium
Anne-Marie Borbely & Jan F. Kreider
Elastoplasticity Theory
Vlado A. Lubarda
Energy Audit of Building Systems: An Engineering Approach
Moncef Krarti
Engineering Experimentation
Euan Somerscales
Entropy Generation Minimization
Adrian Bejan
The Finite Element Method Using MATLAB, 2nd Edition
Young W. Kwon & Hyochoong Bang
Fluid Power Circuits and Controls: Fundamentals and Applications
John S. Cundiff
Fundamentals of Environmental Discharge Modeling
Lorin R. Davis
Heat Transfer in Single and Multiphase Systems
Greg F. Naterer
Introductory Finite Element Method
Chandrakant S. Desai & Tribikram Kundu
Intelligent Transportation Systems: New Principles and Architectures
Sumit Ghosh & Tony Lee
Mathematical & Physical Modeling of Materials Processing Operations
Olusegun Johnson Ilegbusi, Manabu Iguchi & Walter E. Wahnsiedler
Mechanics of Composite Materials, 2nd Edition
Autar K. Kaw
Mechanics of Fatigue
Vladimir V. Bolotin
Mechanics of Solids and Shells: Theories and Approximations
Gerald Wempner & Demosthenes Talaslidis
Mechanism Design: Enumeration of Kinematic Structures According to Function
Lung-Wen Tsai
Multiphase Flow Handbook
Clayton T. Crowe
Nonlinear Analysis of Structures
M. Sathyamoorthy
Optomechatronics: Fusion of Optical and Mechatronic Engineering
Hyungsuck Cho
Practical Inverse Analysis in Engineering
David M. Trujillo & Henry R. Busby
Pressure Vessels: Design and Practice
Somnath Chattopadhyay
Principles of Solid Mechanics
Rowland Richards, Jr.
Thermodynamics for Engineers
Kau-Fui Wong
Vibration and Shock Handbook
Clarence W. de Silva
Viscoelastic Solids
Roderic S. Lakes

MECHANICS OF Composite Materials

SECOND EDITION

AUTAR K. KAW

Taylor & Francis
Taylor & Francis Group
Boca Raton London New York

A CRC title, part of the Taylor & Francis imprint, a member of the Taylor & Francis Group, the academic division of T&F Informa plc.

The cover illustration is an artist's rendition of fiber geometries, cross-sectional views, and crack propagation paths in a composite material. The author gratefully acknowledges and gives his heartfelt thanks to his longtime friend, Dr. Suneet Bahl, for drawing the cover illustration.

Published in 2006 by
CRC Press
Taylor & Francis Group
6000 Broken Sound Parkway NW, Suite 300
Boca Raton, FL 33487-2742

Printed in the United States of America on acid-free paper
10 9 8 7 6 5

International Standard Book Number-10: 0-8493-1343-0 (Hardcover)
International Standard Book Number-13: 978-0-8493-1343-1 (Hardcover)
Library of Congress Card Number 2005049974

Library of Congress Cataloging-in-Publication Data

Kaw, Autar K.
Mechanics of composite materials / Autar K. Kaw.--2nd ed.
p. cm. -- (Mechanical engineering ; v. 29)
Includes bibliographical references and index.
ISBN 0-8493-1343-0 (alk. paper)
1. Composite materials--Mechanical properties. I. Title. II. Mechanical engineering series (Boca Raton, Fla.) ; v. 29

TA418.9.C6K39 2005
620.1'183--dc22 2005049974

Visit the Taylor & Francis Web site at
http://www.taylorandfrancis.com

and the CRC Press Web site at
http://www.crcpress.com

Dedication

To

Sherrie, Candace, Angelie, Chuni, Sushma, Neha, and Trance

and

in memory of my father,

Radha Krishen Kaw,

who gave me the love

of teaching, movies, and music

(necessarily in that order).

There is nothing noble about being superior to another man; the true nobility lies in being superior to your previous self.

Upanishads

Preface to the Second Edition

The first edition of this book was published in 1997, and I am grateful for the response and comments I have received about the book and the accompanying PROMAL software. The changes in the book are mainly a result of comments received from students who used this book in a course or as a self-study.

In this edition, I have added a separate chapter on symmetric and unsymmetric laminated beams. All the other chapters have been updated while maintaining the flow of the content. Key terms and a summary have been added at the end of each chapter. Multiple-choice questions to reinforce the learning from each chapter have been added and are available at the textbook Website: http://www.eng.usf.edu/~kaw/promal/book.html.

Specifically, in Chapter 1, new applications of composite materials have been accommodated. With the ubiquitous presence of the Web, I have annotated articles, videos, and Websites at the textbook Website. In Chapter 2, we have added more examples and derivations have been added. The appendix on matrix algebra has been extended because several engineering departments no longer teach a separate course in matrix algebra. If the reader needs more background knowledge of this subject, he or she can download a free e-book on matrix algebra at http://numericalmethods.eng.usf.edu/ (click on "matrix algebra"). In Chapter 3, derivations are given for the elasticity model of finding the four elastic constants. Two more examples can be found in Chapter 5: design of a pressure vessel and a drive shaft.

The PROMAL program has been updated to include elasticity models in Chapter 3. PROMAL and the accompanying software are available to the eligible buyers of the textbook only at the textbook Website (see the "About the Software" section). The software and the manual will be continually updated.

Preface to the First Edition

Composites are becoming an essential part of today's materials because they offer advantages such as low weight, corrosion resistance, high fatigue strength, faster assembly, etc. Composites are used as materials ranging from making aircraft structures to golf clubs, electronic packaging to medical equipment, and space vehicles to home building. Composites are generating curiosity and interest in students all over the world. They are seeing everyday applications of composite materials in the commercial market, and job opportunities are also increasing in this field. The technology transfer initiative of the U.S. government is opening new and large-scale opportunities for use of advanced composite materials.

Many engineering colleges are offering courses in composite materials as undergraduate technical electives and as graduate-level courses. In addition, as part of their continuing education and retraining, many practicing engineers are participating in workshops and taking short courses in composite materials. The objective of this book is to introduce a senior undergraduate- or graduate-level student to the mechanical behavior of composites. Covering all aspects of the mechanical behavior of composites is impossible to do in one book; also, many aspects require knowledge of advanced graduate study topics such as elasticity, fracture mechanics, and plates and shells theory. Thus, this book emphasizes an overview of composites followed by basic mechanical behavior of composites. Only then will a student form a necessary foundation for further study of topics such as impact, fatigue, fracture mechanics, creep, buckling and vibrations, etc. I think that these topics are important and the interested student has many well-written texts available to follow for that.

This book breaks some traditional rules followed in other textbooks on composites. For example, in the first chapter, composites are introduced in a question–answer format. These questions were raised through my own thought process when I first took a course in composites and then by my students at the University of South Florida, Tampa. Also, this is the first textbook in its field that includes a professional software package. In addition, the book has a format of successful undergraduate books, such as short sections, adequate illustrations, exercise sets with objective questions and numerical problems, reviews wherever necessary, simple language, and many examples.

Chapter 1 introduces basic ideas about composites including why composites are becoming important in today's market. Other topics in Chapter 1 include types of fibers and matrices, manufacturing, applications, recycling, and basic definitions used in the mechanics of composites. In Chapter

2, I start with a review of basic topics of stress, strain, elastic moduli, and strain energy. Then I discuss the mechanical behavior of a single lamina, including concepts about stress–strain relationship for a lamina, stiffness and strength of a lamina, and the stress–strain response due to temperature and moisture change. In Chapter 3, I develop equations for mechanical properties of a lamina such as stiffness, strength, and coefficients of thermal and moisture expansion from individual properties of the constituents (long continuous fibers and matrix) of composites. I introduce experimental characterization of the mechanical properties of a lamina at appropriate places in Chapter 3. Chapter 4 is an extension of Chapter 2, in which the macromechanics of a single lamina are extended to the macromechanics of a laminate. I develop stress–strain equations for a laminate based on individual properties of the laminae that make it. I also discuss stiffness and strength of a laminate and effects of temperature and moisture on residual stresses in a laminate. In Chapter 5, special cases of laminates used in the market are introduced. I develop procedures for analyzing the failure and design of laminated composites. Other mechanical design issues, such as fatigue, environmental effects, and impact, are introduced.

A separate chapter for using the user-friendly software PROMAL is included for supplementing the understanding of Chapter 2 through Chapter 5. Students using PROMAL can instantly conduct pragmatic parametric studies, compare failure theories, and have the information available in tables and graphs instantaneously.

The availability of computer laboratories across the nation allows the instructor to use PROMAL as a teaching tool. Many questions asked by the student can be answered instantly. PROMAL is more than a black box because it shows intermediate results as well. At the end of the course, it will allow students to design laminated composite structures in the classroom. The computer program still maintains the student's need to think about the various inputs to the program to get an optimum design.

You will find this book and software very interesting. I welcome your comments, suggestions, and thoughts about the book and the software at e-mail: promal@eng.usf.edu; and URL: http://www.eng.usf.edu/~kaw/promal/book.html.

Acknowledgments

I acknowledge all the students who have taken the course on composite materials at the University of South Florida since I first taught it in the spring of 1988. Since then, their questions and wish lists have dynamically changed the content of the course.

I would like to thank my talented students — Steven Jourdenais, Brian Shanberg, Franc Urso, Gary Willenbring, and Paula Bond — for their help with building the PROMAL software. PROMAL has been a continuous project since 1988.

I thank my dear friend, Suneet Bahl, who designed yet another unique illustration for the cover for this book. His contribution has been inspirational. I thank J. Ye, J. Meyers, M. Toma, A. Prasad, R. Rodriguez, K. Gangakhedkar, C. Khoe, P. Chalasani, and S. Johnson for drawing the illustrations, proofreading, and checking the examples in the text. Special thanks go again to R. Rodriguez, who painstakingly developed the solutions manual for the book using MATHCAD software.

I would like to thank Sue Britten for helping me in typing the manuscript, especially the equations and the endless loop of revisions and changes. Her effort was very critical in finishing the project on time. I want to thank all the companies that not only sent promotional literature but also made an additional effort to send photographs, videos, slides, design examples, etc. Individual companies whose information has been used in the book are acknowledged for each citation.

A sabbatical granted by the University of South Florida in the fall of 2002 was critical in completing this project. I thank Professor L. Carlsson of Florida Atlantic University, who provided the raw data for some of the figures from his book, *Experimental Characterization of Advanced Composite Materials*. I thank Dr. R.Y. Kim of the University of Dayton Research Institute for providing stress–strain data and photographs for several figures in this book. I want to thank Dr. G.P. Tandon of UDRI for several discussions and references on developing the elasticity models for the elastic moduli of unidirectional composites.

I thank my wife, Sherrie, and our two children, Candace and Angelie, for their support and encouragement during this long project. In their own way, our children have taught me how to be a *good teacher*. I would like to acknowledge my parents, who gave me the opportunities to reach my goals and did that at a great personal sacrifice. I am grateful to my father, who was a role model for my professional career and taught me many things about being a *complete teacher*.

I thank Cindy Carelli and Michael Slaughter, senior editors of Taylor & Francis, and their staff for their support and encouragement. I want to thank Elizabeth Spangenberger, Helena Redshaw, Jessica Vakili, Naomi Lynch, Jonathan Pennell, and their staffs for keeping me updated throughout the production process and giving personal attention to many details, including design, layout, equation editing, etc. of the final product.

I have to thank the authors of *Getting Your Book Published* (Sage Publications) for helping me understand the mechanics of publication and how to create a win–win situation for all the involved parties in this endeavor. I would recommend their book to any educator who is planning to write a textbook.

About the Author

Autar K. Kaw is a professor of mechanical engineering at the University of South Florida, Tampa. Professor Kaw obtained his B.E. (Hons.) degree in mechanical engineering from Birla Institute of Technology and Science, India, in 1981. He received his Ph.D. degree in 1987 and M.S. degree in 1984, both in engineering mechanics from Clemson University, South Carolina. He joined the faculty of the University of South Florida in 1987. He has also been a maintenance engineer (1982) for Ford-Escorts Tractors, India, and a summer faculty fellow (1992) and visiting scientist (1991) at Wright Patterson Air Force Base.

Professor Kaw's main scholarly interests are in the fracture mechanics of composite materials and development of instructional software for engineering education. His research has been funded by the National Science Foundation, Air Force Office of Scientific Research, Florida Department of Transportation, Research and Development Laboratories, Wright Patterson Air Force Base, and Montgomery Tank Lines. He is a fellow of the American Society of Mechanical Engineers (ASME) and a member of the American Society of Engineering Education (ASEE). He has written more than 35 journal papers and developed several software instructional programs for courses such as Mechanics of Composites and Numerical Methods.

Professor Kaw has received the Florida Professor of the Year Award from the Council for Advancement and Support of Education (CASE) and Carnegie Foundation for Advancement of Teaching (CFAT) (2004); Archie Higdon Mechanics Educator Award from the American Society of Engineering Education (ASEE) (2003); Southeastern Section American Society of Engineering Education (ASEE) Outstanding Contributions in Research Award (1996); State of Florida Teaching Incentive Program Award (1994 and 1997); American Society of Engineering Education (ASEE) New Mechanics Educator Award (1992); and Society of Automotive Engineers (SAE) Ralph Teetor Award (1991). At the University of South Florida, he has been awarded the Jerome Krivanek Distinguished Teacher Award (1999); University Outstanding Undergraduate Teaching Award (1990 and 1996); Faculty Honor Guard (1990); and the College of Engineering Teaching Excellence Award (1990 and 1995).

About the Software

Where can I download PROMAL?

You can download PROMAL at http://www.eng.usf.edu/~kaw/promal/book.html. In addition to the restrictions for use given later in this section, only textbook buyers are authorized to download the software.

What is PROMAL?

PROMAL is professionally developed software accompanying this book. Taylor & Francis Group has been given the rights free of charge by the author to supplement this book with this software. PROMAL has five main programs:

1. *Matrix algebra*: Throughout the course of *Mechanics of Composite Materials*, the most used mathematical procedures are based on linear algebra. This feature allows the student to multiply matrices, invert square matrices, and find the solution to a set of simultaneous linear equations. Many students have programmable calculators and access to tools such as MATHCAD to do such manipulations, and we have included this program only for convenience. This program allows the student to concentrate on the fundamentals of the course as opposed to spending time on lengthy matrix manipulations.
2. *Lamina properties database*: In this program, the properties of unidirectional laminae can be added, deleted, updated, and saved. This is useful because these properties can then be loaded into other parts of the program without repeated inputs.
3. *Macromechanical analysis of a lamina*: Using the properties of unidirectional laminae saved in the previously described database, one can find the stiffness and compliance matrices, transformed stiffness and compliance matrices, engineering constants, strength ratios based on four major failure theories, and coefficients of thermal and moisture expansion of angle laminae. These results are then presented in textual, tabular, and graphical forms.
4. *Micromechanics analysis of a lamina*: Using individual elastic moduli, coefficients of thermal and moisture expansion, and specific gravity of fiber and matrix, one can find the elastic moduli and coefficients of thermal and moisture expansion of a unidirectional lamina. Again, the results are available in textual, tabular, and graphical forms.

5. *Macromechanics of a laminate*: Using the properties of the lamina from the database, one can analyze laminated structures. These laminates may be hybrid and unsymmetric. The output includes finding stiffness and compliance matrices, global and local strains, and strength ratios in response to mechanical, thermal, and moisture loads. This program is used for design of laminated structures such as plates and thin pressure vessels at the end of the course.

Who is permitted to use PROMAL?

PROMAL is designed and permitted to be used only as a theoretical–educational tool; it can be used by:

- A university instructor using PROMAL for teaching a formal university-level course in mechanics of composite materials
- A university student using PROMAL to learn about mechanics of composites while enrolled in a formal university-level course in mechanics of composite materials
- A continuing education student using PROMAL to learn about mechanics of composites while enrolled in a formal university-level course in mechanics of composite materials
- A self-study student who has successfully passed a formal university-level course in strength of materials and is using PROMAL while studying the mechanics of composites using a textbook on mechanics of composites

If you or your use of PROMAL does not fall into one of these four categories, you are not permitted to use the PROMAL software.

What is the license agreement to use the software?

Software License

Grant of License: PROMAL is designed and permitted to be used only as a theoretical–educational tool. Also, for using the PROMAL software, the definition of "You" in this agreement should fall into one of four categories.

1. University instructor using PROMAL for teaching a formal university-level course in mechanics of composite materials
2. University student using PROMAL to learn about mechanics of composites while enrolled in a formal university-level course in mechanics of composite materials

3. Continuing education student using PROMAL to learn about mechanics of composites while enrolled in a formal university-level course in mechanics of composite materials
4. Self-study student who has successfully passed a formal university-level course in strength of materials and is using PROMAL while studying the mechanics of composites using a textbook on mechanics of composites

If you or your use of PROMAL does not fall into one of the above four categories, you are not permitted to buy or use the PROMAL software.

Autar K. Kaw and Taylor & Francis Group hereby grant you, and you accept, a nonexclusive and nontransferable license, to use the PROMAL software on the following terms and conditions only: you have been granted an Individual Software License and you may use the Licensed Program on a single personal computer for your own personal use.

Copyright: The software is owned by Autar K. Kaw and is protected by United States copyright laws. A backup copy may be made but all such backup copies are subject to the terms and conditions of this agreement.

Other Restrictions: You may not make or distribute unauthorized copies of the Licensed Program, create by decompilation, or otherwise, the source code of the PROMAL software, or use, copy, modify, or transfer the PROMAL software in whole or in part, except as expressly permitted by this Agreement. If you transfer possession of any copy or modification of the PROMAL software to any third party, your license is automatically terminated. Such termination shall be in addition to and not in lieu of any equitable, civil, or other remedies available to Autar K. Kaw and Taylor & Francis Group.

You acknowledge that all rights (including without limitation, copyrights, patents, and trade secrets) in the PROMAL software (including without limitation, the structure, sequence, organization, flow, logic, source code, object code, and all means and forms of operation of the Licensed Program) are the sole and exclusive property of Autar K. Kaw. By accepting this Agreement, you do not become the owner of the PROMAL software, but you do have the right to use it in accordance with the provision of this Agreement. You agree to protect the PROMAL software from unauthorized use, reproduction, or distribution. You further acknowledge that the PROMAL software contains valuable trade secrets and confidential information belonging to Autar K. Kaw. You may not disclose any component of the PROMAL software, whether or not in machine-readable form, except as expressly provided in this Agreement.

Term: This License Agreement is effective until terminated. This Agreement will also terminate upon the conditions discussed elsewhere in this Agreement, or if you fail to comply with any term or condition of this Agreement. Upon such termination, you agree to destroy the PROMAL software and any copies made of the PROMAL software.

Limited Warranty

This limited warranty is in lieu of all other warranties, expressed or implied, including without limitation, any warranties or merchantability or fitness for a particular purpose. The licensed program is furnished on an "as is" basis and without warranty as to the performance or results you may obtain using the licensed program. The entire risk as to the results or performance, and the cost of all necessary servicing, repair, or correction of the PROMAL software is assumed by you.

In no event will Autar K. Kaw or Taylor & Francis Group be liable to you for any damages whatsoever, including without limitation, lost profits, lost savings, or other incidental or consequential damages arising out of the use or inability to use the PROMAL software even if Autar K. Kaw or Taylor & Francis Group has been advised of the possibility of such damages. **You should not build, design, or analyze any actual structure or component using the results from the PROMAL software**.

This limited warranty gives you specific legal rights. You may have others by operation of law that vary from state to state. If any of the provisions of this agreement are invalid under any applicable statute or rule of law, they are to that extent deemed omitted.

This agreement represents the entire agreement between us and supersedes any proposals or prior agreements, oral or written, and any other communication between us relating to the subject matter of this agreement.

This agreement will be governed and construed as if wholly entered into and performed within the state of Florida.

You acknowledge that you have read this agreement, and agree to be bound by its terms and conditions.

Is there any technical support for the software?

The program is user-friendly and you should not need technical support. However, technical support is available only through e-mail and is free for registered users for 30 days from the day of purchase of this book. Before using technical support, check with your instructor, and study the manual and the home page for PROMAL at http://www.eng.usf.edu/~kaw/

promal/book.html. *At this home page, you can also download upgraded promal.exe files.* Send your questions, comments, and suggestions for future versions by e-mail to promal@eng.usf.edu. I will attempt to include your feedback in the next version of PROMAL.

How do I register the software?

Register by sending an e-mail to promal@eng.usf.edu with "registration" in the subject line and the body with name, university/continuing education affiliation, postal address, e-mail address, telephone number, and how you obtained a copy of the software, i.e., purchase of book, personal copy, site license, continuing education course.
OR
Register by mailing a post card with name, university/continuing education affiliation, address, and e-mail address, telephone number, and how you obtained a copy of the software — i.e., purchase of book, personal copy, site license, continuing education course — to Professor Autar K. Kaw, ENB 118, Mechanical Engineering Department, University of South Florida, Tampa, FL 33620-5350.

What are the requirements of running the program?

The program will generally run on any IBM-PC compatible computer with Microsoft Windows 98 or later, 128 MB of available memory, and a hard disk with 50 MB available, and Microsoft mouse.

Can I purchase a copy of PROMAL separately?

Check the book Website for the latest purchase information for single-copy sales, course licenses, and continuing education course prices.

Contents

1

Introduction to Composite Materials

Chapter Objectives

- Define a composite, enumerate advantages and drawbacks of composites over monolithic materials, and discuss factors that influence mechanical properties of a composite.
- Classify composites, introduce common types of fibers and matrices, and manufacturing, mechanical properties, and applications of composites.
- Discuss recycling of composites.
- Introduce terminology used for studying mechanics of composites.

1.1 Introduction

> You are no longer to supply the people with straw for making bricks; let them go and gather their own straw.
>
> **Exodus 5:7**

Israelites using bricks made of clay and reinforced with straw are an early example of application of composites. The individual constituents, clay and straw, could not serve the function by themselves but did when put together. Some believe that the straw was used to keep the clay from cracking, but others suggest that it blunted the sharp cracks in the dry clay.

Historical examples of composites are abundant in the literature. Significant examples include the use of reinforcing mud walls in houses with bamboo shoots, glued laminated wood by Egyptians (1500 B.C.), and laminated metals in forging swords (A.D. 1800). In the 20th century, modern composites were used in the 1930s when glass fibers reinforced resins. Boats and aircraft were built out of these glass composites, commonly called *fiberglass*. Since the 1970s, application of composites has widely increased due to development of new fibers such as carbon, boron, and aramids,* and new composite systems with matrices made of metals and ceramics.

This chapter gives an overview of composite materials. The question–answer style of the chapter is a suitable way to learn the fundamental aspects of this vast subject. In each section, the questions progressively become more specialized and technical in nature.

What is a composite?

A composite is a structural material that consists of two or more combined constituents that are combined at a macroscopic level and are not soluble in each other. One constituent is called the *reinforcing phase* and the one in which it is embedded is called the *matrix*. The reinforcing phase material may be in the form of fibers, particles, or flakes. The matrix phase materials are generally continuous. Examples of composite systems include concrete reinforced with steel and epoxy reinforced with graphite fibers, etc.

Give some examples of naturally found composites.

Examples include wood, where the lignin matrix is reinforced with cellulose fibers and bones in which the bone-salt plates made of calcium and phosphate ions reinforce soft collagen.

What are advanced composites?

Advanced composites are composite materials that are traditionally used in the aerospace industries. These composites have high performance reinforcements of a thin diameter in a matrix material such as epoxy and aluminum. Examples are graphite/epoxy, Kevlar®†/epoxy, and boron/aluminum composites. These materials have now found applications in commercial industries as well.

Combining two or more materials together to make a composite is more work than just using traditional monolithic metals such as steel and aluminum. What are the advantages of using composites over metals?

Monolithic metals and their alloys cannot always meet the demands of today's advanced technologies. Only by combining several materials can one meet the performance requirements. For example, trusses and benches used in satellites need to be dimensionally stable in space during temperature changes between –256°F (–160°C) and 200°F (93.3°C). Limitations on coefficient of thermal expansion‡ thus are low and may be of the order of ±1 ×

* Aramids are aromatic compounds of carbon, hydrogen, oxygen, and nitrogen.

† Kevlar® is a registered trademark of E.I. duPont deNemours and Company, Inc., Wilimington, DE.

‡ Coefficient of thermal expansion is the change in length per unit length of a material when heated through a unit temperature. The units are in./in./°F and m/m/°C. A typical value for steel is 6.5×10^{-6} in./in.°F (11.7×10^{-6} m/m°C).

10^{-7} in./in./°F ($\pm 1.8 \times 10^{-7}$ m/m/°C). Monolithic materials cannot meet these requirements; this leaves composites, such as graphite/epoxy, as the only materials to satisfy them.

In many cases, using composites is more efficient. For example, in the highly competitive airline market, one is continuously looking for ways to lower the overall mass of the aircraft without decreasing the stiffness* and strength† of its components. This is possible by replacing conventional metal alloys with composite materials. Even if the composite material costs may be higher, the reduction in the number of parts in an assembly and the savings in fuel costs make them more profitable. Reducing one lbm (0.453 kg) of mass in a commercial aircraft can save up to 360 gal (1360 l) of fuel per year;[1] fuel expenses are 25% of the total operating costs of a commercial airline.[2]

Composites offer several other advantages over conventional materials. These may include improved strength, stiffness, fatigue‡ and impact resistance,** thermal conductivity,†† corrosion resistance,‡‡ etc.

How is the mechanical advantage of composite measured?

For example, the axial deflection, *u*, of a prismatic rod under an axial load, *P*, is given by

$$u = \frac{PL}{AE}, \tag{1.1}$$

where

L = length of the rod
E = Young's modulus of elasticity of the material of the rod

Because the mass, *M*, of the rod is given by

$$M = \rho AL, \tag{1.2}$$

where ρ = density of the material of the rod, we have

* Stiffness is defined as the resistance of a material to deflection.

† Strength is defined as the stress at which a material fails.

‡ Fatigue resistance is the resistance to the lowering of mechanical properties such as strength and stiffness due to cyclic loading, such as due to take-off and landing of a plane, vibrating a plate, etc.

** Impact resistance is the resistance to damage and to reduction in residual strength to impact loads, such as a bird hitting an airplane or a hammer falling on a car body.

†† Thermal conductivity is the rate of heat flow across a unit area of a material in a unit time, when the temperature gradient is unity in the direction perpendicular to the area.

‡‡ Corrosion resistance is the resistance to corrosion, such as pitting, erosion, galvanic, etc.

$$M = \frac{PL^2}{u} \frac{1}{E/\rho}. \tag{1.3}$$

This implies that the lightest beam for specified deflection under a specified load is one with the highest (E/ρ) value.

Thus, to measure the mechanical advantage, the (E/ρ) ratio is calculated and is called the *specific modulus* (ratio between the Young's modulus* (E) and the density† (ρ) of the material). The other parameter is called the *specific strength* and is defined as the ratio between the strength (σ_{ult}) and the density of the material (ρ), that is,

$$\text{Specific modulus} = \frac{E}{\rho},$$

$$\text{Specific strength} = \frac{\sigma_{ult}}{\rho}.$$

The two ratios are high in composite materials. For example, the strength of a graphite/epoxy unidirectional composite‡ could be the same as steel, but the specific strength is three times that of steel. What does this mean to a designer? Take the simple case of a rod designed to take a fixed axial load. The rod cross section of graphite/epoxy would be same as that of the steel, but the mass of graphite/epoxy rod would be one third of the steel rod. This reduction in mass translates to reduced material and energy costs. Figure 1.1 shows how composites and fibers rate with other traditional materials in terms of specific strength.[3] Note that the unit of specific strength is inches in Figure 1.1 because specific strength and specific modulus are also defined in some texts as

$$\text{Specific modulus} = \frac{E}{\rho g},$$

$$\text{Specific strength} = \frac{\sigma_{ult}}{\rho g}.$$

where g is the acceleration due to gravity (32.2 ft/s^2 or 9.81 m/s^2).

* Young's modulus of an elastic material is the initial slope of the stress–strain curve.
† Density is the mass of a substance per unit volume.
‡ A unidirectional composite is a composite lamina or rod in which the fibers reinforcing the matrix are oriented in the same direction.

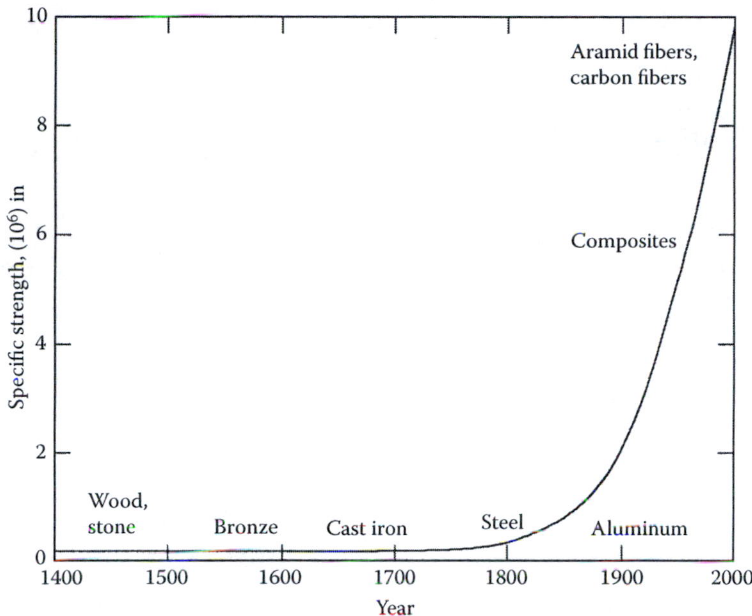

FIGURE 1.1
Specific strength as a function of time of use of materials. (Source: Eager, T.W., Whither advanced materials? *Adv. Mater. Processes*, ASM International, June 1991, 25–29.)

Values of specific modulus and strength are given in Table 1.1 for typical composite fibers, unidirectional composites,* cross-ply† and quasi-isotropic‡ laminated composites, and monolithic metals.

On a first look, fibers such as graphite, aramid, and glass have a specific modulus several times that of metals, such as steel and aluminum. This gives a false impression about the mechanical advantages of composites because they are made not only of fibers, but also of fibers and matrix combined; matrices generally have lower modulus and strength than fibers. Is the comparison of the specific modulus and specific strength parameters of unidirectional composites to metals now fair? The answer is no for two reasons. First, unidirectional composite structures are acceptable only for carrying simple loads such as uniaxial tension or pure bending. In structures with complex requirements of loading and stiffness, composite structures including angle plies will be necessary. Second, the strengths and elastic moduli of unidirectional composites given in Table 1.1 are those in the direction of the fiber. The strength and elastic moduli perpendicular to the fibers are far less.

* A unidirectional laminate is a laminate in which all fibers are oriented in the same direction.

† A cross-ply laminate is a laminate in which the layers of unidirectional lamina are oriented at right angles to each other.

‡ Quasi-isotropic laminate behaves similarly to an isotropic material; that is, the elastic properties are the same in all directions.

TABLE 1.1

Specific Modulus and Specific Strength of Typical Fibers, Composites, and Bulk Metals

Material Units	Specific gravity[a]	Young's modulus (Msi)	Ultimate strength (ksi)	Specific modulus (Msi-in.³/lb)	Specific strength (ksi-in.³/lb)
System of Units: USCS					
Graphite fiber	1.8	33.35	299.8	512.9	4610
Aramid fiber	1.4	17.98	200.0	355.5	3959
Glass fiber	2.5	12.33	224.8	136.5	2489
Unidirectional graphite/epoxy	1.6	26.25	217.6	454.1	3764
Unidirectional glass/epoxy	1.8	5.598	154.0	86.09	2368
Cross-ply graphite/epoxy	1.6	13.92	54.10	240.8	935.9
Cross-ply glass/epoxy	1.8	3.420	12.80	52.59	196.8
Quasi-isotropic graphite/epoxy	1.6	10.10	40.10	174.7	693.7
Quasi-isotropic glass/epoxy	1.8	2.750	10.60	42.29	163.0
Steel	7.8	30.00	94.00	106.5	333.6
Aluminum	2.6	10.00	40.00	106.5	425.8

Material Units	Specific gravity	Young's modulus (GPa)	Ultimate strength (MPa)	Specific modulus (GPa-m³/kg)	Specific strength (MPa-m³/kg)
System of Units: SI					
Graphite fiber	1.8	230.00	2067	0.1278	1.148
Aramid fiber	1.4	124.00	1379	0.08857	0.9850
Glass fiber	2.5	85.00	1550	0.0340	0.6200
Unidirectional graphite/epoxy	1.6	181.00	1500	0.1131	0.9377
Unidirectional glass/epoxy	1.8	38.60	1062	0.02144	0.5900
Cross-ply graphite/epoxy	1.6	95.98	373.0	0.06000	0.2331
Cross-ply glass/epoxy	1.8	23.58	88.25	0.01310	0.0490
Quasi-isotropic graphite/epoxy	1.6	69.64	276.48	0.04353	0.1728
Quasi-isotropic glass/epoxy	1.8	18.96	73.08	0.01053	0.0406
Steel	7.8	206.84	648.1	0.02652	0.08309
Aluminum	2.6	68.95	275.8	0.02652	0.1061

[a] Specific gravity of a material is the ratio between its density and the density of water.

A comparison is now made between popular types of laminates such as cross-ply and quasi-isotropic laminates. Figure 1.2 shows the specific strength plotted as a function of specific modulus for various fibers, metals, and composites.

Are specific modulus and specific strength the only mechanical parameters used for measuring the relative advantage of composites over metals?

No, it depends on the application.[4] Consider compression of a column, where it may fail due to buckling. The Euler buckling formula gives the critical load at which a long column buckles as[5]

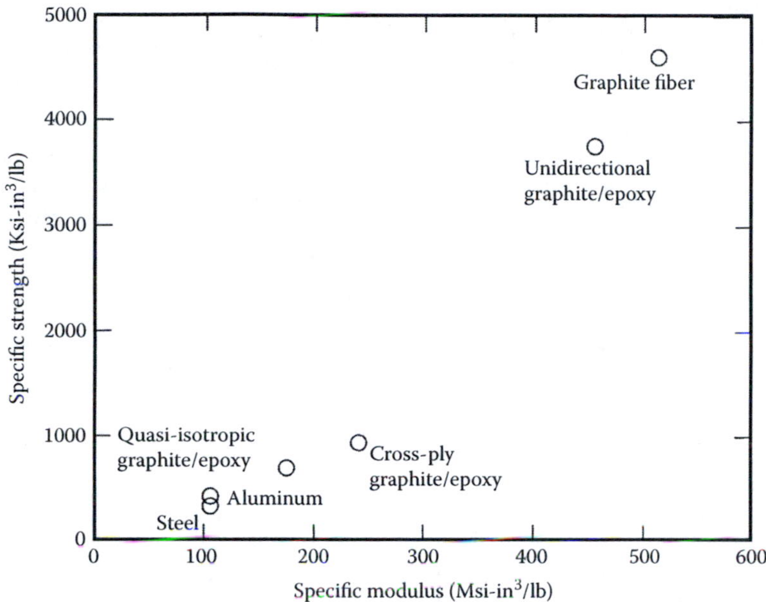

FIGURE 1.2
Specific strength as a function of specific modulus for metals, fibers, and composites.

$$P_{cr} = \frac{\pi^2 EI}{L^2}, \tag{1.4}$$

where

P_{cr} = critical buckling load (lb or N)
E = Young's modulus of column (lb/in.^{2} or N/m^2)
I = second moment of area (in.^{4} or m^4)
L = length of beam (in. or m)

If the column has a circular cross section, the second moment of area is

$$I = \pi \frac{d^4}{64} \tag{1.5}$$

and the mass of the rod is

$$M = \rho \frac{\pi d^2 L}{4}, \tag{1.6}$$

where

M = mass of the beam (lb or kg)
ρ = density of beam (lb/in.^{3} or kg/m^3)
d = diameter of beam (in. or m)

Because the length, L, and the load, P, are constant, we find the mass of the beam by substituting Equation (1.5) and Equation (1.6) in Equation (1.4) as

$$M = \frac{2L^2\sqrt{P_{cr}}}{\sqrt{\pi}}\frac{1}{E^{1/2}/\rho}. \tag{1.7}$$

This means that the lightest beam for specified stiffness is one with the highest value of $E^{1/2}/\rho$.

Similarly, we can prove that, for achieving the minimum deflection in a beam under a load along its length, the lightest beam is one with the highest value of $E^{1/3}/\rho$. Typical values of these two parameters, $E^{1/2}/\rho$ and $E^{1/3}/\rho$ for typical fibers, unidirectional composites, cross-ply and quasi-isotropic laminates, steel, and aluminum are given in Table 1.2. Comparing these numbers with metals shows composites drawing a better advantage for these two parameters. Other mechanical parameters for comparing the performance of composites to metals include resistance to fracture, fatigue, impact, and creep.

Yes, composites have distinct advantages over metals. Are there any drawbacks or limitations in using them?

Yes, drawbacks and limitations in use of composites include:

- High cost of fabrication of composites is a critical issue. For example, a part made of graphite/epoxy composite may cost up to 10 to 15 times the material costs. A finished graphite/epoxy composite part may cost as much as \$300 to \$400 per pound (\$650 to \$900 per kilogram). Improvements in processing and manufacturing techniques will lower these costs in the future. Already, manufacturing techniques such as SMC (sheet molding compound) and SRIM (structural reinforcement injection molding) are lowering the cost and production time in manufacturing automobile parts.
- Mechanical characterization of a composite structure is more complex than that of a metal structure. Unlike metals, composite materials are not isotropic, that is, their properties are not the same in all directions. Therefore, they require more material parameters. For example, a single layer of a graphite/epoxy composite requires *nine*

TABLE 1.2

Specific Modulus Parameters E/ρ, $E^{1/2}/\rho$, and $E^{1/3}/\rho$ for Typical Materials

Material Units	Specific gravity	Young's modulus (Msi)	E/ρ (Msi-in.^{3}/lb)	$E^{1/2}/\rho$ (psi$^{1/2}$-in.^{3}/lb)	$E^{1/3}/\rho$ (psi$^{1/3}$-in.^{3}/lb)
System of Units: USCS					
Graphite fiber	1.8	33.35	512.8	88,806	4,950
Kevlar fiber	1.4	17.98	355.5	83,836	5,180
Glass fiber	2.5	12.33	136.5	38,878	2,558
Unidirectional graphite/epoxy	1.6	26.25	454.1	88,636	5,141
Unidirectional glass/epoxy	1.8	5.60	86.09	36,384	2,730
Cross-ply graphite/epoxy	1.6	13.92	240.8	64,545	4,162
Cross-ply glass/epoxy	1.8	3.42	52.59	28,438	2,317
Quasi-isotropic graphite/epoxy	1.6	10.10	174.7	54,980	3,740
Quasi-isotropic glass/epoxy	1.8	2.75	42.29	25,501	2,154
Steel	7.8	30.00	106.5	19,437	1,103
Aluminum	2.6	10.00	106.5	33,666	2,294

Material Units	Specific gravity	Young's modulus (GPa)	E/ρ (GPa-m^3/kg)	$E^{1/2}/\rho$ (Pa-m^3/kg)	$E^{1/3}/\rho$ (Pa$^{1/3}$-m^3/kg)
System of Units: SI					
Graphite fiber	1.8	230.00	0.1278	266.4	3.404
Kevlar fiber	1.4	124.00	0.08857	251.5	3.562
Glass fiber	2.5	85.00	0.034	116.6	1.759
Unidirectional graphite/epoxy	1.6	181.00	0.1131	265.9	3.535
Unidirectional glass/epoxy	1.8	38.60	0.02144	109.1	1.878
Cross-ply graphite/epoxy	1.6	95.98	0.060	193.6	2.862
Cross-ply glass/epoxy	1.8	23.58	0.0131	85.31	1.593
Quasi-isotropic graphite/epoxy	1.6	69.64	0.04353	164.9	2.571
Quasi-isotropic glass/epoxy	1.8	18.96	0.01053	76.50	1.481
Steel	7.8	206.84	0.02652	58.3	0.7582
Aluminum	2.6	68.95	0.02662	101.0	1.577

stiffness and strength constants for conducting mechanical analysis. In the case of a monolithic material such as steel, one requires only *four* stiffness and strength constants. Such complexity makes structural analysis computationally and experimentally more complicated and intensive. In addition, evaluation and measurement techniques of some composite properties, such as compressive strengths, are still being debated.

- Repair of composites is not a simple process compared to that for metals. Sometimes critical flaws and cracks in composite structures may go undetected.

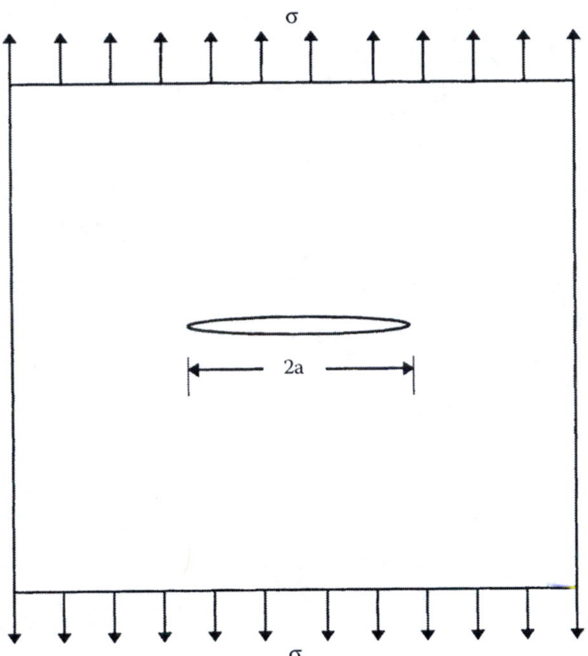

FIGURE 1.3
A uniformly loaded plate with a crack.

- Composites do not have a high combination of strength and fracture toughness* compared to metals. In Figure 1.4, a plot is shown for fracture toughness vs. yield strength for a 1-in. (25-mm) thick material.[3] Metals show an excellent combination of strength and fracture toughness compared to composites. (Note: The transition areas in Figure 1.4 will change with change in the thickness of the specimen.)
- Composites do not necessarily give higher performance in all the properties used for material selection. In Figure 1.5, six primary material selection parameters — strength, toughness, formability,

* In a material with a crack, the value of the stress intensity factor gives the measure of stresses in the crack tip region. For example, for an infinite plate with a crack of length $2a$ under a uniaxial load σ (Figure 1.3), the stress intensity factor is

$$K = \sigma\sqrt{\pi a} \ .$$

If the stress intensity factor at the crack tip is greater than the critical stress intensity factor of the material, the crack will grow. The greater the value of the critical stress intensity factor is, the tougher the material is. The critical stress intensity factor is called the fracture toughness of the material. Typical values of fracture toughness are $23.66 \text{ ksi}\sqrt{\text{in.}}$ $(26 \text{ MPa}\sqrt{\text{m}})$ for aluminum and $25.48 \text{ ksi}\sqrt{\text{in.}}$ $(28 \text{ MPa}\sqrt{\text{m}})$ for steel.

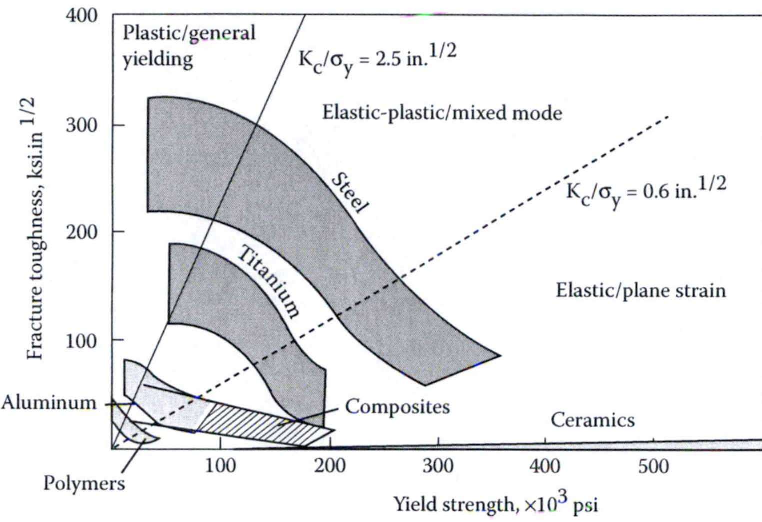

FIGURE 1.4
Fracture toughness as a function of yield strength for monolithic metals, ceramics, and metal–ceramic composites. (Source: Eager, T.W., Whither advanced materials? *Adv. Mater. Processes*, ASM International, June 1991, 25–29.)

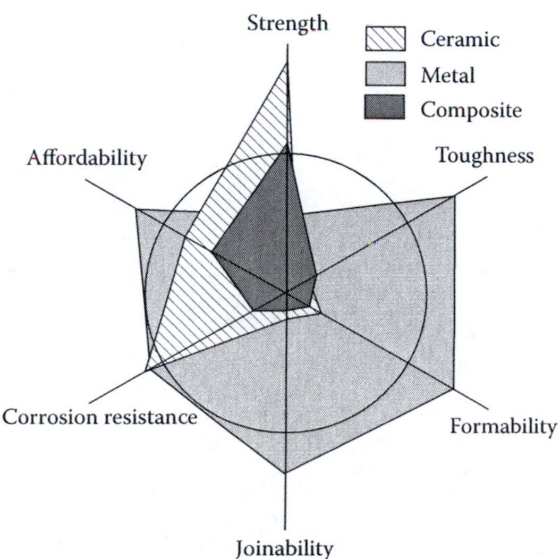

FIGURE 1.5
Primary material selection parameters for a hypothetical situation for metals, ceramics, and metal–ceramic composites. (Source: Eager, T.W., Whither advanced materials? *Adv. Mater. Processes*, ASM International, June 1991, 25–29.)

joinability, corrosion resistance, and affordability — are plotted.[3] If the values at the circumference are considered as the normalized required property level for a particular application, the shaded areas show values provided by ceramics, metals, and metal–ceramic composites. Clearly, composites show better strength than metals, but lower values for other material selection parameters.

Why are fiber reinforcements of a thin diameter?

The main reasons for using fibers of thin diameter are the following:

- Actual strength of materials is several magnitudes lower than the theoretical strength. This difference is due to the inherent flaws in the material. Removing these flaws can increase the strength of the material. As the fibers become smaller in diameter, the chances of an inherent flaw in the material are reduced. A steel plate may have strength of 100 ksi (689 MPa), while a wire made from this steel plate can have strength of 600 ksi (4100 MPa). Figure 1.6 shows how the strength of a carbon fiber increases with the decrease in its diameter.[6]

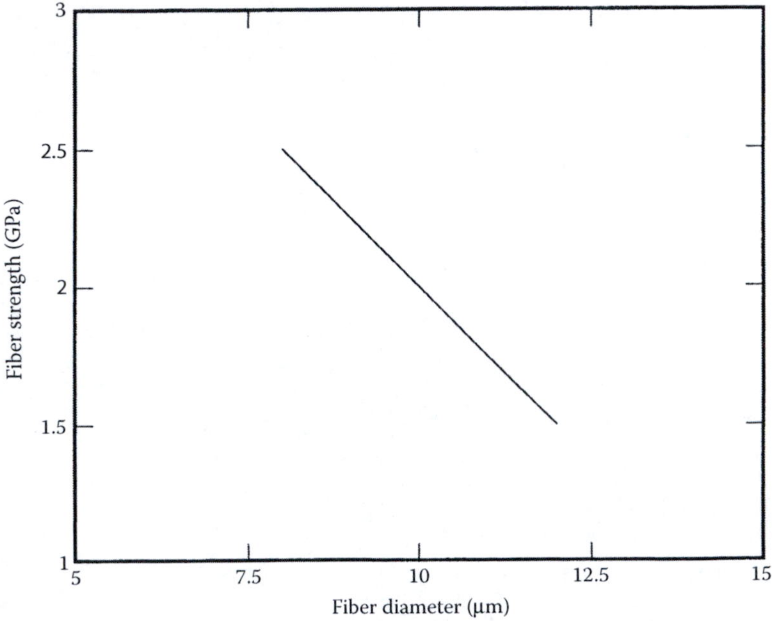

FIGURE 1.6
Fiber strength as a function of fiber diameter for carbon fibers. (Reprinted from Lamotte, E. De, and Perry, A.J., *Fibre Sci. Technol.*, 3, 159, 1970. With permission from Elsevier.)

- For higher ductility* and toughness, and better transfer of loads from the matrix to fiber, composites require larger surface area of the fiber–matrix interface. For the same volume fraction of fibers in a composite, the area of the fiber–matrix interface is inversely proportional to the diameter of the fiber and is proved as follows.

  Assume a lamina consisting of N fibers of diameter D. The fiber–matrix interface area in this lamina is

  $$A_I = N \pi D L. \tag{1.8}$$

  If one replaces the fibers of diameter, D, by fibers of diameter, d, then the number of fibers, n, to keep the fiber volume the same would be

  $$n = N\left(\frac{D}{d}\right)^2. \tag{1.9}$$

  Then, the fiber–matrix interface area in the resulting lamina would be

  $$\begin{aligned} A_{II} &= n \pi d L. \\ &= \frac{N\pi D^2 L}{d} \\ &= \frac{4 \text{ (Volume of fibers)}}{d}. \end{aligned} \tag{1.10}$$

  This implies that, for a fixed fiber volume in a given volume of composite, the area of the fiber–matrix interface is inversely proportional to the diameter of the fiber.

- Fibers able to bend without breaking are required in manufacturing of composite materials, especially for woven fabric composites. Ability to bend increases with a decrease in the fiber diameter and is measured as flexibility. Flexibility is defined as the inverse of bending stiffness and is proportional to the inverse of the product of the elastic modulus of the fiber and the fourth power of its diameter; it can be proved as follows.

  Bending stiffness is the resistance to bending moments. According to the Strength of Materials course, if a beam is subjected to a pure bending moment, M,

* Ductility is the ability of a material to deform without fracturing. It is measured by extending a rod until fracture and measuring the initial (A_i) and final (A_f) cross-sectional area. Then ductility is defined as $R = 1 - (A_f/A_i)$.

$$\frac{d^2v}{dx^2} = \frac{M}{EI}, \tag{1.11}$$

where

v = deflection of the centroidal line (in. or m)
E = Young's modulus of the beam (psi or Pa)
I = second moment of area (in.^{4} or m^4)
x = coordinate along the length of beam (in. or m)

The bending stiffness, then, is *EI* and the flexibility is simply the inverse of *EI*. Because the second moment of area of a cylindrical beam of diameter *d* is

$$I = \frac{\pi d^4}{64}, \tag{1.12}$$

then

$$\text{Flexibility} \propto \frac{1}{Ed^4}. \tag{1.13}$$

For a particular material, unlike strength, the Young's modulus does not change appreciably as a function of its diameter. Therefore, the flexibility for a particular material is inversely proportional to the fourth power of the diameter.

What fiber factors contribute to the mechanical performance of a composite?

Four fiber factors contribute to the mechanical performance of a composite[7]:

- *Length*: The fibers can be long or short. Long, continuous fibers are easy to orient and process, but short fibers cannot be controlled fully for proper orientation. Long fibers provide many benefits over short fibers. These include impact resistance, low shrinkage, improved surface finish, and dimensional stability. However, short fibers provide low cost, are easy to work with, and have fast cycle time fabrication procedures. Short fibers have fewer flaws and therefore have higher strength.
- *Orientation*: Fibers oriented in one direction give very high stiffness and strength in that direction. If the fibers are oriented in more than one direction, such as in a mat, there will be high stiffness and strength in the directions of the fiber orientations. However, for the same volume of fibers per unit volume of the composite, it cannot match the stiffness and strength of unidirectional composites.

- *Shape*: The most common shape of fibers is circular because handling and manufacturing them is easy. Hexagon and square-shaped fibers are possible, but their advantages of strength and high packing factors do not outweigh the difficulty in handling and processing.
- *Material*: The material of the fiber directly influences the mechanical performance of a composite. Fibers are generally expected to have high elastic moduli and strengths. This expectation and cost have been key factors in the graphite, aramids, and glass dominating the fiber market for composites.

What are the matrix factors that contribute to the mechanical performance of composites?

Use of fibers by themselves is limited, with the exceptions of ropes and cables. Therefore, fibers are used as reinforcement to matrices. The matrix functions include binding the fibers together, protecting fibers from the environment, shielding from damage due to handling, and distributing the load to fibers. Although matrices by themselves generally have low mechanical properties compared to those of fibers, the matrix influences many mechanical properties of the composite. These properties include transverse modulus and strength, shear modulus and strength, compressive strength, interlaminar shear strength, thermal expansion coefficient, thermal resistance, and fatigue strength.

Other than the fiber and the matrix, what other factors influence the mechanical performance of a composite?

Other factors include the fiber–matrix interface. It determines how well the matrix transfers the load to the fibers. Chemical, mechanical, and reaction bonding may form the interface. In most cases, more than one type of bonding occurs.

- Chemical bonding is formed between the fiber surface and the matrix. Some fibers bond naturally to the matrix and others do not. Coupling agents* are often added to form a chemical bond.
- The natural roughness or etching of the fiber surface causing interlocking may form a mechanical bond between the fiber and matrix.
- If the thermal expansion coefficient of the matrix is higher than that of the fiber, and the manufacturing temperatures are higher than the operating temperatures, the matrix will radially shrink more than the fiber. This causes the matrix to compress around the fiber.

---

* Coupling agents are compounds applied to fiber surfaces to improve the bond between the fiber and matrix. For example, silane finish is applied to glass fibers to increase adhesion with epoxy matrix.

- Reaction bonding occurs when atoms or molecules of the fiber and the matrix diffuse into each other at the interface. This interdiffusion often creates a distinct interfacial layer, called the interphase, with different properties from that of the fiber or the matrix. Although this thin interfacial layer helps to form a bond, it also forms microcracks in the fiber. These microcracks reduce the strength of the fiber and thus that of the composite.

Weak or cracked interfaces can cause failure in composites and reduce the properties influenced by the matrix. They also allow environmental hazards such as hot gases and moisture to attack the fibers.

Although a strong bond is a requirement in transferring loads from the matrix to the fiber, weak debonding of the fiber–matrix interface is used advantageously in ceramic matrix composites. Weak interfaces blunt matrix cracks and deflect them along the interface. This is the main source of improving toughness of such composites up to five times that of the monolithic ceramics.

What is the world market of composites?

The world market for composites is only 10×10^9 US dollars as compared to more than 450×10^9 US dollars for steel. The annual growth of composites is at a steady rate of 10%. Presently, composite shipments are about 3×10^9 lb annually. Figure 1.7 gives the relative market share of US composite shipments and shows transportation clearly leading in their use. Table 1.3 shows the market share of composites since 1990.

1.2 Classification

How are composites classified?

Composites are classified by the geometry of the reinforcement — particulate, flake, and fibers (Figure 1.8) — or by the type of matrix — polymer, metal, ceramic, and carbon.

- *Particulate* composites consist of particles immersed in matrices such as alloys and ceramics. They are usually isotropic because the particles are added randomly. Particulate composites have advantages such as improved strength, increased operating temperature, oxidation resistance, etc. Typical examples include use of aluminum particles in rubber; silicon carbide particles in aluminum; and gravel, sand, and cement to make concrete.
- *Flake* composites consist of flat reinforcements of matrices. Typical flake materials are glass, mica, aluminum, and silver. Flake compos-

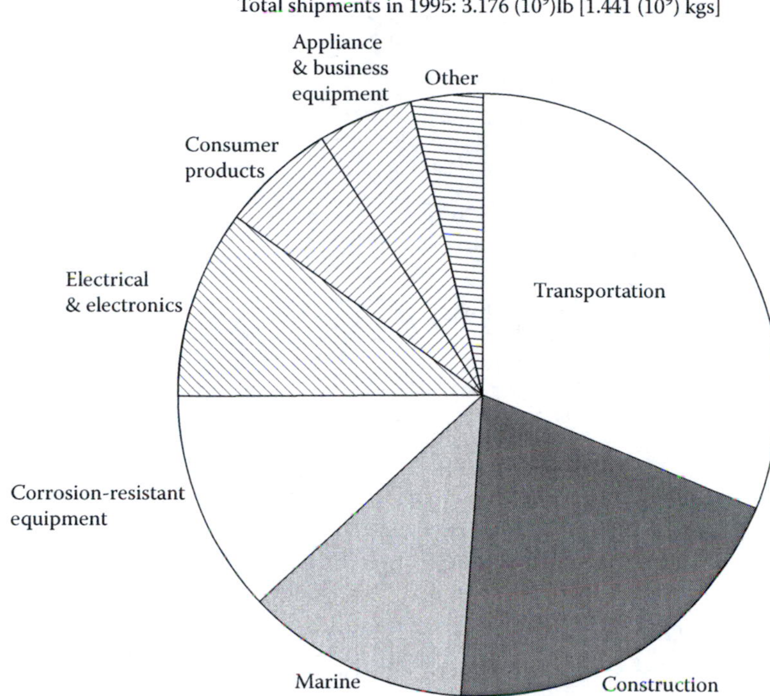

FIGURE 1.7
Approximate shipments of polymer-based composites in 1995. (Source: Data used in figure published with permission of the SPI, Inc.; http://www.socplas.org.)

TABLE 1.3

U.S. Composites Shipment in 10^6 lb, Including Reinforced Thermoset and Thermoplastic Resin Composites, Reinforcements, and Fillers

Markets	1990	1991	1992	1993	1994	1995
Aircraft/aerospace/military	39	38.7	32.3	25.4	24.2	24.0
Appliance/business equipment	153	135.2	143.2	147.5	160.7	166.5
Construction	468	420.0	483.0	530.0	596.9	626.9
Consumer products	165	148.7	162.2	165.7	174.8	183.6
Corrosion-resistant equipment	350	355.0	332.3	352.0	376.3	394.6
Electrical/electronic	241	231.1	260.0	274.9	299.3	315.1
Marine	375	275.0	304.4	319.3	363.5	375.1
Transportation	705	682.2	750.0	822.1	945.6	984.0
Other	79	73.8	83.4	89.3	101.8	106.6
TOTAL	**2575**	**2360**	**2551**	**2726**	**3043.1**	**3176.4**

Source: Published with permission of the SPI, Inc.

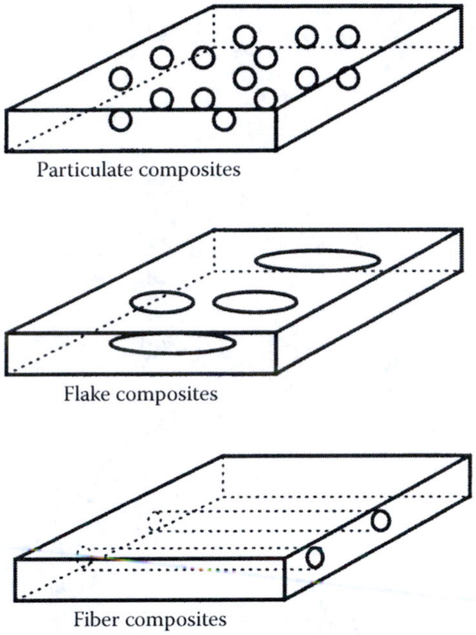

FIGURE 1.8
Types of composites based on reinforcement shape.

ites provide advantages such as high out-of-plane flexural modulus,* higher strength, and low cost. However, flakes cannot be oriented easily and only a limited number of materials are available for use.

- *Fiber* composites consist of matrices reinforced by short (discontinuous) or long (continuous) fibers. Fibers are generally anisotropic† and examples include carbon and aramids. Examples of matrices are resins such as epoxy, metals such as aluminum, and ceramics such as calcium–alumino silicate. Continuous fiber composites are emphasized in this book and are further discussed in this chapter by the types of matrices: polymer, metal, ceramic, and carbon. The fundamental units of continuous fiber matrix composite are unidirectional or woven fiber laminas. Laminas are stacked on top of each other at various angles to form a multidirectional laminate.
- *Nanocomposites* consist of materials that are of the scale of nanometers (10^{-9} m). The accepted range to be classified as a nanocomposite is that one of the constituents is less than 100 nm. At this scale, the

* Out of plane flexural stiffness is the resistance to deflection under bending that is out of the plane, such as bending caused by a heavy stone placed on a simply supported plate.

† Anisotropic materials are the opposite of isotropic materials like steel and aluminum; they have different properties in different directions. For example, the Young's modulus of a piece of wood is higher (different) in the direction of the grain than in the direction perpendicular to the grain. In comparison, a piece of steel has the same Young's modulus in all directions.

properties of materials are different from those of the bulk material. Generally, advanced composite materials have constituents on the microscale (10^{-6} m). By having materials at the nanometer scale, most of the properties of the resulting composite material are better than the ones at the microscale. Not all properties of nanocomposites are better; in some cases, toughness and impact strength can decrease.

Applications of nanocomposites include packaging applications for the military in which nanocomposite films show improvement in properties such as elastic modulus, and transmission rates for water vapor, heat distortion, and oxygen.[8]

Body side molding of the 2004 Chevrolet Impala is made of olefin-based nanocomposites.[9] This reduced the weight of the molding by 7% and improved its surface quality. General Motors™ currently uses 540,000 lb of nanocomposite materials per year.

Rubber containing just a few parts per million of metal conducts electricity in harsh conditions just like solid metal. Called Metal Rubber®, it is fabricated molecule by molecule by a process called electrostatic self-assembly. Awaited applications of the Metal Rubber include artificial muscles, smart clothes, flexible wires, and circuits for portable electronics.[10]

1.2.1 Polymer Matrix Composites

What are the most common advanced composites?

The most common advanced composites are polymer matrix composites (PMCs) consisting of a polymer (e.g., epoxy, polyester, urethane) reinforced by thin diameter fibers (e.g., graphite, aramids, boron). For example, graphite/epoxy composites are approximately five times stronger than steel on a weight-for-weight basis. The reasons why they are the most common composites include their low cost, high strength, and simple manufacturing principles.

What are the drawbacks of polymer matrix composites?

The main drawbacks of PMCs include low operating temperatures, high coefficients of thermal and moisture expansion,* and low elastic properties in certain directions.

What are the typical mechanical properties of some polymer matrix composites? Compare these properties with metals.

Table 1.4 gives typical mechanical properties of common polymer matrix composites.

* Some materials, such as polymers, absorb or deabsorb moisture that results in dimensional changes. The coefficient of moisture expansion is the change in length per unit length per unit mass of moisture absorbed per unit mass of the substance.

TABLE 1.4

Typical Mechanical Properties of Polymer Matrix Composites and Monolithic Materials

Property	Units	Graphite/ epoxy	Glass/ epoxy	Steel	Aluminum
System of units: USCS					
Specific gravity	—	1.6	1.8	7.8	2.6
Young's modulus	Msi	26.25	5.598	30.0	10.0
Ultimate tensile strength	ksi	217.6	154.0	94.0	40.0
Coefficient of thermal expansion	μin./in./°F	0.01111	4.778	6.5	12.8
System of units: SI					
Specific gravity	—	1.6	1.8	7.8	2.6
Young's modulus	GPa	181.0	38.6	206.8	68.95
Ultimate tensile strength	MPa	150.0	1062	648.1	275.8
Coefficient of thermal expansion	μm/m/°C	0.02	8.6	11.7	23

TABLE 1.5

Typical Mechanical Properties of Fibers Used in Polymer Matrix Composites

Property	Units	Graphite	Aramid	Glass	Steel	Aluminum
System of units: USCS						
Specific gravity	—	1.8	1.4	2.5	7.8	2.6
Young's modulus	Msi	33.35	17.98	12.33	30	10.0
Ultimate tensile strength	ksi	299.8	200.0	224.8	94	40.0
Axial coefficient of thermal expansion	μin./in./°F	–0.722	–2.778	2.778	6.5	12.8
System of units: SI						
Specific gravity	—	1.8	1.4	2.5	7.8	2.6
Young's modulus	GPa	230	124	85	206.8	68.95
Ultimate tensile strength	MPa	2067	1379	1550	648.1	275.8
Axial coefficient of thermal expansion	μm/m/°C	–1.3	–5	5	11.7	23

Give names of various fibers used in advanced polymer composites.

The most common fibers used are glass, graphite, and Kevlar. Typical properties of these fibers compared with bulk steel and aluminum are given in Table 1.5.

Give a description of the glass fiber.

Glass is the most common fiber used in polymer matrix composites. Its advantages include its high strength, low cost, high chemical resistance, and good insulating properties. The drawbacks include low elastic modulus,

TABLE 1.6

Comparison of Properties of E-Glass and S-Glass

Property	Units	E-Glass	S-Glass
System of units: USCS			
Specific gravity	—	2.54	2.49
Young's modulus	Msi	10.5	12.4
Ultimate tensile strength	ksi	500	665
Coefficient of thermal expansion	μin./in./°F	2.8	3.1
System of units: SI			
Specific gravity	—	2.54	2.49
Young's modulus	GPa	72.40	85.50
Ultimate tensile strength	MPa	3447	4585
Coefficient of thermal expansion	μm/m/°C	5.04	5.58

TABLE 1.7

Chemical Composition of E-Glass and S-Glass Fibers

	% Weight	
Material	**E-Glass**	**S-Glass**
Silicon oxide	54	64
Aluminum oxide	15	25
Calcium oxide	17	0.01
Magnesium oxide	4.5	10
Boron oxide	8	0.01
Others	1.5	0.8

poor adhesion to polymers, high specific gravity, sensitivity to abrasion (reduces tensile strength), and low fatigue strength.

Types: The main types are E-glass (also called "fiberglass") and S-glass. The "E" in E-glass stands for electrical because it was designed for electrical applications. However, it is used for many other purposes now, such as decorations and structural applications. The "S" in S-glass stands for higher content of silica. It retains its strength at high temperatures compared to E-glass and has higher fatigue strength. It is used mainly for aerospace applications. Some property differences are given in Table 1.6.

The difference in the properties is due to the compositions of E-glass and S-glass fibers. The main elements in the two types of fibers are given in Table 1.7.

Other types available commercially are C-glass ("C" stands for corrosion) used in chemical environments, such as storage tanks; R-glass used in structural applications such as construction; D-glass (dielectric) used for applications requiring low dielectric constants, such as radomes; and A-glass (appearance) used to improve surface appearance. Combination types such

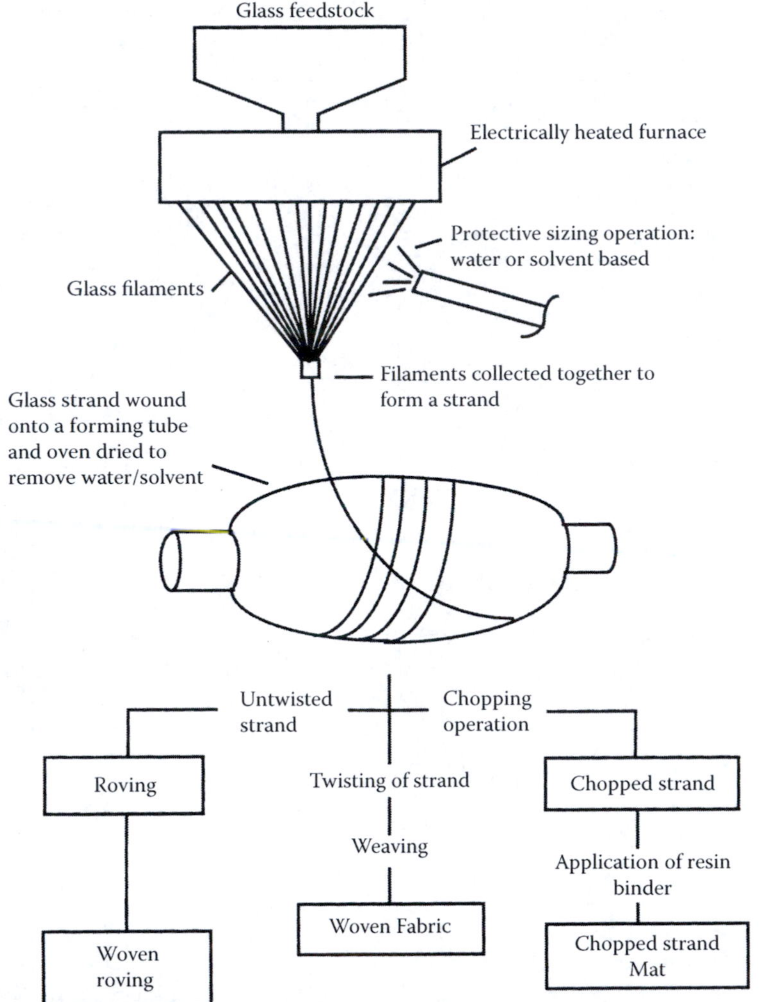

FIGURE 1.9
Schematic of manufacturing glass fibers and available glass forms. (From Bishop, W., in *Advanced Composites*, Partridge, I.K., Ed., Kluwer Academic Publishers, London, 1990, Figure 4, p. 177. Reproduced with kind permission of Springer.)

as E-CR glass ("E-CR" stands for electrical and corrosion resistance) and AR glass (alkali resistant) also exist.

Manufacturing: Glass fibers are made generally by drawing from a melt[11] as shown in Figure 1.9. The melt is formed in a refractory furnace at about 2550°F (1400°C) from a mixture that includes sand, limestone, and alumina. The melt is stirred and maintained at a constant temperature. It passes through as many as 250 heated platinum alloy nozzles of about 394 μin. (10 μm) diameter, where it is drawn into filaments of needed size at high speeds of about 361 mi/h (25 m/s). These fibers are sprayed with an organic sizing solution before they are drawn. The sizing solution is a mixture of binders, lubricants, and coupling and antistatic agents; binders allow filaments to be packed in strands, lubricants prevent abrasion of filaments, and coupling agents give better adhesion between the inorganic glass fiber and the organic matrix.

Fibers are then drawn into strands and wound on a forming tube. Strands are groups of more than 204 filaments. The wound array of strands is then removed and dried in an oven to remove any water or sizing solutions. The glass strand can then be converted into several forms as shown in Figure 1.9. Different forms of various fibers are shown in Figure 1.10.

Give a description of graphite fibers.

Graphite fibers are very common in high-modulus and high-strength applications such as aircraft components, etc. The advantages of graphite fibers include high specific strength and modulus, low coefficient of thermal expansion, and high fatigue strength. The drawbacks include high cost, low impact resistance, and high electrical conductivity.

Manufacturing: Graphite fibers have been available since the late 1800s. However, only since the early 1960s has the manufacturing of graphite fibers taken off. Graphite fibers are generally manufactured from three precursor materials: rayon, polyacrylonitrile (PAN), and pitch. PAN is the most popular precursor and the process to manufacture graphite fibers from it is given next (Figure 1.11).

PAN fibers are first stretched five to ten times their length to improve their mechanical properties and then passed through three heating processes. In the first process, called stabilization, the fiber is passed through a furnace between 392 and 572°F (200 and 300°C) to stabilize its dimensions during the subsequent high-temperature processes. In the second process, called carbonization, it is pyrolized* in an inert atmosphere of nitrogen or argon between 1832 and 2732°F (1000 and 1500°C). In the last process, called graphitization, it is heat treated above 4532°F (2500°C). The graphitization yields a microstructure that is more graphitic than that produced by carbonization. The fibers may also be subjected to tension in the last two heating processes to develop fibers with a higher degree of orientation.

At the end of this three-step heat treatment process, the fibers are surface treated to develop fiber adhesion and increase laminar shear strength when they are used in composite structures. They are then collected on a spool.

Properties: Table 1.8 gives properties of graphite fibers obtained from two different precursors.

Are carbon and graphite the same?

No,[7] they are different. Carbon fibers have 93 to 95% carbon content, but graphite has more than 99% carbon content. Also, carbon fibers are produced

* Pyrolysis is defined as the decomposition of a complex organic substance to one of a simpler structure by means of heat.

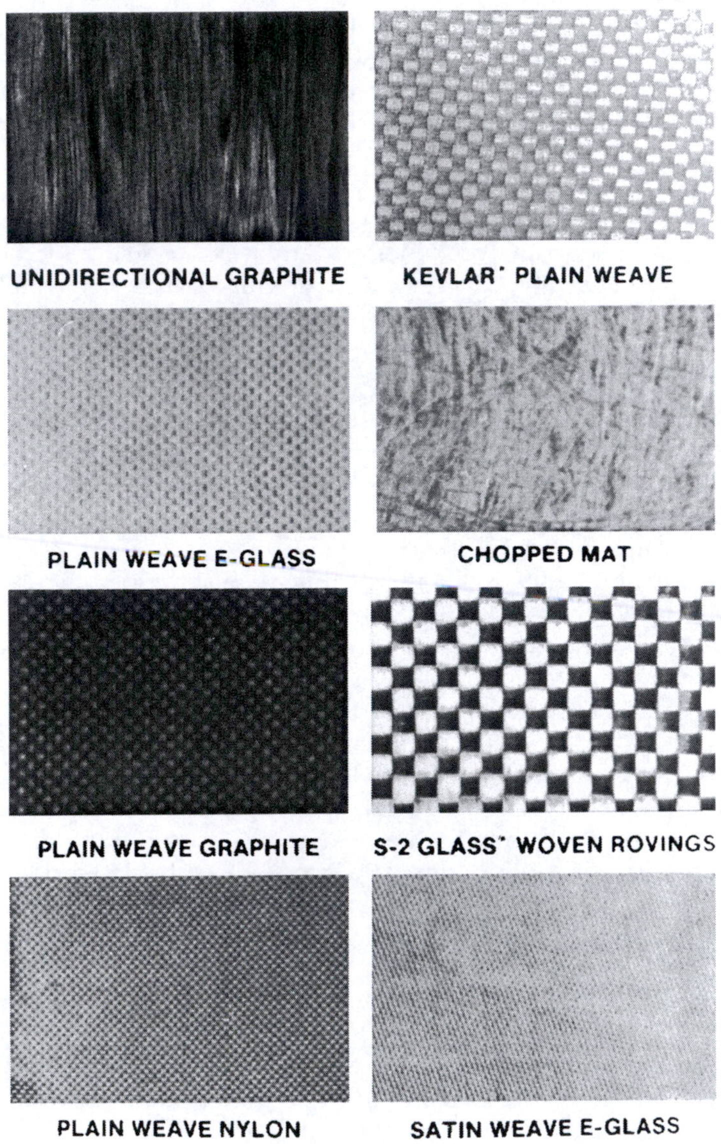

FIGURE 1.10
Forms of available fibers. (Graphic courtesy of M.C. Gill Corporation, http://www.mcgillcorp.com.)

at 2400°F (1316°C), and graphite fibers are typically produced in excess of 3400°F (1900°C).

Give a description of the aramid fiber.

An aramid fiber is an aromatic organic compound made of carbon, hydrogen, oxygen, and nitrogen. Its advantages are low density, high tensile

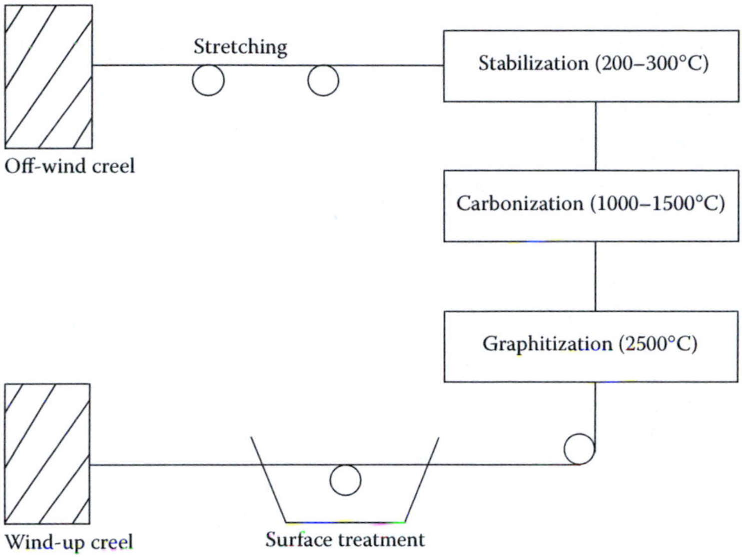

FIGURE 1.11
Stages of manufacturing a carbon fiber from PAN-based precursors.

TABLE 1.8
Mechanical Properties of Two Typical Graphite Fibers

Property	Units	PITCH	PAN
System of units: USCS			
Specific gravity	—	1.99	1.78
Young's modulus	Msi	55	35
Ultimate tensile strength	ksi	250	500
Axial coefficient of thermal expansion	μin/in/°F	−0.3	−0.7
System of units: SI			
Specific gravity	—	1.99	1.78
Young's modulus	GPa	379.2	241.3
Ultimate tensile strength	MPa	1723	3447
Axial coefficient of thermal expansion	μm/m/°C	−0.54	−1.26

strength, low cost, and high impact resistance. Its drawbacks include low compressive properties and degradation in sunlight.

Types: The two main types of aramid fibers are Kevlar 29®* and Kevlar 49®†. Both types of Kevlar fibers have similar specific strengths, but Kevlar 49 has a higher specific stiffness. Kevlar 29 is mainly used in bulletproof

* Kevlar 29 is a registered trademark of E.I. duPont deNemours and Company, Inc., Wilmington, DE.

† Kevlar 49 is a registered trademark of E.I. duPont deNemours and Company, Inc., Wilmington, DE.

TABLE 1.9

Properties of Kevlar Fibers

Property	Units	Kevlar 29	Kevlar 49
System of units: USCS			
Specific gravity	—	1.44	1.48
Young's modulus	Msi	9	19
Ultimate tensile strength	ksi	525	525
Axial coefficient of thermal expansion	μin./in./°F	–1.111	–1.111
System of units: SI			
Specific gravity	—	1.44	1.48
Young's modulus	GPa	62.05	131.0
Ultimate tensile strength	MPa	3620	3620
Axial coefficient of thermal expansion	μm/m/°C	–2	–2

vests, ropes, and cables. High performance applications in the aircraft industry use Kevlar 49. Table 1.9 gives the relative properties of Kevlar 29 and Kevlar 49.

Manufacturing: The fiber is produced by making a solution of proprietary polymers and strong acids such as sulfuric acid. The solution is then extruded into hot cylinders at 392°F (200°C), washed, and dried on spools. The fiber is then stretched and drawn to increase its strength and stiffness.

Give names of various polymers used in advanced polymer composites.

These polymers include epoxy, phenolics, acrylic, urethane, and polyamide.

Why are there so many resin systems in advanced polymer composites?

Each polymer has its advantages and drawbacks in its use[12]:

- Polyesters: The advantages are low cost and the ability to be made translucent; drawbacks include service temperatures below 170°F (77°C), brittleness, and high shrinkage* of as much as 8% during curing.
- Phenolics: The advantages are low cost and high mechanical strength; drawbacks include high void content.
- Epoxies: The advantages are high mechanical strength and good adherence to metals and glasses; drawbacks are high cost and difficulty in processing.

* Shrinkage in resins is found by measuring the density of the resin before and after cross-linking. If ρ is the density before cross-linking and ρ' is the density after cross-linking. The percent shrinkage is defined as shrinkage = $(\rho' - \rho)/\rho' \times 100$.

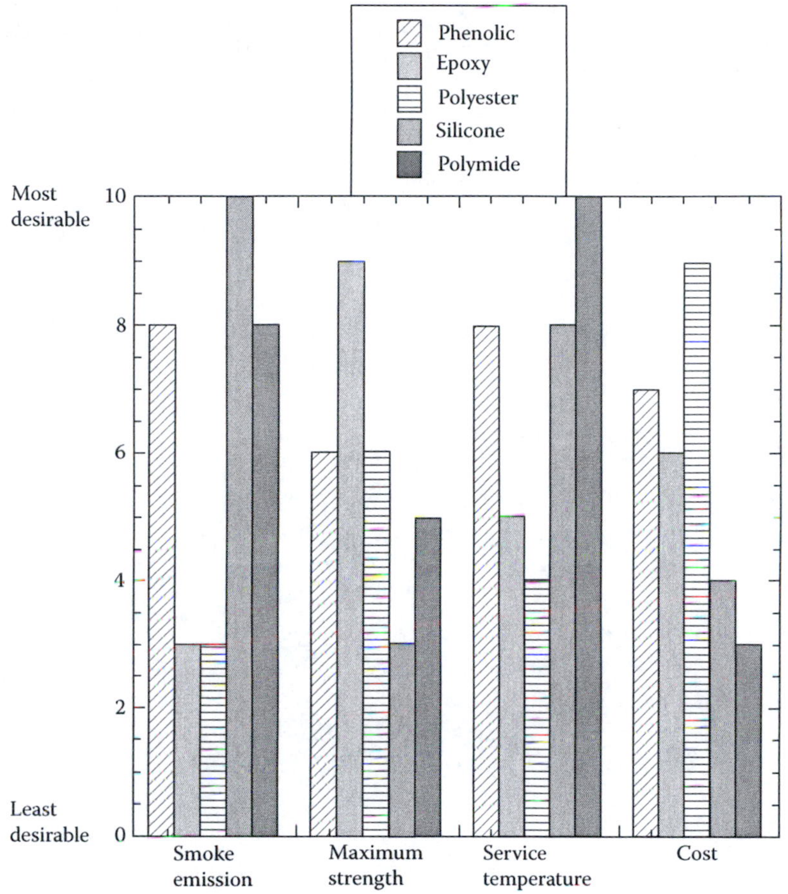

FIGURE 1.12
Comparison of performance of several common matrices used in polymer matrix composites. (Graphic courtesy of M.C. Gill Corporation, http://www.mcgillcorp.com.)

As can be seen, each of the resin systems has its advantages and drawbacks. The use of a particular system depends on the application. These considerations include mechanical strength, cost, smoke emission, temperature excursions, etc. Figure 1.12 shows the comparison of five common resins based on smoke emission, strength, service temperature, and cost.[12]

Give a description of epoxy.

Epoxy resins are the most commonly used resins. They are low molecular weight organic liquids containing epoxide groups. Epoxide has three members in its ring: one oxygen and two carbon atoms. The reaction of epichlorohydrin with phenols or aromatic amines makes most epoxies. Hardeners,*

* Hardeners are substances that are added to polymers for aiding in curing of composites.

TABLE 1.10

Room Temperature Properties of a Typical Epoxy

Property	Units	Value
System of units: USCS		
Specific gravity	—	1.28
Young's modulus	Msi	0.55
Ultimate tensile strength	ksi	12.0
System of units: SI		
Specific gravity	—	1.28
Young's modulus	GPa	3.792
Ultimate tensile strength	MPa	82.74

plasticizers,* and fillers† are also added to produce epoxies with a wide range of properties of viscosity, impact, degradation, etc. The room temperature properties of a typical epoxy are given in Table 1.10.

Epoxy is the most common type of matrix material. Why?

Although epoxy is costlier than other polymer matrices, it is the most popular PMC matrix. More than two-thirds of the polymer matrices used in aerospace applications are epoxy based. The main reasons why epoxy is the most used polymer matrix material are

- High strength
- Low viscosity and low flow rates, which allow good wetting of fibers and prevent misalignment of fibers during processing
- Low volatility during cure
- Low shrink rates, which reduce the tendency of gaining large shear stresses of the bond between epoxy and its reinforcement
- Available in more than 20 grades to meet specific property and processing requirements

Polymers are classified as thermosets and thermoplastics. What is the difference between the two? Give some examples of both.

Thermoset polymers are insoluble and infusible after cure because the chains are rigidly joined with strong covalent bonds; thermoplastics are formable at high temperatures and pressure because the bonds are weak and

---

* Plasticizers are lubricants that improve the toughness, flexibility, processibility, and ductility of polymers. This improvement is generally at the expense of lower strength.

† Fillers are ingredients added to enhance properties such as strength, surface texture, and ultraviolet absorption of a polymer, and to lower the cost of polymers. Typical examples include chopped fabric and wood flour.

of the van der Waals type. Typical examples of thermoset include epoxies, polyesters, phenolics, and polyamide; typical examples of thermoplastics include polyethylene, polystyrene, polyether–ether–ketone (PEEK), and polyphenylene sulfide (PPS). The differences between thermosets and thermoplastics are given in the following table.[13]

Thermoplastics	Thermoset
Soften on heating and pressure, and thus easy to repair	Decompose on heating
High strains to failure	Low strains to failure
Indefinite shelf life	Definite shelf life
Can be reprocessed	Cannot be reprocessed
Not tacky and easy to handle	Tacky
Short cure cycles	Long cure cycles
Higher fabrication temperature and viscosities have made it difficult to process	Lower fabrication temperature
Excellent solvent resistance	Fair solvent resistance

What are prepregs?

Prepregs are a ready-made tape composed of fibers in a polymer matrix (Figure 1.13). They are available in standard widths from 3 to 50 in. (76 to 1270 mm). Depending on whether the polymer matrix is thermoset or thermoplastic, the tape is stored in a refrigerator or at room temperature, respectively. One can lay these tapes manually or mechanically at various orientations to make a composite structure. Vacuum bagging and curing under high pressures and temperatures may follow.

Figure 1.14 shows the schematic of how a prepreg is made.[14] A row of fibers is passed through a resin bath. The resin-impregnated fibers are then

FIGURE 1.13
Boron/epoxy prepreg tape. (Photo courtesy of Specialty Materials, Inc., http://www.specmaterials.com.)

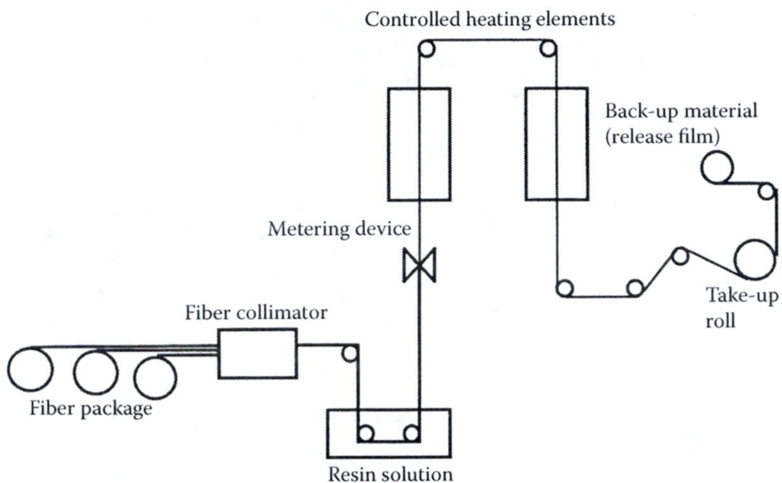

FIGURE 1.14
Schematic of prepreg manufacturing. (Reprinted from Mallick, P.K., *Fiber-Reinforced Composites: Materials, Manufacturing, and Design*, Marcel Dekker, Inc., New York, Chap. 2, 1988, p. 62. Courtesy of CRC Press, Boca Raton, FL.)

heated to advance the curing reaction from A-stage* to the B-stage. A release film is now wound over a take-up roll and backed with a release film. The release film keeps the prepregs from sticking to each other during storage.

Give examples of how a polymer matrix composite is manufactured.

Techniques of manufacturing a polymer matrix composite include *filament winding* (used generally for making pipes and tanks to handle chemicals), *autoclave forming* (used to make complex shapes and flat panels for structures in which low void content and high quality are important), and *resin transfer molding* (used extensively in the automotive industry because short production runs are necessary).

Filament winding: Fibers are impregnated with a resin by drawing them through an in-line resin bath (wet winding) (Figure 1.16) or prepregs (dry winding) are wound over a mandrel. Wet winding is inexpensive and lets one control the properties of the composite. Dry winding is cleaner, but more expensive and thus quite uncommon.

---

* Thermosetting resins have three curing stages: A, B, and C (Figure 1.15).

Resins are manufactured in the A-stage, in which the resin may be solid or liquid but is able to flow if heat is applied. The A-stage is also called the completely uncured stage.

The B-stage is the middle stage of the reaction of a thermosetting resin used when prepregs are manufactured. This stage allows easy processing and handling of composite layers, such as graphite/epoxy.

The C-stage is the final stage in the reaction of a thermosetting resin. This stage is accomplished when a composite structure is made out of composite layers. Heat and pressure may be applied at the B-stage to cure the resin completely. This stage results in irreversible hardening and insolubility.

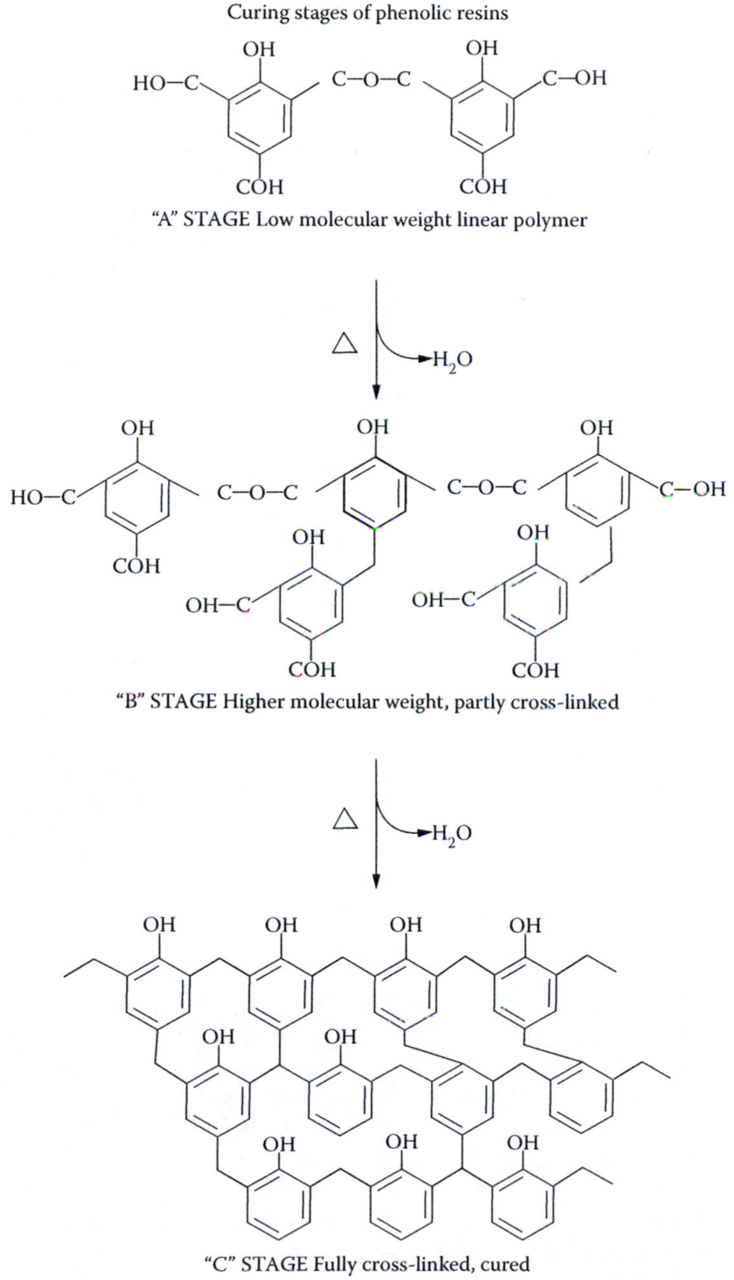

Curing stages of phenolic resins

"A" STAGE Low molecular weight linear polymer

"B" STAGE Higher molecular weight, partly cross-linked

"C" STAGE Fully cross-linked, cured

FIGURE 1.15
Curing stages of phenolic resins. (Graphic courtesy of M.C. Gill Corporation, http://www.mcgillcorp.com.)

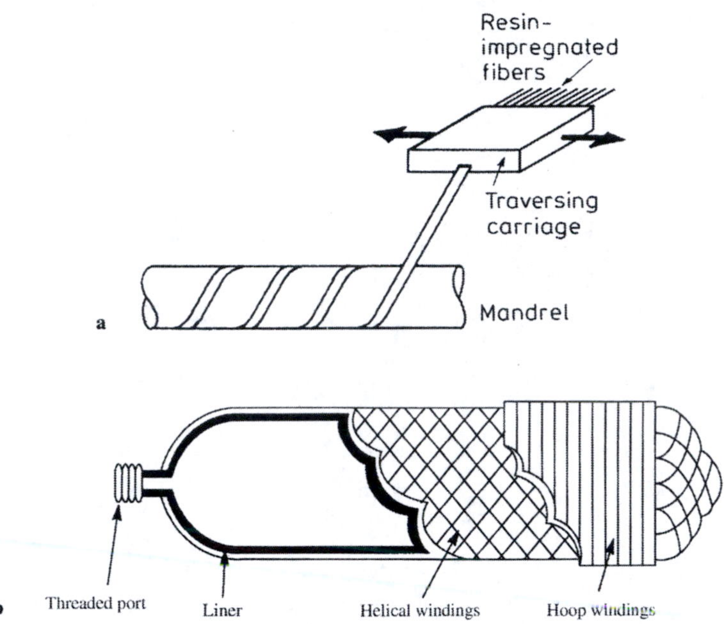

FIGURE 1.16
(a) Filament winding process; (b) filament wound pressure vessel with liner. (From Chawla, K.K., *Composite Materials — Science and Engineering*, Springer–Verlag, 1998. Reprinted by permission of Springer–Verlag.)

Depending on the desired properties of the product, winding patterns such as hoop, helical, and polar can be developed. The product is then cured with or without heat and pressure. Depending on the application, mandrels are made of wood, aluminum, steel, plaster, or salts. For example, steel mandrels are chosen for manufacturing large quantities of open-ended cylinders, and low-melting alloys or water-soluble salts are used for closed-ended cylinders so that one can easily remove the mandrel.

Autoclave forming: This method of manufacturing is used with composites available as prepregs. First, a peel ply made out of nylon or cellophane coated with Teflon* is placed on the mold.† Teflon is used for easy removal of the part and the peel ply achieves a desired finish that is smooth and wrinkle free. Replacing Teflon by mold releasing powders and liquids can also accomplish removal of the part. Prepregs of the required number are laid up one ply at a time by automated means or by hand. Each ply is pressed to remove any entrapped air and wrinkles. The lay-up is sealed at the edges to form a vacuum seal.

* Teflon is a registered trademark of E.I. duPont deNemours and Company, Inc., Wilmington, DE.

† Mold: a structure around or in which the composite forms a desired shape. Molds are female and male. If the composite part is in the mold, it is called a female mold; if it is made around the mold, it is called a male mold. (See in Figure 1.17 the male mold that was used in making a human-powered submarine.)

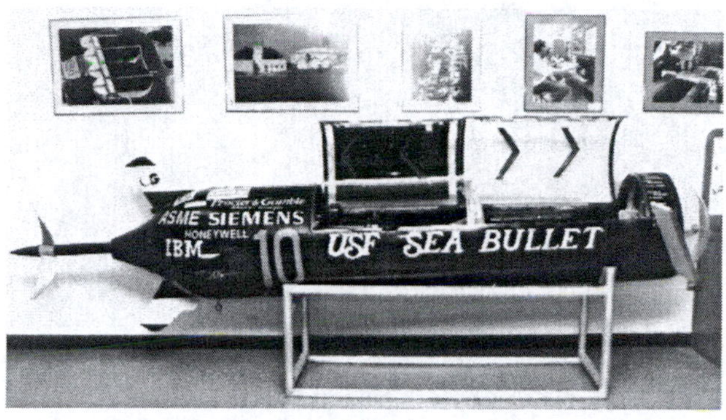

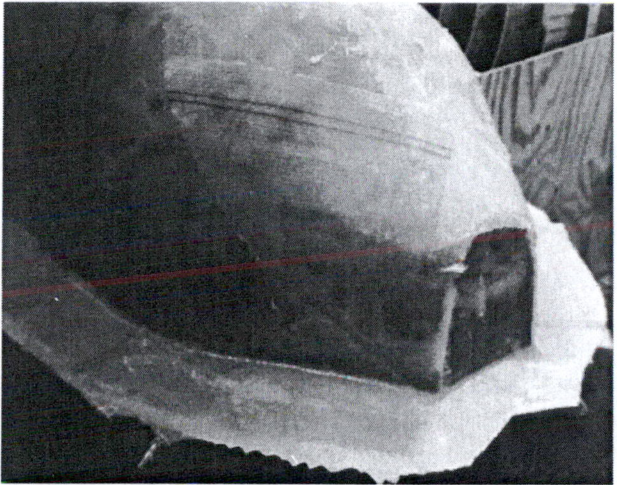

FIGURE 1.17
Human-powered submarine and its mold. (Courtesy of Professor G.H. Besterfield and student section of ASME, University of South Florida, Tampa.)

Now one establishes the bleeder system to get rid of the volatiles and excess resin during the heating and vacuum process that follows later. The bleeder system consists of several bleeder sheets made of glass cloth. These are placed on the edges and the top of the lay-up.

Then, vacuum connections are placed over the bleeders and the lay-up is bagged. A partial vacuum is developed to smooth the bag surface. The whole assembly is put in an autoclave (Figure 1.18), where heat and pressure are applied with an inert gas such as nitrogen. The vacuum system is kept functioning to remove volatiles during the cure cycle and to keep the part conformed to the mold. The cure cycle may last more than 5 h.

Resin transfer molding (RTM) (also called liquid molding): A low viscosity resin such as polyester or epoxy resin is injected under low pressure into a closed mold that contains the fiber preform. The resin flow is stopped and the part is allowed to cure. The cure is done at room temperature or at elevated temperatures. The latter is done if the part is to be used for high-temperature

FIGURE 1.18
Autoclave used for processing polymer matrix composites. (Photo courtesy of ACP Composites, MN, http://www.acp_composites.com.)

application. The advantages of RTM are that it is less expensive than hand lay-up, can be automated, and does not need refrigerated storage for prepregs. Major drawbacks include the capital expense required for having two molds instead of one.[15]

Give typical applications of polymer matrix composites.

Applications of polymer matrix composites range from tennis racquets to the space shuttle. Rather than enumerating only the areas in which polymer-based composites are used, a few examples have been taken from each industry. Emphasis has been placed on why a composite material is the material of choice.

Aircraft: The military aircraft industry has mainly led the use of polymer composites. The percentage of structural weight of composites that was less than 2% in F-15s in the 1970s has increased to about 30% on the AV-8B in the 1990s. In both cases, the weight reduction over metal parts was more than 20%.

In commercial airlines, the use of composites has been conservative because of safety concerns. Use of composites is limited to secondary structures such as rudders and elevators made of graphite/epoxy for the Boeing 767 and landing gear doors made of Kevlar–graphite/epoxy. Composites are also used in panels and floorings of airplanes. Some examples of using composites in the primary structure are the all-composite Lear Fan 2100 plane and the tail fin of the Airbus A310-300. In the latter case, the tail fin consists of graphite/epoxy and aramid honeycomb. It not only reduced the weight of the tail fin by 662 lb (300 kg) but also reduced the number of parts from 2000 to 100. Skins of aircraft engine cowls shown in Figure 1.19 are also made of polymer matrix composites for reducing weight.[16]

With increasing competition in model airplane flying, the weight of composite materials has been reduced. Figure 1.20 shows a World War II model airplane with fuselage made of glass/epoxy, wings made of balsa-wood

FIGURE 1.19
Aircraft engine cowling. (Photo provided courtesy of Alliant Techsystems, Inc.)

FIGURE 1.20
Model BF109 WWII German fighter plane using glass/epoxy-molded fuselage and wing spars of graphite/epoxy. (Photo courtesy of Russell A. Lepré, Tampa, FL.)

FIGURE 1.21
The BELL™ V-22 Osprey in combat configuration. (Courtesy of Bell Helicopter Textron Inc.)

facings/Styrofoam core sandwich construction, and wingspars made of graphite/epoxy.[17]

Helicopters and tiltrotors (Figure 1.21) use graphite/epoxy and glass/epoxy rotor blades that not only increase the life of blades by more than 100% over metals but also increase the top speeds.

Space: Two factors make composites the material of choice in space applications: high specific modulus and strength, and dimensional stability during large changes in temperature in space. Examples include the Graphite/epoxy-honeycomb payload bay doors in the space shuttle (Figure 1.22). Weight savings[7] over conventional metal alloys translate to higher payloads that cost as much as $1000/lb ($2208/kg). Also, for the space shuttles, graphite/epoxy was chosen primarily for weight savings and for small mechanical and thermal deflections concerning the remote manipulator arm, which deploys and retrieves payloads.

Figure 1.23 shows a high-gain antenna for the space station that uses sandwiches made of graphite/epoxy facings with an aluminum honeycomb core. Antenna ribs and struts in satellite systems use graphite/epoxy for their high specific stiffness and its ability to meet the dimensional stability[16] requirements due to large temperature excursions in space.

In June 2004, Paul G. Allen and Scaled Composites[18] launched the first privately manned vehicle, called SpaceshipOne, beyond the Earth's atmosphere (Figure 1.24). The spaceship reached a record-breaking altitude of approximately 62 miles (100 km). SpaceshipOne is constructed from graphite-epoxy composite materials; a *trowel-on* ablative thermal protection layer[19] protects its hotter sections.

Sporting goods: Graphite/epoxy is replacing metals in golf club shafts mainly to decrease the weight and use the saved weight in the head. This increase in the head weight has improved driving distances by more than 25 yards (23 m).

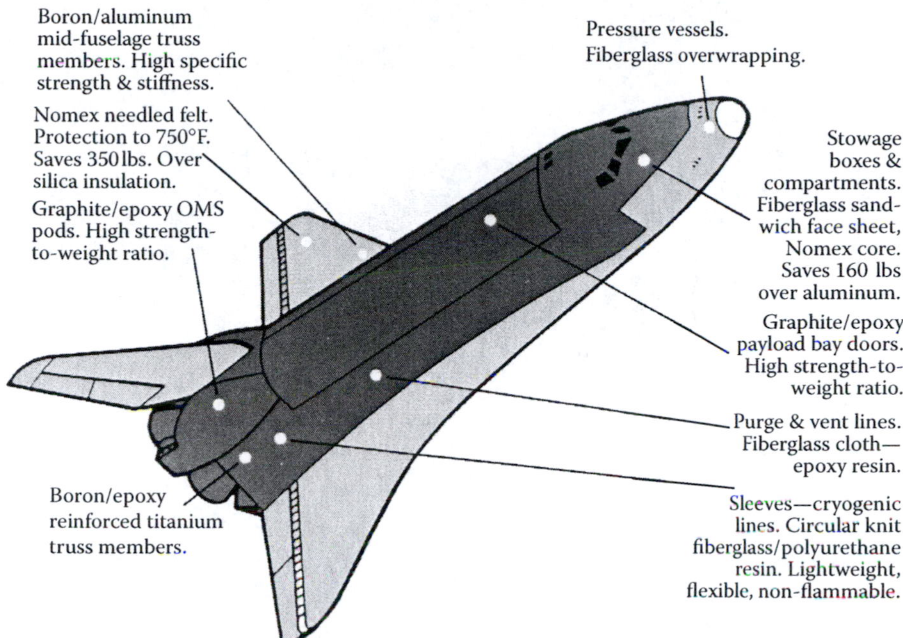

FIGURE 1.22
Use of composites in the space shuttle. (Graphic courtesy of M.C. Gill Corporation, http://www.mcgillcorp.com.)

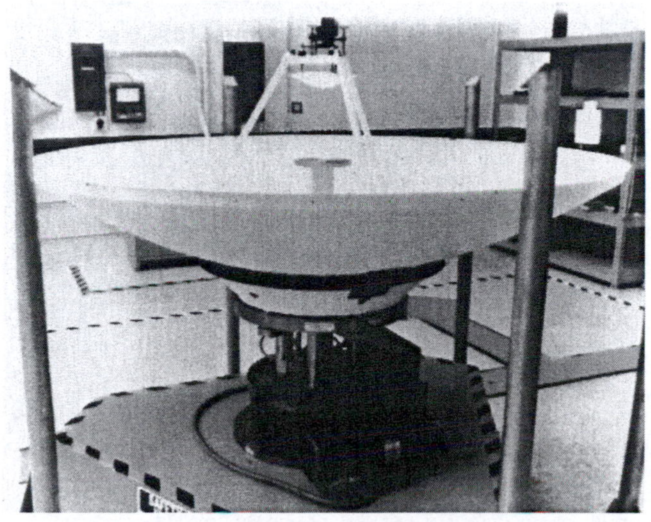

FIGURE 1.23
High-gain antenna for space station. (Photo provided courtesy of Alliant Techsystems, Inc.)

FIGURE 1.24
First privately manned vehicle (SpaceShipOne) to go beyond the Earth's atmosphere. (Photo provided courtesy of Scaled Composites, http://www.scaled.com.)

Bicycles use hybrid construction of graphite/epoxy composites wound on an aluminum tubing or chopped S-glass reinforced urethane foam. The graphite/epoxy composite increases the specific modulus of the tube and decreases the mass of the frame by 25%. Composites also allow frames to consist of one piece, which improves fatigue life and avoids stress concentration* found in metallic frames at their joints. Bicycle wheels made of carbon–polymide composites offer low weight and better impact resistance than aluminum.

Tennis and racquetball rackets with graphite/epoxy frames are now commonplace. The primary reasons for using composites are that they improve the torsional rigidity of the racquet and reduce risk of elbow injury due to vibration damping.† Ice hockey sticks are now manufactured out of hybrids such as Kevlar–glass/epoxy. Kevlar is added for durability and stiffness. Ski poles made of glass/polyester composites have higher strength, flexibility, and lower weight than conventional ski poles. This reduces stress and impact on upper body joints as the skier plants his poles.

Medical devices: Applications here include the use of glass–Kevlar/epoxy lightweight face masks for epileptic patients. Artificial portable lungs are made of graphite–glass/epoxy so that a patient can be mobile. X-ray tables made of graphite/epoxy facing sandwiches are used for their high stiffness, light weight, and transparency to radiation. The latter feature allows the

* If a loaded machine element has a discontinuity, the stresses are different at the discontinuity. The ratio between the stresses at the discontinuity and the nominal stress is defined as the stress concentration factor. For example, in a plate with a small hole, the stress concentration factor is three at the edge of the hole.

† Vibration damping is the ability of a material to dissipate energy during vibration. Damping of composites is higher than that of conventional metals such as steel and aluminum. Damping of composites depends on fiber volume fraction, orientation, constituent properties, and stacking sequence. Damping in composites is measured by calculating the ratio of energy dissipated to the energy stored.[20]

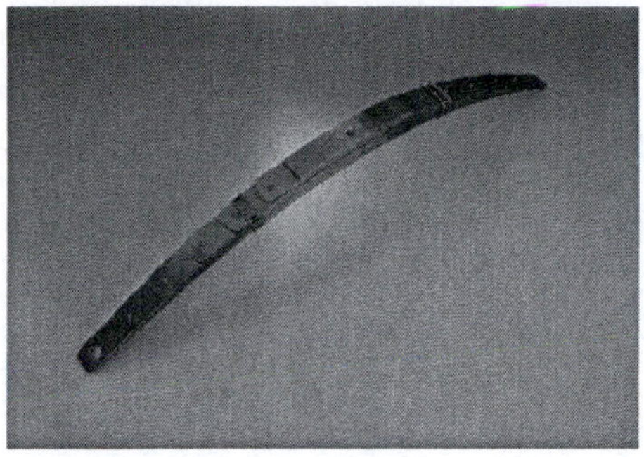

FIGURE 1.25
Rear fiberglass monosprings for Corvettes. (Photo courtesy of Vette Brakes and Products, St. Petersburg, FL, http://www.vbandp.com.)

patient to stay on one bed for an operation as well as x-rays and be subjected to a lower dosage of radiation.

Marine: The application of fiberglass in boats is well known. Hybrids of Kevlar–glass/epoxy are now replacing fiberglass for improved weight savings, vibration damping, and impact resistance. Kevlar–epoxy by itself would have poor compression properties.

Housings made of metals such as titanium to protect expensive oceanographic research instruments during explorations of sea wrecks are cost prohibitive. These housings are now made out of glass/epoxy and sustain pressures as high as 10 ksi (69 MPa) and extremely corrosive conditions.

Bridges made of polymer composite materials are gaining wide acceptance due to their low weight, corrosion resistance, longer life cycle, and limited earthquake damage. Although bridge components made of composites may cost $5/lb as opposed to components made of steel, reinforced concrete may only cost $0.30 to $1.00 per pound; the former weighs 80% less than the latter. Also, by lifetime costs, fewer composite bridges need to be built than traditional bridges.[21]

Automotive: The fiberglass body of the Corvette® comes to mind when considering automotive applications of polymer matrix composites. In addition, the Corvette has glass/epoxy composite leaf springs (Figure 1.25) with a fatigue life of more than five times that of steel. Composite leaf springs also give a smoother ride than steel leaf springs and give more rapid response to stresses caused by road shock. Moreover, composite leaf springs offer less chance of catastrophic failure, and excellent corrosion resistance.[22] By weight, about 8% of today's automobile parts are made of composites, including bumpers, body panels, and doors. However, since 1981, the average engine horsepower has increased by 84%, while average vehicle weight has increased by more than 20%. To overcome the increasing weight but also maintain the safety of modern vehicles, some estimate that carbon composite bodies will reduce the weight by 50%.[23]

Commercial: Fiber-reinforced polymers have many other commercial applications too. Examples include mops with pultruded fiberglass handles (Fig-

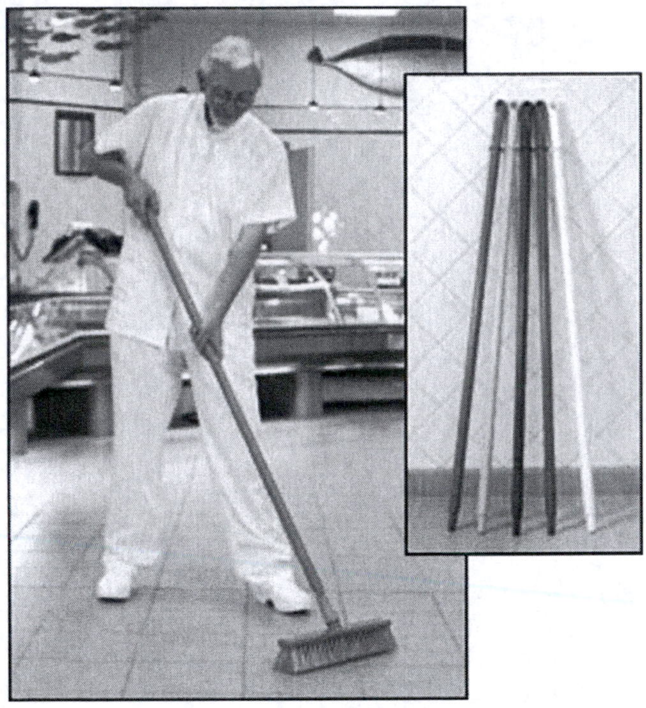

FIGURE 1.26
Fiberglass mop handle. (Photo courtesy of RTP Company, MN.)

ure 1.26). Some brooms used in pharmaceutical factories have handles that have no joints or seams; the surfaces are smooth and sealed. This keeps the bacteria from staying and growing. To have a handle that also is strong, rigid, and chemically and heat resistant, the material of choice is glass-fiber-reinforced polypropylene.[24] Other applications include pressure vessels for applications such as chemical plants. Garden tools (Figure 1.27)[25] can be made lighter than traditional metal tools and thus are suitable for children and people with physically challenged hands. Figure 1.27 shows the Powergear® Fiskars® anvil lopper. The handles of the lopper are made of Nyglass® composite, making it extremely lightweight and durable.

1.2.2 Metal Matrix Composites

What are metal matrix composites?

Metal matrix composites (MMCs), as the name implies, have a metal matrix. Examples of matrices in such composites include aluminum, magnesium, and titanium. Typical fibers include carbon and silicon carbide. Metals are mainly reinforced to increase or decrease their properties to suit the needs of design. For example, the elastic stiffness and strength of metals can be increased, and large coefficients of thermal expansion and thermal

FIGURE 1.27
Strong, efficient, and lightweight Fiskars Powergear anvil lopper. (Photo courtesy of Fiskars Brands, Inc.)

and electric conductivities of metals can be reduced, by the addition of fibers such as silicon carbide.

What are the advantages of metal matrix composites?

Metal matrix composites are mainly used to provide advantages over monolithic metals such as steel and aluminum. These advantages include higher specific strength and modulus by reinforcing low-density metals, such as aluminum and titanium; lower coefficients of thermal expansion by reinforcing with fibers with low coefficients of thermal expansion, such as graphite; and maintaining properties such as strength at high temperatures.

MMCs have several advantages over polymer matrix composites. These include higher elastic properties; higher service temperature; insensitivity to moisture; higher electric and thermal conductivities; and better wear, fatigue, and flaw resistances. The drawbacks of MMCs over PMCs include higher processing temperatures and higher densities.

Do any properties degrade when metals are reinforced with fibers?

Yes, reinforcing metals with fibers may reduce ductility and fracture toughness.[26] Ductility of aluminum is 48% and it can decrease to below 10% with

TABLE 1.11

Typical Mechanical Properties of Metal Matrix Composites

Property	Units	SiC/ aluminum	Graphite/ aluminum	Steel	Aluminum
System of units: USCS					
Specific gravity	—	2.6	2.2	7.8	2.6
Young's modulus	Msi	17	18	30	10
Ultimate tensile strength	ksi	175	65	94	34
Coefficient of thermal expansion	μin./in./°F	6.9	10	6.5	12.8
System of units: SI					
Specific gravity	—	2.6	2.2	7.8	2.6
Young's modulus	GPa	117.2	124.1	206.8	68.95
Ultimate tensile strength	MPa	1206	448.2	648.1	234.40
Coefficient of thermal expansion	μm/m/°C	12.4	18	11.7	23

simple reinforcements of silicon carbide whiskers. The fracture toughness of aluminum alloys is 18.2 to 36.4 ksi$\sqrt{\text{in.}}$ (20 to 40 MPa$\sqrt{\text{m}}$) and it reduces by 50% or more when reinforced with silicon fibers.

What are the typical mechanical properties of some metal matrix composites? Compare the properties with metals.

Typical mechanical properties of MMCs are given in Table 1.11.

Show one process of how metal matrix composites are manufactured.

Fabrication methods for MMCs are varied. One method of manufacturing them is diffusion bonding[26] (Figure 1.28), which is used in manufacturing boron/aluminum composite parts (Figure 1.29). A fiber mat of boron is placed between two thin aluminum foils about 0.002 in. (0.05 mm) thick. A polymer binder or an acrylic adhesive holds the fibers together in the mat. Layers of these metal foils are stacked at angles as required by the design. The laminate is first heated in a vacuum bag to remove the binder. The laminate is then hot pressed with a temperature of about 932°F (500°C) and pressure of about 5 ksi (35 MPa) in a die to form the required machine element.

What are some of the applications of metal matrix composites?

Metal matrix composites applications are

- *Space*: The space shuttle uses boron/aluminum tubes to support its fuselage frame. In addition to decreasing the mass of the space shuttle by more than 320 lb (145 kg), boron/aluminum also reduced the thermal insulation requirements because of its low thermal con-

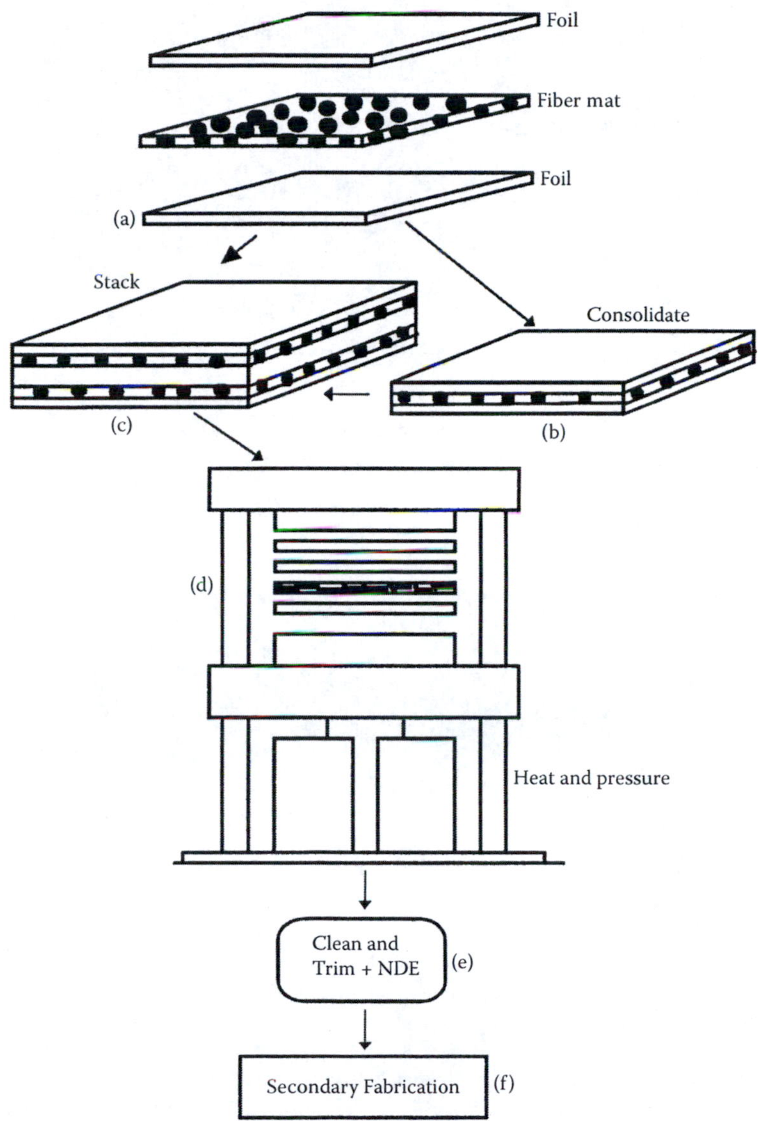

FIGURE 1.28
Schematic of diffusion bonding for metal matrix composites. (Reproduced with permission from Matthews, F.L. and Rawlings, R.D., *Composite Materials: Engineering and Science*, Chapman & Hall, London, 1994, Figure 3.1, p. 81. Copyright CRC Press, Boca Raton, FL.)

ductivity. The mast of the Hubble Telescope uses carbon-reinforced aluminum.

- *Military*: Precision components of missile guidance systems demand dimensional stability — that is, the geometries of the components cannot change during use.[27] Metal matrix composites such as SiC/aluminum composites satisfy this requirement because they have

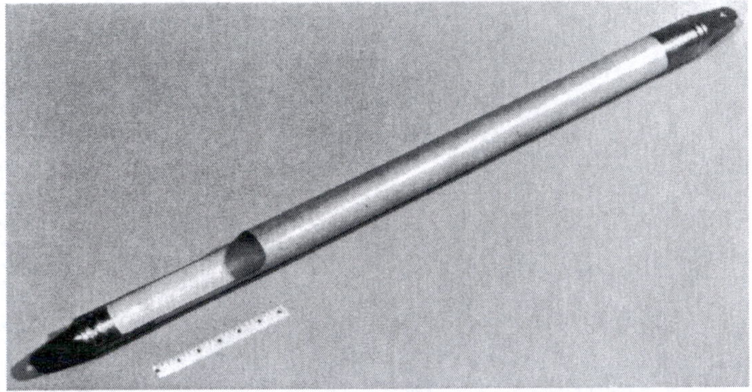

FIGURE 1.29
Boron/aluminum component made from diffusion bonding. (Photo courtesy of Specialty Materials, Inc., http://www.specmaterials.com.)

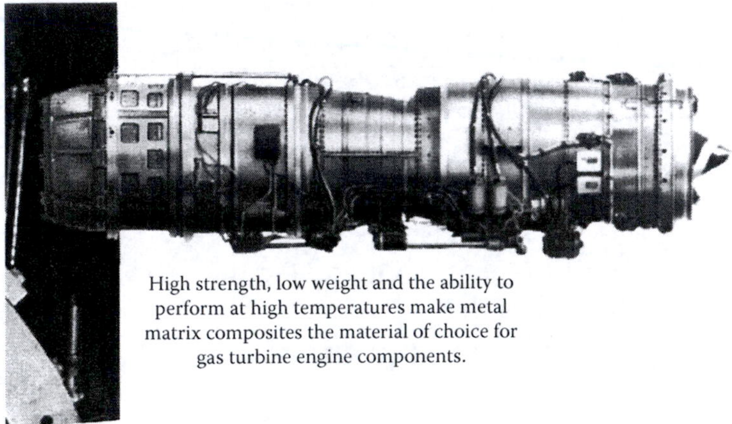
High strength, low weight and the ability to
perform at high temperatures make metal
matrix composites the material of choice for
gas turbine engine components.

FIGURE 1.30
Gas turbine engine components made of metal matrix composites. (Photo courtesy of Specialty Materials, Inc., http://www.specmaterials.com.)

high microyield strength.* In addition, the volume fraction of SiC can be varied to have a coefficient of thermal expansion compatible with other parts of the system assembly.

- *Transportation*: Metal matrix composites are finding use now in automotive engines that are lighter than their metal counterparts. Also, because of their high strength and low weight, metal matrix composites are the material of choice for gas turbine engines (Figure 1.30).

* Microyield strength is a major design parameter for elements that are required to be dimensionally stable. It is defined as the stress required to create a plastic (residual) strain of 1×10^{-6} or 1 μm.

TABLE 1.12

Typical Fracture Toughness of Monolithic Materials and Ceramic Matrix Composites

Material	Fracture toughness (MPa $\sqrt{\text{m}}$)	Fracture toughness (ksi $\sqrt{\text{in.}}$)
Epoxy	3	2.73
Aluminum alloys	35	31.85
Silicon carbide	3	2.73
SiC/Al_2O_3	27	24.6
SiC/SiC	30	27.3

1.2.3 Ceramic Matrix Composites

What are ceramic matrix composites?

Ceramic matrix composites (CMCs) have a ceramic matrix such as alumina calcium alumino silicate reinforced by fibers such as carbon or silicon carbide.

What are the advantages of ceramic matrix composites?

Advantages of CMCs include high strength, hardness, high service temperature limits* for ceramics, chemical inertness, and low density. However, ceramics by themselves have low fracture toughness. Under tensile or impact loading, they fail catastrophically. Reinforcing ceramics with fibers, such as silicon carbide or carbon, increases their fracture toughness (Table 1.12) because it causes gradual failure of the composite. This combination of a fiber and ceramic matrix makes CMCs more attractive for applications in which high mechanical properties and extreme service temperatures are desired.

What are the typical mechanical properties of some ceramic matrix composites? Compare them with metals.

Typical mechanical properties of ceramic matrix composites are given in Table 1.13.

Show one process of how ceramic matrix composites are manufactured.

One of the most common methods to manufacture ceramic matrix composites is called the hot pressing method.[28] Glass fibers in continuous tow are passed through slurry consisting of powdered matrix material, solvent such as alcohol, and an organic binder (Figure 1.31). The tow is then wound on a drum and dried to form prepreg tapes. The prepreg tapes can now be stacked to make a required laminate. Heating at about 932°F (500°C) burns out the binder. Hot pressing at high temperatures in excess of 1832°F (1000°C) and pressures of 1 to 2 ksi (7 to 14 MPa) follows this.

* Current service temperatures limits are 750°F (400°C) for polymers, 1800°F (1000ºC) for metals and their alloys, and 2700°F (1500°C) for ceramics.

TABLE 1.13

Typical Mechanical Properties of Some Ceramic Matrix Composites

Property	Units	SiC/LAS	SiC/CAS	Steel	Aluminum
System of units: USCS					
Specific gravity	—	2.1	2.5	7.8	2.6
Young's modulus	Msi	13	17.55	30.0	10.0
Ultimate tensile strength	ksi	72	58.0	94.0	34.0
Coefficient of thermal expansion	μin./in./°F	2	2.5	6.5	12.8
System of units: SI					
Specific gravity	—	2.1	2.5	7.8	2.6
Young's modulus	GPa	89.63	121	206.8	68.95
Ultimate tensile strength	MPa	496.4	400	648.1	234.4
Coefficient of thermal expansion	μm/m/°C	3.6	4.5	11.7	23

What are the applications of ceramic matrix composites?

Ceramic matrix composites are finding increased application in high-temperature areas in which metal and polymer matrix composites cannot be used. This is not to say that CMCs are not attractive otherwise, especially considering their high strength and modulus, and low density. Typical applications include cutting tool inserts in oxidizing and high-temperature environments. Textron Systems Corporation® has developed fiber-reinforced ceramics with SCS™ monofilaments for future aircraft engines (Figure 1.32).

1.2.4 Carbon–Carbon Composites

What are carbon–carbon composites?

Carbon–carbon composites use carbon fibers in a carbon matrix. These composites are used in very high-temperature environments of up to 6000°F (3315°C), and are 20 times stronger and 30% lighter than graphite fibers.[29]

What are the advantages of carbon–carbon composites?

Carbon is brittle and flaw sensitive like ceramics. Reinforcement of a carbon matrix allows the composite to fail gradually and also gives advantages such as ability to withstand high temperatures, low creep at high temperatures, low density, good tensile and compressive strengths, high fatigue resistance, high thermal conductivity, and high coefficient of friction. Drawbacks include high cost, low shear strength, and susceptibility to oxidations at high temperatures. Typical properties of carbon–carbon composites are given in Table 1.14.

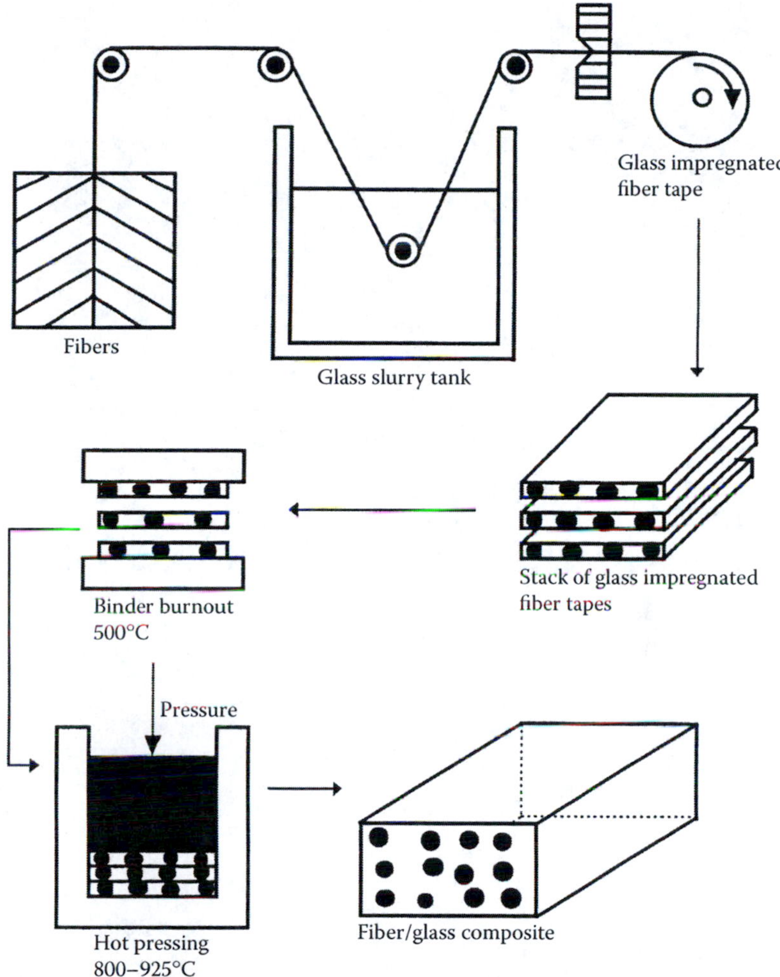

FIGURE 1.31
Schematic of slurry infiltration process for ceramic matrix composites. (From Chawla, K.K., *Science and Business Media from Ceramics Matrix Composites*, Kluwer Academic Publishers, London, 1993, Figure 4.1, p. 128. Reproduced with permission of Springer–Verlag.)

Give a typical method of processing a carbon–carbon composite.

A typical method for manufacturing carbon–carbon composites is called low-pressure carbonization[29] and is shown in Figure 1.33. A graphite cloth is taken, impregnated by resin (such as phenolic, pitch, and furfuryl ester), and laid up in layers. It is laid in a mold, cured, and trimmed. The part is then pyrolized, converting the phenolic resin to graphite. The composite is then impregnated by furfuryl alcohol. The process drives off the resin and any volatiles. The process is repeated three or four times until the level of porosity is reduced to an acceptable level. Each time, this process increases

FIGURE 1.32
Ceramic matrix composites for high temperature and oxidation resistant application. (Photo courtesy of Specialty Materials, Inc., http://www.specmaterials.com.)

TABLE 1.14
Typical Mechanical Properties of Carbon–Carbon Matrix Composites

Property	Units	C–C	Steel	Aluminum
System of units: USCS				
Specific gravity	—	1.68	7.8	2.6
Young's modulus	Msi	1.95	30	10
Ultimate tensile strength	ksi	5.180	94	40
Coefficient of thermal expansion	μin./in./°F	1.11	6.5	12.8
System of units: SI				
Specific gravity	—	1.68	7.8	2.6
Young's modulus	GPa	13.5	206.8	68.95
Ultimate strength	MPa	35.7	648.1	234.4
Coefficient of thermal expansion	μm/m/°C	2.0	11.7	23

its modulus and strength. Because carbon–carbon composites oxidize at temperatures as low as 842°F (450°C), an outer layer of silicon carbide may be deposited.[30]

What are the applications of carbon–carbon composites?

The main uses of carbon–carbon composites are the following:

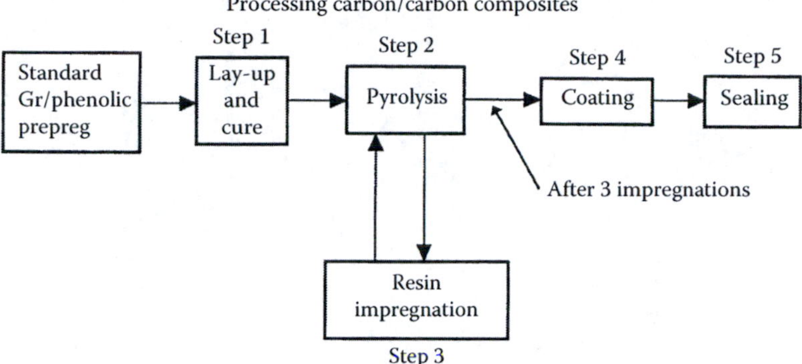

FIGURE 1.33
Schematic of processing carbon–carbon composites. (Reprinted with permission from Klein, A.J., *Adv. Mater. Processes*, 64–68, November 1986, ASM International.)

- *Space shuttle nose cones*: As the shuttle enters Earth's atmosphere, temperatures as high as 3092°F (1700°C) are experienced. Carbon–carbon composite is a material of choice for the nose cone because it has the lowest overall weight of all ablative* materials; high thermal conductivity to prevent surface cracking; high specific heat† to absorb large heat flux; and high thermal shock resistance to low temperatures in space of –238°F (–150°C) to 3092°F (1700°C) due to re-entry. Also, the carbon–carbon nose remains undamaged and can be reused many times.
- *Aircraft brakes*: The carbon–carbon brakes shown in Figure 1.34 cost \$440/lb (\$970/kg), which is several times more than their metallic counterpart; however, the high durability (two to four times that of steel), high specific heat (2.5 times that of steel), low braking distances and braking times (three-quarters that of berylium), and large weight savings of up to 990 lb (450 kg) on a commercial aircraft such as Airbus A300-B2K and A300-B4 are attractive.[29] As mentioned earlier, 1 lb (0.453 kg) weight savings on a full-service commercial aircraft can translate to fuel savings of about 360 gal/year (1360 L/year). Other advantages include reduced inventory due to longer endurance of carbon brakes.
- *Mechanical fasteners*: Fasteners needed for high temperature applications are made of carbon–carbon composites because they lose little strength at high temperatures.

* Ablative materials absorb heat through pyrolysis at or near the exposed surfaces.
† Specific heat is the amount of heat required to heat a unit mass of a substance through a unit temperature.

FIGURE 1.34
Sectioned carbon–carbon brake from Airbus A320. (From Savage, G., *Science and Business Media from Carbon–Carbon Composites*, Kluwer Academic Publishers, London, 1993, Figure 9.2, p. 325. Reproduced with kind permission of Springer–Verlag.)

1.3 Recycling Fiber-Reinforced Composites

What types of processes are used for recycling of composites?

The two main processes are called chemical and mechanical processes. Why is recycling of composites complex?

This is because of the many variables in material types — thermoset vs. thermoplastics, long vs. short fibers, glass vs. carbon, etc.

What are the various steps in mechanical recycling of short fiber-reinforced composites?

These are shredding, separation, washing, grinding, drying, and extrusion.

Where are mechanically recycled short-fiber composites used?

The recycled material is available in powder or fiber form. Powder form is reused as paste for sheet-molding compounds, and the fiber form is used for reinforcement in bulk-molding compounds. One cannot use too much of these as replacements because the impact resistance and electrical properties degrade after about 20% content. Products from recycled plastics are limited to fences and benches.

Why is chemical recycling not as popular as mechanical recycling?

Chemical processing is very costly. Processes such as pyrolysis (decomposing materials in an oxygen-free atmosphere) produce many gases, and hydrogenation gives high filler content. However, General Motors has adapted pyrolysis to recycle composite automobile parts. Gases and oils are recovered, and the residues are used as fillers in concrete and roof shingles.

One other problem is the chlorine content. The scrap needs to be dehalogenated after separation, especially if carbon fibers were used as reinforcement. Glass fibers in recycled composites also pose the problem of low compressive strength of the new material.

What can one do if the different types of composites cannot be separated?

Incineration or use as fuel may be the only solution because metals, thermosets, and thermoplastics may be mixed, and they may be soiled with toxic materials. The fuel value* of polymer matrix composites is around 5000 BTU/lb (11,622 kJ/kg). This is about half the value for coal.

Which chemical process shows the most promise?

Incineration offers the most promise. Its advantages include minimal cost, high-volume reduction, and no residual material. It is also feasible for low scrap volume.

1.4 Mechanics Terminology

How is a composite structure analyzed mechanically?

A composite material consists of two or more constituents; thus, the analysis and design of such materials is different from that for conventional materials such as metals. The approach to analyze the mechanical behavior of composite structures is as follows (Figure 1.35).

1. Find the average properties of a composite ply from the individual properties of the constituents. Properties include stiffness, strength, thermal, and moisture expansion coefficients. Note that average properties are derived by considering the ply to be homogeneous. At this level, one can optimize for the stiffness and strength requirements of a lamina. This is called the *micromechanics* of a lamina.

* Fuel value is the heat transferred when the products of complete combustion of a fuel are cooled to the initial temperature of air and fuel. Units of fuel value are Btu/lbm and J/kg. Typical fuel value for lignite coal is 7000 Btu/lbm.

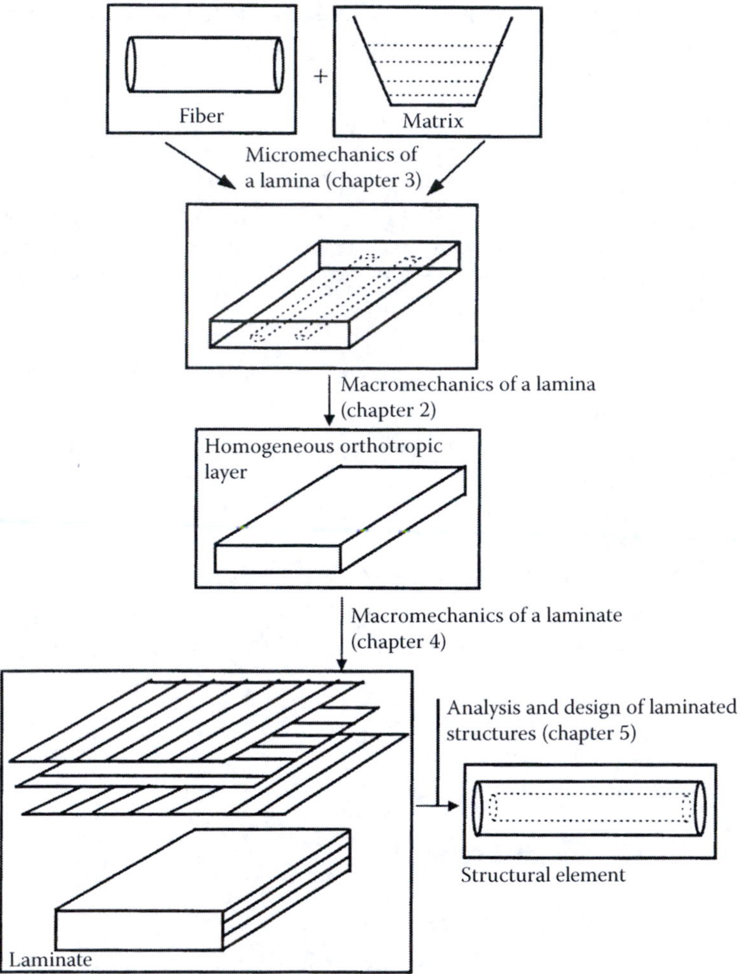

FIGURE 1.35
Schematic of analysis of laminated composites.

2. Develop the stress–strain relationships for a unidirectional/bidirectional lamina. Loads may be applied along the principal directions of symmetry of the lamina or off-axis. Also, one develops relationships for stiffness, thermal and moisture expansion coefficients, and strengths of angle plies. Failure theories of a lamina are based on stresses in the lamina and strength properties of a lamina. This is called the *macromechanics* of a lamina.

A structure made of composite materials is generally a laminate structure made of various laminas stacked on each other. Knowing the macromechanics of a single lamina, one develops the macromechanics of a laminate. Stiffness, strengths, and thermal and moisture expansion coefficients can be found for the whole laminate. Laminate failure is based on stresses and application of failure theories to each ply. This knowledge of analysis of composites can then eventually form the basis for the mechanical design of structures made of composites.

Several terms are defined to develop the fundamentals of the mechanical behavior of composites. These include the following.

What is an isotropic body?

An isotropic material has properties that are the same in all directions. For example, the Young's modulus of steel is the same in all directions.

What is a homogeneous body?

A homogeneous body has properties that are the same at all points in the body. A steel rod is an example of a homogeneous body. However, if one heats this rod at one end, the temperature at various points on the rod would be different. Because Young's modulus of steel varies with temperature, one no longer has a homogeneous body. The body is still isotropic because the properties at a particular point are still identical in all directions.

Are composite materials isotropic and/or homogeneous?

Most composite materials are neither isotropic nor homogeneous. For example, consider epoxy reinforced with long glass fibers. If one chooses a location on the glass fiber, the properties are different from a location on the epoxy matrix. This makes the composite material nonhomogeneous (not homogeneous). Also, the stiffness in the direction parallel to the fibers is higher than in the direction perpendicular to the fibers and thus the properties are not independent of the direction. This makes the composite material anisotropic (not isotropic).

What is an anisotropic material?

At a point in an anisotropic material, material properties are different in all directions.

What is a nonhomogeneous body?

A nonhomogeneous or inhomogeneous body has material properties that are a function of the position on the body.

What is a lamina?

A lamina (also called a ply or layer) is a single flat layer of unidirectional fibers or woven fibers arranged in a matrix.

What is a laminate?

A laminate is a stack of plies of composites. Each layer can be laid at various orientations and can be made up of different material systems.

What is a hybrid laminate?

Hybrid composites contain more than one fiber or one matrix system in a laminate. The main four types of hybrid laminates follow.

- *Interply hybrid laminates* contain plies made of two or more different composite systems. Examples include car bumpers made of glass/epoxy layers to provide torsional rigidity and graphite/epoxy to give stiffness. The combinations also lower the cost of the bumper.
- *Intraply hybrid composites* consist of two or more different fibers used in the same ply. Examples include golf clubs that use graphite and aramid fibers. Graphite fibers provide the torsional rigidity and the aramid fibers provide tensile strength and toughness.
- An *interply–intraply hybrid* consists of plies that have two or more different fibers in the same ply and distinct composite systems in more than one ply.
- *Resin hybrid laminates* combine two or more resins instead of combining two or more fibers in a laminate. Generally, one resin is flexible and the other one is rigid. Tests have proven that these resin hybrid laminates can increase shear and work of fracture properties by more than 50% over those of all-flexible or all-rigid resins.[31]

1.5 Summary

This chapter introduced advanced composite materials and enumerated the advantages and drawbacks of composite materials over monolithic materials. Fiber and matrix factors were discussed to understand their influence on mechanical properties of the composites. The classification of the composites based on the matrix materials — polymer, metal, and ceramics — was discussed. In addition, carbon–carbon composites were also examined. The manufacturing and mechanical properties and application of composites were described. Discussion also covered the recycling of composite materials as well as the terminology used in studying the mechanics of composite materials.

Key Terms

Composite
Advanced composite materials
Specific modulus
Specific strength

Material selection
Fiber factors
Matrix factors
Classification of composites
Polymer matrix composites
Resins
Prepregs
Thermosets
Thermoplastics
Autoclave
Resin transfer molding
Metal matrix composites
Diffusion bonding
Ceramic matrix composites
Carbon–carbon composites
Recycling
Isotropic body
Anisotropic body
Homogeneous body
Nonhomogeneous body
Lamina
Laminate
Hybrid laminate

Exercise Set

1.1 What is a composite?

1.2 Why did Israelites reinforce clay with straw?

1.3 Give a brief historical review of composites.

1.4 Give four examples of naturally found composites. What are the constituents of these natural composites?

1.5 Airbus A-300 saved 300 kg of mass by making the tailfin out of advanced composites. Estimate in gallons the amount of fuel saved per year.

1.6 Give the definitions and units of the following in the SI and USCS system of units:

Coefficient of thermal expansion

Coefficient of moisture expansion

Thermal conductivity

Young's modulus

Ultimate strength
Poisson's ratio
Specific modulus
Specific strength
Density
Specific gravity
Ductility
Fracture toughness
Specific heat

1.7 Draw the graphs for the ratios $E^{1/2}/\rho$ vs. $\sigma^{1/2}_{ult}/\rho$ for materials in Table 1.1.

1.8 A lamina consists of 100 fibers of 10-μm diameter. The fibers are 10 mm long. Find the interfacial area. What is the increase in the interfacial area if the diameter of the fiber is reduced to 5 μm and the total volume of fibers is kept constant?

1.9 Compare the flexibility of a 0.01-in. diameter steel wire to a 0.02-in. diameter aluminum wire. The Young's modulus of steel is 30 Msi and that of aluminum is 10 Msi.

1.10 What are the limitations of modern composites?

1.11 Enumerate six primary material selection parameters that are used in evaluating the use of a particular material.

1.12 How are composites classified?

1.13 Compare the specific modulus, specific strength, and coefficient of thermal expansion coefficient of a pitch based graphite fiber, Kevlar 49, and S-glass.

1.14 Describe one manufacturing method of polymer matrix composites other than those given in Chapter 1.

1.15 Why is epoxy the most popular resin?

1.16 Find ten applications of polymer matrix components other than those given in Chapter 1.

1.17 Give the advantages and drawbacks of metal matrix composites over polymer matrix composites.

1.18 Find three applications of metal matrix composites other than those given in Chapter 1.

1.19 Describe one manufacturing method of metal matrix composites other than given in Chapter 1.

1.20 Find three applications of ceramic matrix composites other than those given in Chapter 1.

1.21 Describe one manufacturing method of ceramic matrix composites other than those given in Chapter 1.

1.22 Find three applications of carbon matrix composites other than those given in Chapter 1.

1.23 Describe one manufacturing method of carbon matrix composites other than those given in Chapter 1.

1.24 Give the upper limit of operating temperatures of polymer, metal, ceramic, and carbon matrix composites.

1.25 Define the following:

Isotropic body

Homogeneous body

Anisotropic body

Nonhomogeneous body

Micromechanics

Macromechanics

Lamina

Laminate

1.26 Give an example of a:

Homogeneous body that is not isotropic

Nonhomogeneous body that is isotropic

1.27 Do all properties of composites always improve over their individual constituents? Give examples.

1.28 How are hybrid composites classified?

References

1. Mack, J., Advanced polymer composites, *Mater. Edge*, 18, January 1988.
2. Meetham, G.W., Design considerations for aerospace applications in *Handbook of Polymer–Fiber Composites*, Jones, F.R., Ed, Longman Scientific and Technical, Essex, England, Chap. 5, 1994.
3. Eager, T.W., Whither advanced materials? *Adv. Mater. Processes*, 25, June 1991.
4. Ashby, M.F., On engineering properties of materials, *Acta Metallurgica*, 37, 1273, 1989.
5. Buchanan, G.R., *Mechanics of Materials*, HRW Inc., New York, 1988.
6. Lamotte, E. De and Perry, A.J., Diameter and strain-rate dependence of the ultimate tensile strength and Young's modulus of carbon fiber, *Fiber Sci. Technol.*, 3, 159, 1970.
7. Schwartz, M.M., *Composite Materials Handbook*, McGraw–Hill, New York, 1984.
8. Polymer nanocomposites for packaging applications, see http://www.natick.army.mil/soldier/media/fact/food/PolyNano.htm, last accessed August 31, 2004.
9. GM, GMability advanced technology GM to use nanocomposites on highest volume car, see http://www.gm.com/company/gmability/adv_tech/100_news/nanocomposites_012704.html, last accessed August 31, 2004.

10. Allen, L., A limber future, *Popular Sci.*, 36, August 2004.
11. Partridge, I.K., *Advanced Composites*, Elsevier Applied Science, New York, 1989.
12. Cooks, G., Composite resins for the 90s, *M.C. Gill Doorway*, 7, 27, Spring, 1990.
13. Hergenrother, P.M. and Johnston, N.J., *Polymer Mater. Sci. Eng. Proc.*, 59, 697, 1988.
14. Mallick, P.K., *Fiber-Reinforced Composites Materials, Manufacturing, and Design*, Marcell Dekker, Inc., New York, Chap. 2, 1988.
15. McConnell, V.P. and Stover, D., Advanced composites: performance materials of choice for innovative products, *High-Performance Composites Sourcebook*, 8, November 1995.
16. Composite structures, Alliant Techsystems, Magna, UT.
17. Lepré, R.A., personal communication, 1995.
18. Scaled composites, see http://www.scaled.com/, last accessed September 21, 2004.
19. How stuff works, see http://science.howstuffworks.com, last accessed September 21, 2004.
20. Sun, C.T. and Lu, Y.P., *Vibration Damping of Structural Elements*, Prentice Hall, Englewood Cliffs, NJ, 1995.
21. Ashley, S., Bridging the cost gap with composites, *Mech. Eng.*, 118, 76, 1996.
22. Bursel, J.S., Composite Springs, Inc. technical bulletin, October 1990, St. Petersburg, FL.
23. Neil, D., Our driving conundrum, *Popular Sci.*, 62, September 2004.
24. Clean handling, *Mech. Eng.*, 26–27, September 2004.
25. Fiskars introduces two new PowerGear pruners, press release, April 1, 2004, see http://www.fiskars.com/en_US/press_release.do?num=8&res=6, last accessed September 28, 2004.
26. Matthews, F.L. and Rawlings, R.D., *Composites, Materials, Engineering and Science*, Springer–Verlag, New York, Chap. 3, 1987.
27. Niskanen, P. and Mohn, W.R., Versatile metal-matrix composites, *Adv. Mat. Processes*, 3, 39, 1988.
28. Chawla, K.K., *Ceramic Matrix Composites*, Chapman & Hall, London, Chap. 4, 1993.
29. Klein, A.J., Carbon/carbon composites, *Adv. Mater. Processes*, 64, November 1986.
30. Strife, J.R. and Sheehan, J.E., Ceramic coatings for carbon–carbon composites, *Ceramic Bull.*, 67, 369, 1988.
31. Sheppard, L.M., On the road with composites, *Adv. Mater. Processes*, 36, December 1986.

General References

Chung, D.D.L., *Carbon Fiber Composites*, Butterworth–Heinemann, Boston, 1994.

Gill, R.M., *Carbon Fibers in Composite Materials*, Butterworth and Co., London, 1972.

Geier, M.H., *Quality Handbook for Composite Materials*, Chapman & Hall, London, 1994.

Holloway, L., *Polymer Composites for Civil and Structural Engineering*, Blackie Academic and Professional, London, 1993.

Jones, F.R. Ed., *Handbook of Polymer–Fiber Composites*, Longman Scientific and Technical, Essex, 1994.

Lubin, G., 1982, *Handbook of Advanced Composites*, 2nd ed., Van Nostrand Reinhold, New York.

Phillips, L.N., Ed., *Design with Advanced Composite Materials*, Springer–Verlag, New York, 1989.

Powell, P.C., *Engineering with Fiber–Polymer Laminates*, Chapman & Hall, London, 1994.

Savage, G., *Carbon–Carbon Composites*, Chapman & Hall, London, 1993.

Vinson, J.R. and Chou, T., *Composite Materials and Their Use in Structures*, John Wiley & Sons, New York, 1975.

Vinson, J.R. and Sierakowski, R.L., *The Behavior of Structures Composed of Composite Materials*, Martinus Nighoff Publishers, Dordrecht, 1986.

Video References

Advanced Composites in Manufacturing, Society of Manufacturing Engineering, Dearborn, MI, 1986.

New Materials, Films for the Humanities and Sciences, 1989.

The Light Stuff, Coronet Film and Video, Northbrook, IL, 1988.

Tooling for Composites, Society of Manufacturing Engineers, Dearborn, MI, 1989.

2

Macromechanical Analysis of a Lamina

Chapter Objectives

- Review definitions of stress, strain, elastic moduli, and strain energy.
- Develop stress–strain relationships for different types of materials.
- Develop stress–strain relationships for a unidirectional/bidirectional lamina.
- Find the engineering constants of a unidirectional/bidirectional lamina in terms of the stiffness and compliance parameters of the lamina.
- Develop stress–strain relationships, elastic moduli, strengths, and thermal and moisture expansion coefficients of an angle ply based on those of a unidirectional/bidirectional lamina and the angle of the ply.

2.1 Introduction

A lamina is a thin layer of a composite material that is generally of a thickness on the order of 0.005 in. (0.125 mm). A laminate is constructed by stacking a number of such laminae in the direction of the lamina thickness (Figure 2.1). Mechanical structures made of these laminates, such as a leaf spring suspension system in an automobile, are subjected to various loads, such as bending and twisting. The design and analysis of such laminated structures demands knowledge of the stresses and strains in the laminate. Also, design tools, such as failure theories, stiffness models, and optimization algorithms, need the values of these laminate stresses and strains.

However, the building blocks of a laminate are single lamina, so understanding the mechanical analysis of a lamina precedes understanding that of a laminate. A lamina is unlike an isotropic homogeneous material. For example, if the lamina is made of isotropic homogeneous fibers and an

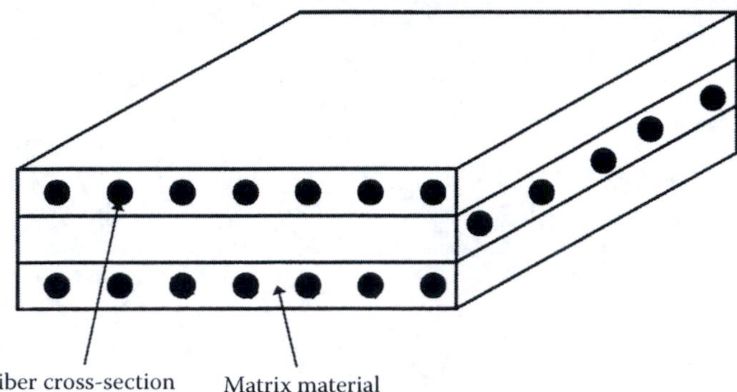

FIGURE 2.1
Typical laminate made of three laminae.

isotropic homogeneous matrix, the stiffness of the lamina varies from point to point depending on whether the point is in the fiber, the matrix, or the fiber–matrix interface. Accounting for these variations will make any kind of mechanical modeling of the lamina very complicated. For this reason, the macromechanical analysis of a lamina is based on average properties and considering the lamina to be homogeneous. Methods to find these average properties based on the individual mechanical properties of the fiber and the matrix, as well as the content, packing geometry, and shape of fibers are discussed in Chapter 3.

Even with the homogenization of a lamina, the mechanical behavior is still different from that of a homogeneous isotropic material. For example, take a square plate of length and width w and thickness t out of a large isotropic plate of thickness t (Figure 2.2) and conduct the following experiments.

Case A: Subject the square plate to a pure normal load P in direction 1. Measure the normal deformations in directions 1 and 2, δ_{1A} and δ_{2A}, respectively.

Case B: Apply the same pure normal load P as in case A, but now in direction 2. Measure the normal deformations in directions 1 and 2, δ_{1B} and δ_{2B}, respectively.

Note that

$$\delta_{1A} = \delta_{2B}\,,$$
$$\delta_{2A} = \delta_{1B}\,. \tag{2.1a,b}$$

However, taking a unidirectional square plate (Figure 2.3) of the same dimensions $w \times w \times t$ out of a large composite lamina of thickness t and conducting the same case A and B experiments, note that the deformations

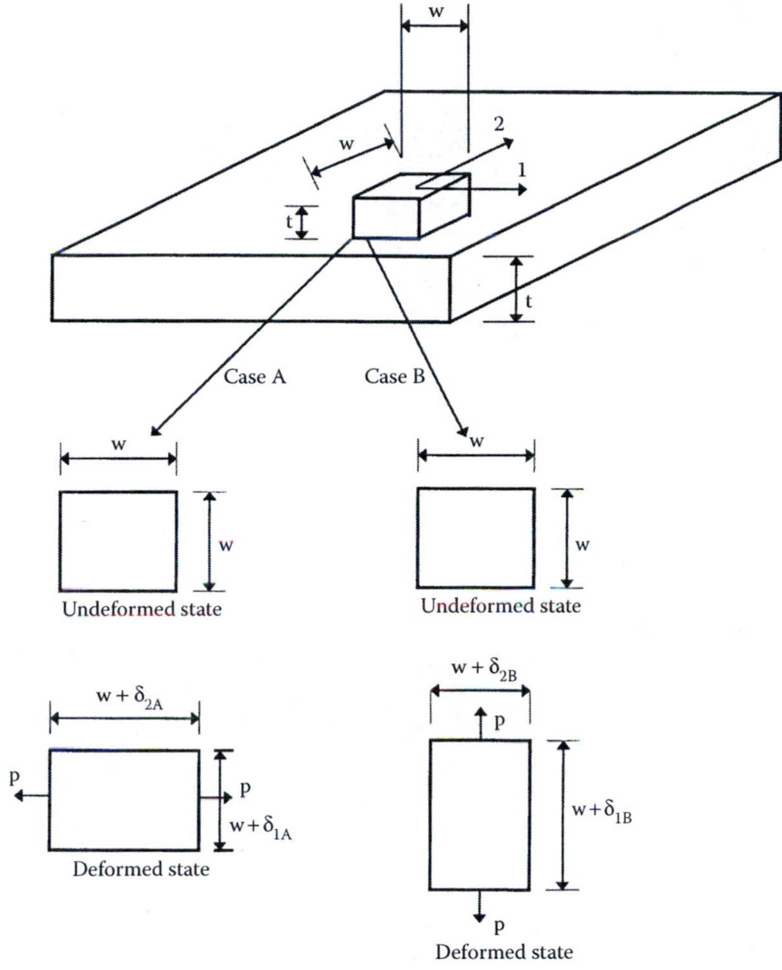

FIGURE 2.2
Deformation of square plate taken from an isotropic plate under normal loads.

$$\delta_{1A} \neq \delta_{2B}\,,$$
$$\delta_{2A} \neq \delta_{1B}\,. \tag{2.2a,b}$$

because the stiffness of the unidirectional lamina in the direction of fibers is much larger than the stiffness in the direction perpendicular to the fibers. Thus, the mechanical characterization of a unidirectional lamina will require more parameters than it will for an isotropic lamina.

Also, note that if the square plate (Figure 2.4) taken out of the lamina has fibers at an angle to the sides of the square plate, the deformations will be different for different angles. In fact, the square plate would not only have

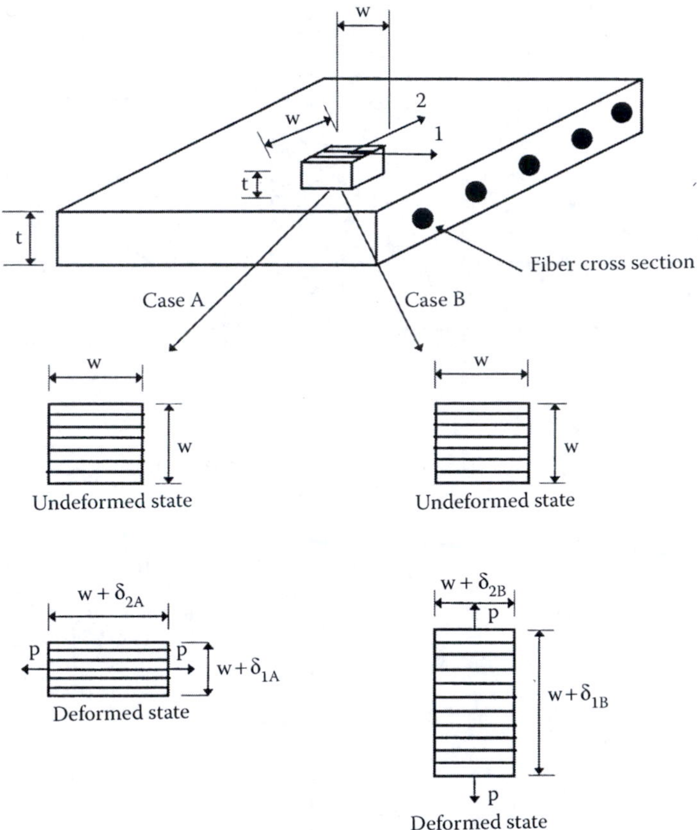

FIGURE 2.3
Deformation of a square plate taken from a unidirectional lamina with fibers at zero angle under normal loads.

deformations in the normal directions but would also distort. This suggests that the mechanical characterization of an angle lamina is further complicated.

Mechanical characterization of materials generally requires costly and time-consuming experimentation and/or theoretical modeling. Therefore, the goal is to find the minimum number of parameters required for the mechanical characterization of a lamina.

Also, a composite laminate may be subjected to a temperature change and may absorb moisture during processing and operation. These changes in temperature and moisture result in residual stresses and strains in the laminate. The calculation of these stresses and strains in a laminate depends on the response of each lamina to these two environmental parameters. In this chapter, the stress–strain relationships based on temperature change and moisture content will also be developed for a single lamina. The effects of temperature and moisture on a laminate are discussed later in Chapter 4.

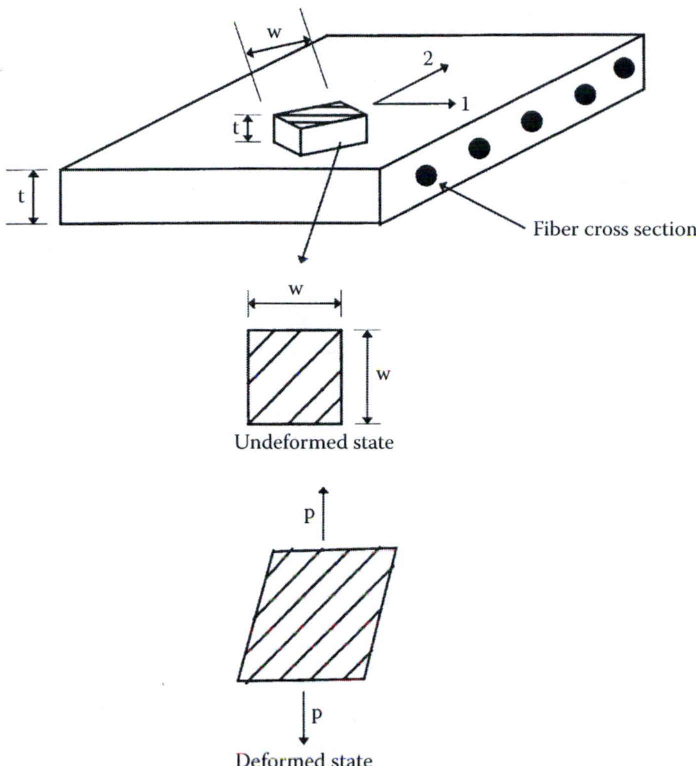

FIGURE 2.4
Deformation of a square plate taken from a unidirectional lamina with fibers at an angle under normal loads.

2.2 Review of Definitions

2.2.1 Stress

A mechanical structure takes external forces, which act upon a body as surface forces (for example, bending a stick) and body forces (for example, the weight of a standing vertical telephone pole on itself). These forces result in internal forces inside the body. Knowledge of the internal forces at all points in the body is essential because these forces need to be less than the strength of the material used in the structure. Stress, which is defined as the intensity of the load per unit area, determines this knowledge because the strengths of a material are intrinsically known in terms of stress.

Imagine a body (Figure 2.5) in equilibrium under various loads. If the body is cut at a cross-section, forces will need to be applied on the cross-sectional area so that it maintains equilibrium as in the original body. At any cross-

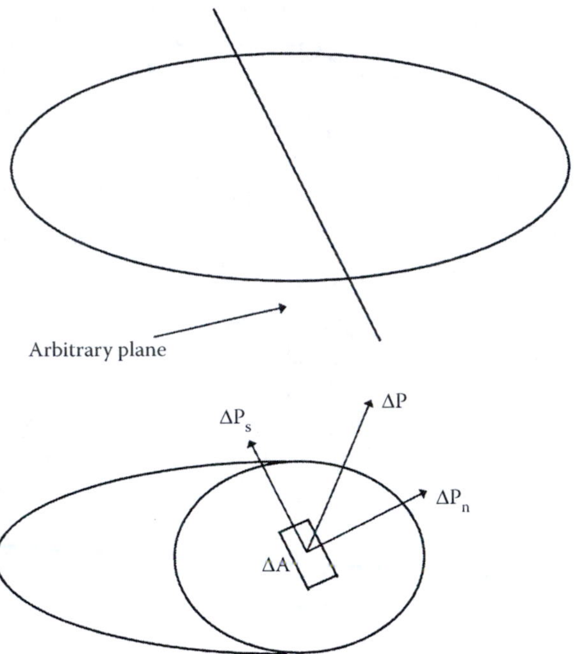

FIGURE 2.5
Stresses on an infinitesimal area on an arbitrary plane.

section, a force ΔP is acting on an area of ΔA. This force vector has a component normal to the surface, ΔP_n, and one parallel to the surface, ΔP_s. The definition of stress then gives

$$\sigma_n = \lim_{\Delta A \to 0} \frac{\Delta P_n}{\Delta A},$$

$$\tau_s = \lim_{\Delta A \to 0} \frac{\Delta P_s}{\Delta A}. \tag{2.3a,b}$$

The component of the stress normal to the surface, σ_n, is called the normal stress and the stress parallel to the surface, τ_s, is called the shear stress. If one takes a different cross-section through the same point, the stress remains unchanged but the two components of stress, normal stress, σ_n, and shear stress, τ_s, will change. However, it has been proved that a complete definition of stress at a point only needs use of any three mutually orthogonal coordinate systems, such as a Cartesian coordinate system.

Take the right-hand coordinate system x–y–z. Take a cross-section parallel to the yz-plane in the body as shown in Figure 2.6. The force vector ΔP acts

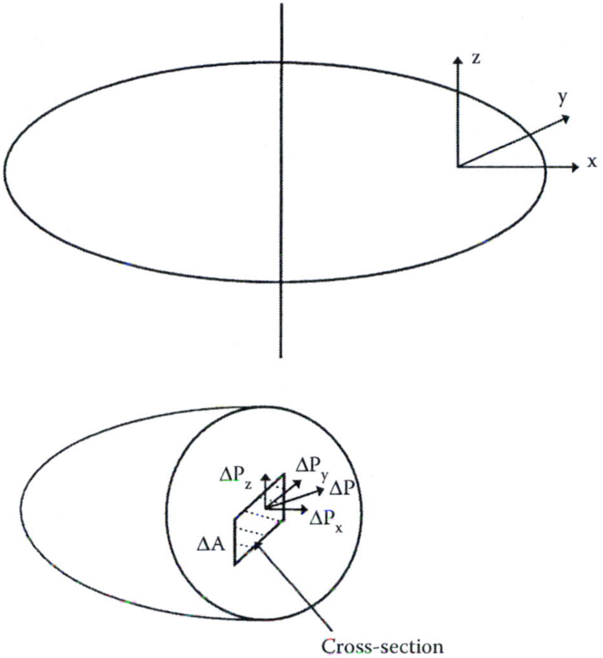

FIGURE 2.6
Forces on an infinitesimal area on the y–z plane.

on an area ΔA. The component ΔP_x is normal to the surface. The force vector ΔP_s is parallel to the surface and can be further resolved into components along the y and z axes: ΔP_y and ΔP_z. The definition of the various stresses then is

$$\sigma_x = \lim_{\Delta A \to 0} \frac{\Delta P_x}{\Delta A}$$

$$\tau_{xy} = \lim_{\Delta A \to 0} \frac{\Delta P_y}{\Delta A},$$

$$\tau_{xz} = \lim_{\Delta A \to 0} \frac{\Delta P_z}{\Delta A}. \quad (2.4a\text{–}c)$$

Similarly, stresses can be defined for cross-sections parallel to the xy and xz planes. For defining all these stresses, the stress at a point is defined generally by taking an infinitesimal cuboid in a right-hand coordinate system

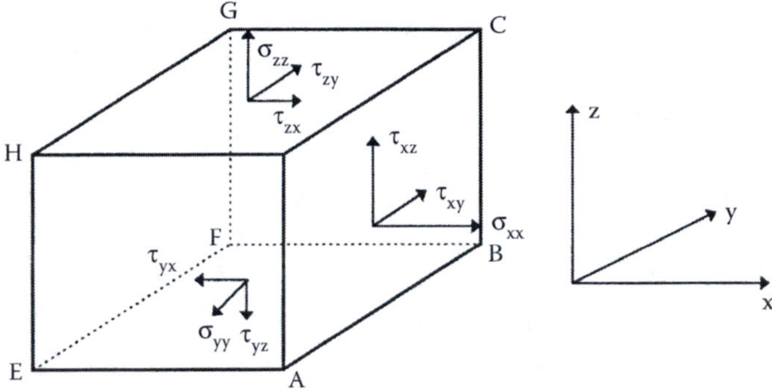

FIGURE 2.7
Stresses on an infinitesimal cuboid.

and finding the stresses on each of its faces. Nine different stresses act at a point in the body as shown in Figure 2.7. The six shear stresses are related as

$$\tau_{xy} = \tau_{yx} ,$$

$$\tau_{yz} = \tau_{zy} ,$$

$$\tau_{zx} = \tau_{xz} . \tag{2.5a–c}$$

The preceding three relations are found by equilibrium of moments of the infinitesimal cube. There are thus six independent stresses. The stresses σ_x, σ_y, and σ_z are normal to the surfaces of the cuboid and the stresses τ_{yz}, τ_{zx}, and τ_{xy} are along the surfaces of the cuboid.

A tensile normal stress is positive, and a compressive normal stress is negative. A shear stress is positive, if its direction and the direction of the normal to the face on which it is acting are both in positive or negative direction; otherwise, the shear stress is negative.

2.2.2 Strain

Similar to the need for knowledge of forces inside a body, knowing the deformations because of the external forces is also important. For example, a piston in an internal combustion engine may not develop larger stresses than the failure strengths, but its excessive deformation may seize the engine. Also, finding stresses in a body generally requires finding deformations. This is because a stress state at a point has six components, but there are only three force-equilibrium equations (one in each direction).

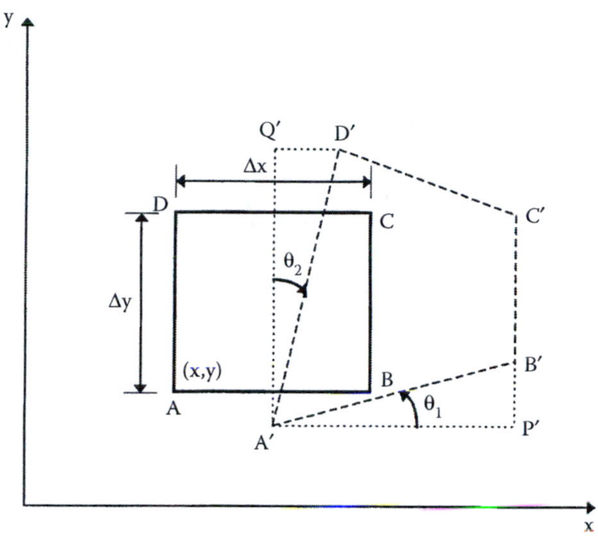

FIGURE 2.8
Normal and shearing strains on an infinitesimal area in the x–y plane.

The knowledge of deformations is specified in terms of strains — that is, the relative change in the size and shape of the body. The strain at a point is also defined generally on an infinitesimal cuboid in a right-hand coordinate system. Under loads, the lengths of the sides of the infinitesimal cuboid change. The faces of the cube also get distorted. The change in length corresponds to a normal strain and the distortion corresponds to the shearing strain. Figure 2.8 shows the strains on one of the faces, *ABCD*, of the cuboid.

The strains and displacements are related to each other. Take the two perpendicular lines *AB* and *AD*. When the body is loaded, the two lines become *A′B′* and *A′D′*. Define the displacements of a point (x,y,z) as

$u = u(x,y,z)$ = displacement in x-direction at point (x,y,z)

$v = v(x,y,z)$ = displacement in y-direction at point (x,y,z)

$w = w(x,y,z)$ = displacement in z-direction at point (x,y,z)

The normal strain in the x-direction, ε_x, is defined as the change of length of line *AB* per unit length of *AB* as

$$\varepsilon_x = \lim_{AB \to 0} \frac{A'B' - AB}{AB}, \tag{2.6}$$

where

$$A'B' = \sqrt{(A'P')^2 + (B'P')^2},$$

$$= \sqrt{[\Delta x + u(x+\Delta x, y) - u(x,y)]^2 + [v(x+\Delta x, y) - v(x,y)]^2},$$

$$AB = \Delta x. \tag{2.7a,b}$$

Substituting the preceding expressions of Equation (2.7) in Equation (2.6),

$$\varepsilon_x = \lim_{\Delta x \to 0} \left\{ \left[1 + \frac{u(x+\Delta x, y) - u(x,y)}{\Delta x} \right]^2 + \left[\frac{v(x+\Delta x, y) - v(x,y)}{\Delta x} \right]^2 \right\}^{1/2} - 1 .$$

Using definitions of partial derivatives

$$\varepsilon_x = \left[\left(1 + \frac{\partial u}{\partial x} \right)^2 + \left(\frac{\partial v}{\partial x} \right)^2 \right]^{1/2} - 1$$

$$\varepsilon_x = \frac{\partial u}{\partial x}, \tag{2.8}$$

because

$$\frac{\partial u}{\partial x} << 1 ,$$

$$\frac{\partial v}{\partial x} << 1 ,$$

for small displacements.

The normal strain in the y-direction, ε_y is defined as the change in the length of line AD per unit length of AD as

$$\varepsilon_y = \lim_{AD \to 0} \frac{A'D' - AD}{AD}, \tag{2.9}$$

where

$$A'D' = \sqrt{(A'Q')^2 + (Q'D')^2},$$

$$A'D' = \sqrt{[\Delta y + v(x, y + \Delta y) - v(x, y)]^2 + [u(x, y + \Delta y) - u(x, y)]^2},$$

$$AD = \Delta y. \tag{2.10a,b}$$

Substituting the preceding expressions of Equation (2.10) in Equation (2.9),

$$\varepsilon_y = \lim_{\Delta y \to 0} \left\{ \left[1 + \frac{v(x, y + \Delta y) - v(x, y)\}}{\Delta y} \right]^2 + \left[\frac{u(x, y + \Delta y) - u(x, y)}{\Delta y} \right]^2 \right\}^{1/2} - 1 .$$

Using definitions of partial derivatives,

$$\varepsilon_y = \left[\left(1 + \frac{\partial v}{\partial y} \right)^2 + \left(\frac{\partial u}{\partial y} \right)^2 \right]^{1/2} - 1$$

$$\varepsilon_y = \frac{\partial v}{\partial y}, \tag{2.11}$$

because

$$\frac{\partial u}{\partial y} << 1,$$

$$\frac{\partial v}{\partial y} << 1,$$

for small displacements.

A normal strain is positive if the corresponding length increases; a normal strain is negative if the corresponding length decreases.

The shearing strain in the x–y plane, γ_{xy} is defined as the change in the angle between sides *AB* and *AD* from 90°. This angular change takes place by the inclining of sides *AB* and *AD*. The shearing strain is thus defined as

$$\gamma_{xy} = \theta_1 + \theta_2, \tag{2.12}$$

where

$$q_1 = \lim_{AB \to 0} \frac{P'B'}{A'P'},$$

$$P'B' = v(x + \Delta x, y) - v(x, y),$$

$$A'P' = u(x + \Delta x, y) + \Delta x - u(x, y), \tag{2.13a–c}$$

$$\theta_2 = \lim_{AD \to 0} \frac{Q'D'}{A'Q'},$$

$$Q'D' = u(x, y + \Delta y) - u(x, y),$$

$$A'Q' = v(x, y + \Delta y) + \Delta y - v(x, y). \tag{2.14a–c}$$

Substituting Equation (2.13) and Equation (2.14) in Equation (2.12),

$$\gamma_{xy} = \lim_{\substack{\Delta x \to 0 \\ \Delta y \to 0}} \frac{\dfrac{v(x + \Delta x, y) - v(x, y)}{\Delta x}}{\dfrac{u(x + \Delta x, y) + \Delta x - u(x, y)}{\Delta x}} + \frac{\dfrac{u(x, y + \Delta y) - u(x, y)}{\Delta y}}{\dfrac{v(x, y + \Delta y) + \Delta y - v(x, y)}{\Delta y}}$$

$$= \frac{\dfrac{\partial v}{\partial x}}{1 + \dfrac{\partial u}{\partial x}} + \frac{\dfrac{\partial u}{\partial y}}{1 + \dfrac{\partial u}{\partial y}}$$

$$= \frac{\partial v}{\partial x} + \frac{\partial u}{\partial y}, \tag{2.15}$$

because

$$\frac{\partial u}{\partial x} << 1,$$

$$\frac{\partial v}{\partial y} << 1,$$

for small displacements.

The shearing strain is positive when the angle between the sides *AD* and *AB* decreases; otherwise, the shearing strain is negative.

The definitions of the remaining normal and shearing strains can be found by noting the change in size and shape of the other sides of the infinitesimal cuboid in Figure 2.7 as

$$\gamma_{yz} = \frac{\partial v}{\partial z} + \frac{\partial w}{\partial y},$$

$$\gamma_{zx} = \frac{\partial w}{\partial x} + \frac{\partial u}{\partial z},$$

$$\varepsilon_z = \frac{\partial w}{\partial z}. \qquad (2.16a\text{–}c)$$

Example 2.1

A displacement field in a body is given by

$u = 10^{-5}(x^2 + 6y + 7xy)$
$v = 10^{-5}(yz)$
$w = 10^{-5}(xy + yz^2)$

Find the state of strain at $(x,y,z) = (1,2,3)$.

Solution

From Equation (2.8),

$$\epsilon_x = \frac{\partial u}{\partial x}$$

$$= \frac{\partial}{\partial x}\left(10^{-5}\left(x^2 + 6y + 7xz\right)\right)$$

$$= 10^{-5}\left(2x + 7z\right)$$

$$= 10^{-5}\left(2 \times 1 + 7 \times 3\right)$$

$$= 2.300 \times 10^{-4}.$$

From Equation (2.11),

$$\epsilon_y = \frac{\partial v}{\partial y}$$

$$= \frac{\partial}{\partial y}\left(10^{-5}\left(yz\right)\right)$$

$$= 10^{-5}\left(z\right)$$

$$= 10^{-5}\left(3\right)$$

$$= 3.000 \times 10^{-5} .$$

From Equation (2.16c),

$$\epsilon_z = \frac{\partial w}{\partial z}$$

$$= \frac{\partial}{\partial z}\left(10^{-5}\left(xy + yz^2\right)\right)$$

$$= 10^{-5}\left(2yz\right)$$

$$= 10^{-5}(2 \times 2 \times 3)$$

$$= 1.2 \times 10^{-4} .$$

From Equation (2.15),

$$\gamma_{xy} = \frac{\partial u}{\partial y} + \frac{\partial v}{\partial x}$$

$$= \frac{\partial}{\partial y}\left(10^{-5}\left(x^2 + 6y + 7xz\right)\right) + \frac{\partial}{\partial x}\left(10^{-5}\left(yz\right)\right)$$

$$= 10^{-5}(6) + 10^{-5}(0)$$

$$= 6.000 \times 10^{-5}\ .$$

From Equation (2.16a),

$$\gamma_{yz} = \frac{\partial v}{\partial z} + \frac{\partial w}{\partial y}$$

$$= \frac{\partial}{\partial z}\left(10^{-5}(yz)\right) + \frac{\partial}{\partial y}\left(10^{-5}(xy + yz^2)\right)$$

$$= 10^{-5}(y) + 10^{-5}(x + z^2)$$

$$= 10^{-5}(2) + 10^{-5}(1 + 3^2)$$

$$= 1.2 \times 10^{-4}\ .$$

From Equation (2.16b),

$$\gamma_{zx} = \frac{\partial w}{\partial x} + \frac{\partial u}{\partial z}$$

$$= \frac{\partial}{\partial x}\left(10^{-5}(xy + yz^2)\right) + \frac{\partial}{\partial z}\left(10^{-5}(x^2 + 6y + 7xz)\right)$$

$$= 10^{-5}(y) + 10^{-5}(7x)$$

$$= 10^{-5}(2) + 10^{-5}(7 \times 1)$$

$$= 9.000 \times 10^{-5}\ .$$

2.2.3 Elastic Moduli

As mentioned in Section 2.2.2, three equilibrium equations are insufficient for defining all six stress components at a point. For a body that is linearly

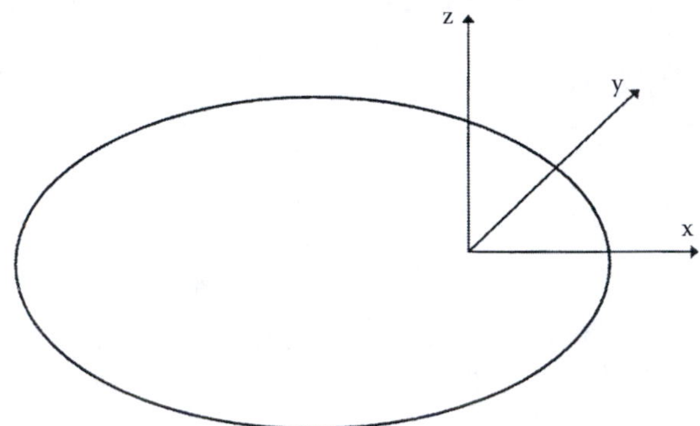

FIGURE 2.9
Cartesian coordinates in a three-dimensional body.

elastic and has small deformations, stresses and strains at a point are related through six simultaneous linear equations called Hooke's law. Note that 15 unknown parameters are at a point: six stresses, six strains, and three displacements. Combined with six simultaneous linear equations of Hooke's law, six strain-displacement relations — given by Equation (2.8), Equation (2.11), Equation (2.15), and Equation (2.16) — and three equilibrium equations give 15 equations for the solution of 15 unknowns.[1] Because strain-displacement and equilibrium equations are differential equations, they are subject to knowing boundary conditions for complete solutions.

For a linear isotropic material in a three-dimensional stress state, the Hooke's law stress–strain relationships at a point in an x–y–z orthogonal system (Figure 2.9) in matrix form are

$$\begin{bmatrix} \varepsilon_x \\ \varepsilon_y \\ \varepsilon_z \\ \gamma_{yz} \\ \gamma_{zx} \\ \gamma_{xy} \end{bmatrix} = \begin{bmatrix} \frac{1}{E} & -\frac{\nu}{E} & -\frac{\nu}{E} & 0 & 0 & 0 \\ -\frac{\nu}{E} & \frac{1}{E} & -\frac{\nu}{E} & 0 & 0 & 0 \\ -\frac{\nu}{E} & -\frac{\nu}{E} & \frac{1}{E} & 0 & 0 & 0 \\ 0 & 0 & 0 & \frac{1}{G} & 0 & 0 \\ 0 & 0 & 0 & 0 & \frac{1}{G} & 0 \\ 0 & 0 & 0 & 0 & 0 & \frac{1}{G} \end{bmatrix} \begin{bmatrix} \sigma_x \\ \sigma_y \\ \sigma_z \\ \tau_{yz} \\ \tau_{zx} \\ \tau_{xy} \end{bmatrix}, \tag{2.17}$$

$$\begin{bmatrix} \sigma_x \\ \sigma_y \\ \sigma_z \\ \tau_{yz} \\ \tau_{zx} \\ \tau_{xy} \end{bmatrix} = \begin{bmatrix} \frac{E(1-\nu)}{(1-2\nu)(1+\nu)} & \frac{\nu E}{(1-2\nu)(1+\nu)} & \frac{\nu E}{(1-2\nu)(1+\nu)} & 0 & 0 & 0 \\ \frac{\nu E}{(1-2\nu)(1+\nu)} & \frac{E(1-\nu)}{(1-2\nu)(1+\nu)} & \frac{\nu E}{(1-2\nu)(1+\nu)} & 0 & 0 & 0 \\ \frac{\nu E}{(1-2\nu)(1+\nu)} & \frac{\nu E}{(1-2\nu)(1+\nu)} & \frac{E(1-\nu)}{(1-2\nu)(1+\nu)} & 0 & 0 & 0 \\ 0 & 0 & 0 & G & 0 & 0 \\ 0 & 0 & 0 & 0 & G & 0 \\ 0 & 0 & 0 & 0 & 0 & G \end{bmatrix} \begin{bmatrix} \varepsilon_x \\ \varepsilon_y \\ \varepsilon_z \\ \gamma_{yz} \\ \gamma_{zx} \\ \gamma_{xy} \end{bmatrix}, \tag{2.18}$$

where ν is the Poisson's ratio. The shear modulus G is a function of two elastic constants, E and ν, as

$$G = \frac{E}{2(1+\nu)}. \tag{2.19}$$

The 6 × 6 matrix in Equation (2.17) is called the compliance matrix [S] of an isotropic material. The 6 × 6 matrix in Equation (2.18), obtained by inverting the compliance matrix in Equation (2.17), is called the stiffness matrix [C] of an isotropic material.

2.2.4 Strain Energy

Energy is defined as the capacity to do work. In solid, deformable, elastic bodies under loads, the work done by external loads is stored as recoverable strain energy. The strain energy stored in the body per unit volume is then defined as

$$W = \frac{1}{2}(\sigma_x \varepsilon_x + \sigma_y \varepsilon_y + \sigma_z \varepsilon_z + \tau_{xy}\gamma_{xy} + \tau_{yz}\gamma_{yz} + \tau_{zx}\gamma_{zx}). \tag{2.20}$$

Example 2.2

Consider a bar of cross-section A and length L (Figure 2.10). A uniform tensile load P is applied to the two ends of the rod; find the state of stress and strain, and strain energy per unit volume of the body. Assume that the rod is made of a homogeneous isotropic material of Young's modulus, E.

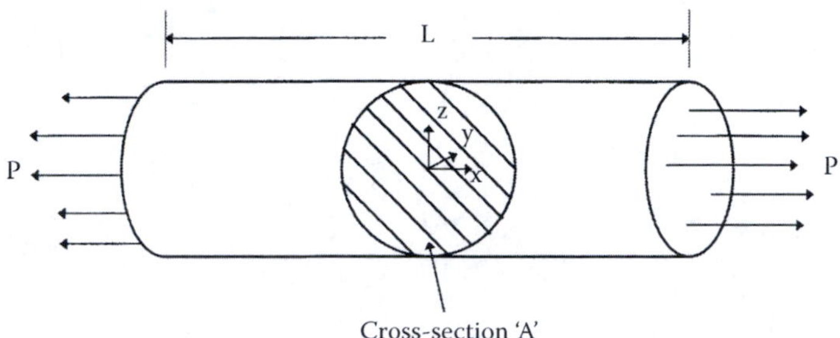

FIGURE 2.10
Cylindrical rod under uniform uniaxial load, *P*.

Solution

The stress state at any point is given by

$$\sigma_x = \frac{P}{A}, \sigma_y = 0, \sigma_z = 0, \tau_{yz} = 0, \tau_{zx} = 0, \tau_{xy} = 0. \tag{2.21}$$

If the circular rod is made of an isotropic, homogeneous, and linearly elastic material, then the stress–strain at any point is related as

$$\begin{bmatrix} \varepsilon_x \\ \varepsilon_y \\ \varepsilon_z \\ \gamma_{yz} \\ \gamma_{zx} \\ \gamma_{xy} \end{bmatrix} = \begin{bmatrix} \frac{1}{E} & -\frac{\nu}{E} & -\frac{\nu}{E} & 0 & 0 & 0 \\ -\frac{\nu}{E} & \frac{1}{E} & -\frac{\nu}{E} & 0 & 0 & 0 \\ -\frac{\nu}{E} & -\frac{\nu}{E} & \frac{1}{E} & 0 & 0 & 0 \\ 0 & 0 & 0 & \frac{1}{G} & 0 & 0 \\ 0 & 0 & 0 & 0 & \frac{1}{G} & 0 \\ 0 & 0 & 0 & 0 & 0 & \frac{1}{G} \end{bmatrix} \begin{bmatrix} \frac{P}{A} \\ 0 \\ 0 \\ 0 \\ 0 \\ 0 \end{bmatrix}, \tag{2.22}$$

$$\begin{gathered} \varepsilon_x = \frac{P}{AE}, \varepsilon_y = -\frac{\nu P}{AE}, \varepsilon_z = -\frac{\nu P}{AE}, \\ \gamma_{yz} = 0, \gamma_{zx} = 0, \gamma_{xy} = 0. \end{gathered} \tag{2.23}$$

The strain energy stored per unit volume in the rod, per Equation (2.20), is

$$W=\frac{1}{2}\left[\left(\frac{P}{A}\right)\left(\frac{P}{AE}\right)+(0)\left(-\frac{\nu P}{AE}\right)+(0)\left(-\frac{\nu P}{AE}\right)+(0)(0)+(0)(0)+(0)(0)\right]$$

$$=\frac{1}{2}\frac{P^2}{A^2E}$$

$$=\frac{1}{2}\frac{\sigma_x^2}{E}. \tag{2.24}$$

2.3 Hooke's Law for Different Types of Materials

The stress–strain relationship for a general material that is not linearly elastic and isotropic is more complicated than Equation (2.17) and Equation (2.18). Assuming linear and elastic behavior for a composite is acceptable; however, assuming it to be isotropic is generally unacceptable. Thus, the stress–strain relationships follow Hooke's law, but the constants relating stress and strain are more in number than seen in Equation (2.17) and Equation (2.18). The most general stress–strain relationship is given as follows for a three-dimensional body in a 1–2–3 orthogonal Cartesian coordinate system:

$$\begin{bmatrix}\sigma_1\\\sigma_2\\\sigma_3\\\tau_{23}\\\tau_{31}\\\tau_{12}\end{bmatrix}=\begin{bmatrix}C_{11}&C_{12}&C_{13}&C_{14}&C_{15}&C_{16}\\C_{21}&C_{22}&C_{23}&C_{24}&C_{25}&C_{26}\\C_{31}&C_{32}&C_{33}&C_{34}&C_{35}&C_{36}\\C_{41}&C_{42}&C_{43}&C_{44}&C_{45}&C_{46}\\C_{51}&C_{52}&C_{53}&C_{54}&C_{55}&C_{56}\\C_{61}&C_{62}&C_{63}&C_{64}&C_{65}&C_{66}\end{bmatrix}\begin{bmatrix}\varepsilon_1\\\varepsilon_2\\\varepsilon_3\\\gamma_{23}\\\gamma_{31}\\\gamma_{12}\end{bmatrix}, \tag{2.25}$$

where the 6×6 [C] matrix is called the stiffness matrix. The stiffness matrix has 36 constants.

What happens if one changes the system of coordinates from an orthogonal system 1–2–3 to some other orthogonal system, 1′–2′–3′? Then, new stiffness and compliance constants will be required to relate stresses and strains in the new coordinate system 1′–2′–3′. However, the new stiffness and compliance matrices in the 1′–2′–3′ system will be a function of the stiffness and compliance matrices in the 1–2–3 system and the angle between the axes of the 1′–2′–3′system and the 1–2–3 system.

Inverting Equation (2.25), the general strain–stress relationship for a three-dimensional body in a 1–2–3 orthogonal Cartesian coordinate system is

$$\begin{bmatrix} \varepsilon_1 \\ \varepsilon_2 \\ \varepsilon_3 \\ \gamma_{23} \\ \gamma_{31} \\ \gamma_{12} \end{bmatrix} = \begin{bmatrix} S_{11} & S_{12} & S_{13} & S_{14} & S_{15} & S_{16} \\ S_{21} & S_{22} & S_{23} & S_{24} & S_{25} & S_{26} \\ S_{31} & S_{32} & S_{33} & S_{34} & S_{35} & S_{36} \\ S_{41} & S_{42} & S_{43} & S_{44} & S_{45} & S_{46} \\ S_{51} & S_{52} & S_{53} & S_{54} & S_{55} & S_{56} \\ S_{61} & S_{62} & S_{63} & S_{64} & S_{65} & S_{66} \end{bmatrix} \begin{bmatrix} \sigma_1 \\ \sigma_2 \\ \sigma_3 \\ \tau_{23} \\ \tau_{31} \\ \tau_{12} \end{bmatrix}. \tag{2.26}$$

In the case of an isotropic material, relating the preceding strain–stress equation to Equation (2.17), one finds that the compliance matrix is related directly to engineering constants as

$$S_{11} = \frac{1}{E} = S_{22} = S_{33}$$

$$S_{12} = -\frac{\nu}{E} = S_{13} = S_{21} = S_{23} = S_{31} = S_{32}\,, \tag{2.27}$$

$$S_{44} = \frac{1}{G} = S_{55} = S_{66}\,,$$

and S_{ij}, other than in the preceding, are zero.

It can be shown that the 36 constants in Equation (2.25) actually reduce to 21 constants due to the symmetry of the stiffness matrix [C] as follows. The stress–strain relationship (2.25) can also be written as

$$\sigma_i = \sum_{j=1}^{6} C_{ij}\varepsilon_j, \quad i = 1 \ldots 6\,, \tag{2.28}$$

where, in a contracted notation,

$$\sigma_4 = \tau_{23},\ \sigma_5 = \tau_{31},\ \sigma_6 = \tau_{12},$$

$$\varepsilon_4 = \gamma_{23},\ \varepsilon_5 = \gamma_{31},\ \varepsilon_6 = \gamma_{12}. \tag{2.29a–f}$$

The strain energy in the body per unit volume, per Equation (2.20), is expressed as

$$W = \frac{1}{2}\sum_{i=1}^{6} \sigma_i \varepsilon_i. \tag{2.30}$$

Substituting Hooke's law, Equation (2.28), in Equation (2.30),

$$W = \frac{1}{2}\sum_{i=1}^{6}\sum_{j=1}^{6} C_{ij}\varepsilon_j \varepsilon_i. \tag{2.31}$$

Now, by partial differentiation of Equation (2.31),

$$\frac{\partial W}{\partial \varepsilon_i \partial \varepsilon_j} = C_{ij}, \tag{2.32}$$

and

$$\frac{\partial W}{\partial \varepsilon_j \partial \varepsilon_i} = C_{ji}. \tag{2.33}$$

Because the differentiation does not necessarily need to be in either order,

$$C_{ij} = C_{ji}. \tag{2.34}$$

Equation (2.34) can also be proved by realizing that

$$\sigma_i = \frac{\partial W}{\partial \varepsilon_i}.$$

Thus, only 21 independent elastic constants are in the general stiffness matrix [C] of Equation (2.25). This also implies that only 21 independent constants are in the general compliance matrix [S] of Equation (2.26).

2.3.1 Anisotropic Material

The material that has 21 independent elastic constants at a point is called an anisotropic material. Once these constants are found for a particular point, the stress and strain relationship can be developed at that point. Note that

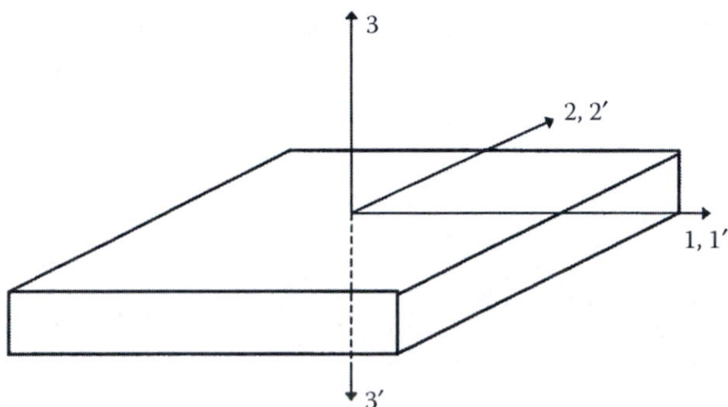

FIGURE 2.11
Transformation of coordinate axes for 1–2 plane of symmetry for a monoclinic material.

these constants can vary from point to point if the material is nonhomogeneous. Even if the material is homogeneous (or assumed to be), one needs to find these 21 elastic constants analytically or experimentally. However, many natural and synthetic materials do possess material symmetry — that is, elastic properties are identical in directions of symmetry because symmetry is present in the internal structure. Fortunately, this symmetry reduces the number of the independent elastic constants by zeroing out or relating some of the constants within the 6 × 6 stiffness [C] and 6 × 6 compliance [S] matrices. This simplifies the Hooke's law relationships for various types of elastic symmetry.

2.3.2 Monoclinic Material

If, in one plane of material symmetry* (Figure 2.11), for example, direction 3 is normal to the plane of material symmetry, then the stiffness matrix reduces to

$$[C]=\begin{bmatrix} C_{11} & C_{12} & C_{13} & 0 & 0 & C_{16} \\ C_{12} & C_{22} & C_{23} & 0 & 0 & C_{26} \\ C_{13} & C_{23} & C_{33} & 0 & 0 & C_{36} \\ 0 & 0 & 0 & C_{44} & C_{45} & 0 \\ 0 & 0 & 0 & C_{45} & C_{55} & 0 \\ C_{16} & C_{26} & C_{36} & 0 & 0 & C_{66} \end{bmatrix}. \tag{2.35}$$

as

* Material symmetry implies that the material and its mirror image about the plane of symmetry are identical.

$$C_{14}=0, C_{15}=0, C_{24}=0, C_{25}=0, C_{34}=0, C_{35}=0, C_{46}=0, C_{56}=0.$$

The direction perpendicular to the plane of symmetry is called the *principal direction*. Note that there are 13 independent elastic constants. Feldspar is an example of a monoclinic material.

The compliance matrix correspondingly reduces to

$$[S]=\begin{bmatrix} S_{11} & S_{12} & S_{13} & 0 & 0 & S_{16} \\ S_{12} & S_{22} & S_{23} & 0 & 0 & S_{26} \\ S_{13} & S_{23} & S_{33} & 0 & 0 & S_{36} \\ 0 & 0 & 0 & S_{44} & S_{45} & 0 \\ 0 & 0 & 0 & S_{45} & S_{55} & 0 \\ S_{16} & S_{26} & S_{36} & 0 & 0 & S_{66} \end{bmatrix}. \quad (2.36)$$

Modifying an excellent example[2] of demonstrating the meaning of elastic symmetry for a monoclinic material given, consider a cubic element of Figure 2.12 taken out of a monoclinic material, in which 3 is the direction perpendicular to the 1–2 plane of symmetry. Apply a normal stress, σ_3, to the element. Then using the Hooke's law Equation (2.26) and the compliance matrix (Equation 2.36) for the monoclinic material, one gets

$$\varepsilon_1 = S_{13}\sigma_3$$

$$\varepsilon_2 = S_{23}\sigma_3$$

$$\varepsilon_3 = S_{33}\sigma_3$$

$$\gamma_{23} = 0$$

$$\gamma_{31} = 0$$

$$\gamma_{12} = S_{36}\sigma_3 \,. \quad (2.37a–f)$$

The cube will deform in all directions as determined by the normal strain equations. The shear strains in the 2–3 and 3–1 plane are zero, showing that the element will not change shape in those planes. However, it will change

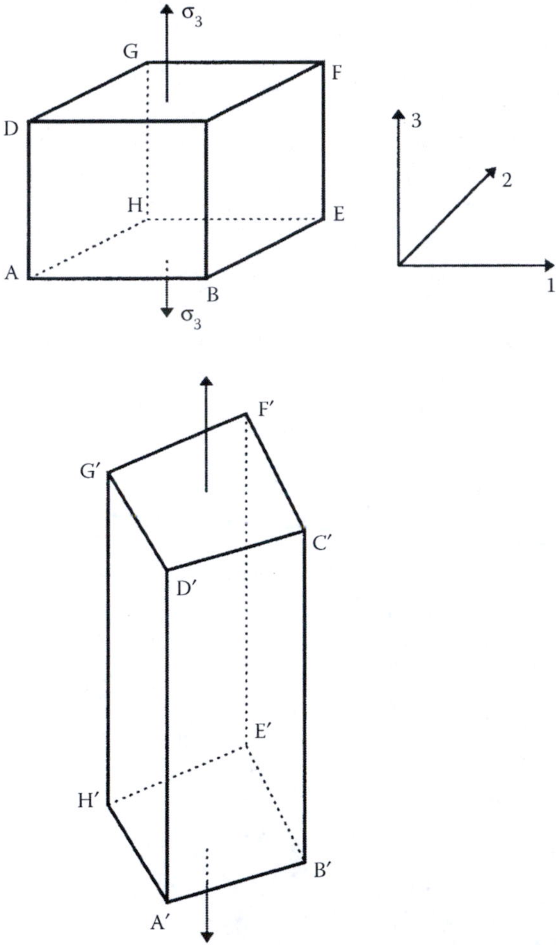

FIGURE 2.12
Deformation of a cubic element made of monoclinic material.

shape in the 1–2 plane. Thus, the faces *ABEH* and *CDFG* perpendicular to the 3 direction will change from rectangles to parallelograms, while the other four faces *ABCD*, *BEFC*, *GFEH*, and *AHGD* will stay as rectangles. This is unlike anisotropic behavior, in which all faces will be deformed in shape, and also unlike isotropic behavior, in which all faces will remain undeformed in shape.

2.3.3 Orthotropic Material (Orthogonally Anisotropic)/Specially Orthotropic

If a material has three mutually perpendicular planes of material symmetry, then the stiffness matrix is given by

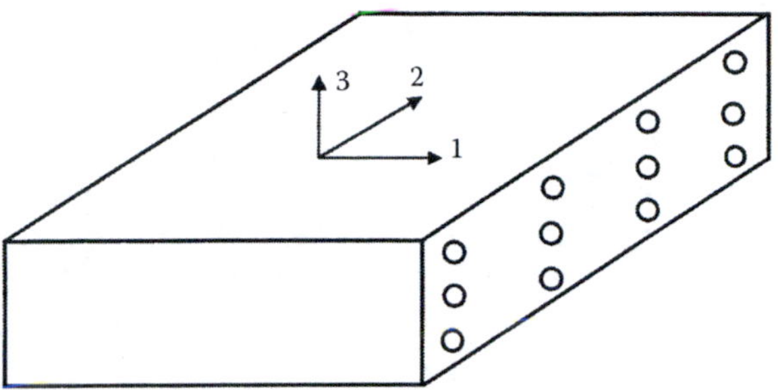

FIGURE 2.13
A unidirectional lamina as a monoclinic material with fibers, arranged in a rectangular array.

$$[C] = \begin{bmatrix} C_{11} & C_{12} & C_{13} & 0 & 0 & 0 \\ C_{12} & C_{22} & C_{23} & 0 & 0 & 0 \\ C_{13} & C_{23} & C_{33} & 0 & 0 & 0 \\ 0 & 0 & 0 & C_{44} & 0 & 0 \\ 0 & 0 & 0 & 0 & C_{55} & 0 \\ 0 & 0 & 0 & 0 & 0 & C_{66} \end{bmatrix}. \tag{2.38}$$

The preceding stiffness matrix can be derived by starting from the stiffness matrix [C] for the monoclinic material (Equation 2.35). With two more planes of symmetry, it gives

$$C_{16} = 0, C_{26} = 0, C_{36} = 0, C_{45} = 0 .$$

Three mutually perpendicular planes of material symmetry also imply three mutually perpendicular planes of elastic symmetry. Note that nine independent elastic constants are present. This is a commonly found material symmetry unlike anisotropic and monoclinic materials. Examples of an orthotropic material include a single lamina of continuous fiber composite, arranged in a rectangular array (Figure 2.13), a wooden bar, and rolled steel.

The compliance matrix reduces to

$$[S] = \begin{bmatrix} S_{11} & S_{12} & S_{13} & 0 & 0 & 0 \\ S_{12} & S_{22} & S_{23} & 0 & 0 & 0 \\ S_{13} & S_{23} & S_{33} & 0 & 0 & 0 \\ 0 & 0 & 0 & S_{44} & 0 & 0 \\ 0 & 0 & 0 & 0 & S_{55} & 0 \\ 0 & 0 & 0 & 0 & 0 & S_{66} \end{bmatrix}. \tag{2.39}$$

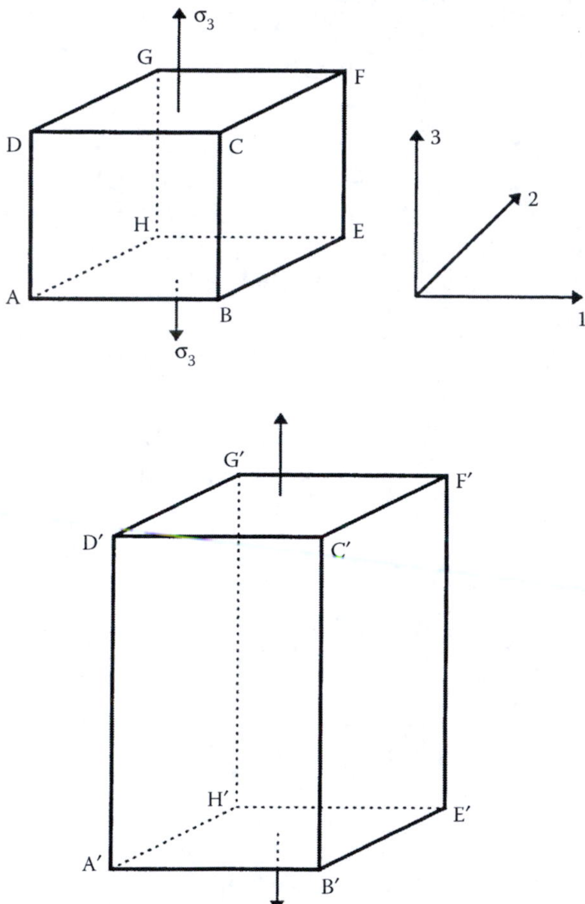

FIGURE 2.14
Deformation of a cubic element made of orthotropic material.

Demonstrating the meaning of elastic symmetry for an orthotropic material is similar to the approach taken for a monoclinic material (Section 2.3.2). Consider a cubic element (Figure 2.14) taken out of the orthotropic material, where 1, 2, and 3 are the principal directions or 1–2, 2–3, and 3–1 are the three mutually orthogonal planes of symmetry. Apply a normal stress, σ_3, to the element. Then, using the Hooke's law Equation (2.26) and the compliance matrix (Equation 2.39) for the orthotropic material, one gets

$$\varepsilon_1 = S_{13}\sigma_3$$

$$\varepsilon_2 = S_{23}\sigma_3$$

$$\varepsilon_3 = S_{33}\sigma_3$$

$$\gamma_{23} = 0 \qquad (2.40a\text{–}f)$$

$$\gamma_{31} = 0$$

$$\gamma_{12} = 0.$$

The cube will deform in all directions as determined by the normal strain equations. However, the shear strains in all three planes (1–2, 2–3, and 3–1) are zero, showing that the element will not change shape in those planes. Thus, the cube will not deform in shape under any normal load applied in the principal directions. This is unlike the monoclinic material, in which two out of the six faces of the cube changed shape.

A cube made of isotropic material would not change its shape either; however, the normal strains, ε_1 and ε_2, will be different in an orthotropic material and identical in an isotropic material.

2.3.4 Transversely Isotropic Material

Consider a plane of material isotropy in one of the planes of an orthotropic body. If direction 1 is normal to that plane (2–3) of isotropy, then the stiffness matrix is given by

$$[C] = \begin{bmatrix} C_{11} & C_{12} & C_{12} & 0 & 0 & 0 \\ C_{12} & C_{22} & C_{23} & 0 & 0 & 0 \\ C_{12} & C_{23} & C_{22} & 0 & 0 & 0 \\ 0 & 0 & 0 & \dfrac{C_{22}-C_{23}}{2} & 0 & 0 \\ 0 & 0 & 0 & 0 & C_{55} & 0 \\ 0 & 0 & 0 & 0 & 0 & C_{55} \end{bmatrix}. \qquad (2.41)$$

Transverse isotropy results in the following relations:

$$C_{22} = C_{33}, C_{12} = C_{13}, C_{55} = C_{66}, C_{44} = \frac{C_{22} - C_{23}}{2}.$$

Note the five independent elastic constants. An example of this is a thin unidirectional lamina in which the fibers are arranged in a square array or

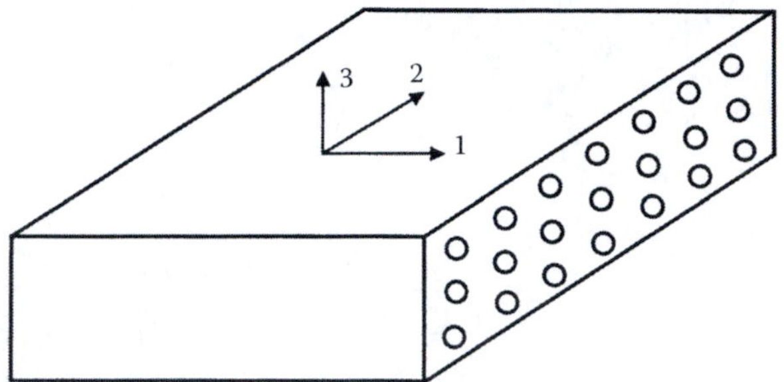

FIGURE 2.15
A unidirectional lamina as a transversely isotropic material with fibers arranged in a square array.

a hexagonal array. One may consider the elastic properties in the two directions perpendicular to the fibers to be the same. In Figure 2.15, the fibers are in direction 1, so plane 2–3 will be considered as the plane of isotropy.

The compliance matrix reduces to

$$[S] = \begin{bmatrix} S_{11} & S_{12} & S_{12} & 0 & 0 & 0 \\ S_{12} & S_{22} & S_{23} & 0 & 0 & 0 \\ S_{12} & S_{23} & S_{22} & 0 & 0 & 0 \\ 0 & 0 & 0 & 2(S_{22}-S_{23}) & 0 & 0 \\ 0 & 0 & 0 & 0 & S_{55} & 0 \\ 0 & 0 & 0 & 0 & 0 & S_{55} \end{bmatrix}. \quad (2.42)$$

2.3.5 Isotropic Material

If all planes in an orthotropic body are identical, it is an isotropic material; then, the stiffness matrix is given by

$$[C] = \begin{bmatrix} C_{11} & C_{12} & C_{12} & 0 & 0 & 0 \\ C_{12} & C_{11} & C_{12} & 0 & 0 & 0 \\ C_{12} & C_{12} & C_{11} & 0 & 0 & 0 \\ 0 & 0 & 0 & \frac{C_{11}-C_{12}}{2} & 0 & 0 \\ 0 & 0 & 0 & 0 & \frac{C_{11}-C_{12}}{2} & 0 \\ 0 & 0 & 0 & 0 & 0 & \frac{C_{11}-C_{12}}{2} \end{bmatrix}. \quad (2.43)$$

Isotropy results in the following additional relationships:

$$C_{11} = C_{22}, C_{12} = C_{23}, C_{66} = \frac{C_{22} - C_{23}}{2} = \frac{C_{11} - C_{12}}{2}.$$

This also implies infinite principal planes of symmetry. Note the two independent constants. This is the most common material symmetry available. Examples of isotropic bodies include steel, iron, and aluminum. Relating Equation (2.43) to Equation (2.18) shows that

$$C_{11} = \frac{E(1-\nu)}{(1-2\nu)(1+\nu)},$$

$$C_{12} = \frac{\nu E}{(1-2\nu)(1+\nu)}. \quad (2.44a\text{–}b)$$

Note that

$$\frac{C_{11} - C_{12}}{2}$$

$$= \frac{1}{2}\left[\frac{E(1-\nu)}{(1-2\nu)(1+\nu)} - \frac{\nu E}{(1-2\nu)(1+\nu)}\right]$$

$$= \frac{E}{2(1+\nu)}$$

$$= G.$$

The compliance matrix reduces to

$$[S] = \begin{bmatrix} S_{11} & S_{12} & S_{12} & 0 & 0 & 0 \\ S_{12} & S_{11} & S_{12} & 0 & 0 & 0 \\ S_{12} & S_{12} & S_{11} & 0 & 0 & 0 \\ 0 & 0 & 0 & 2(S_{11} - S_{12}) & 0 & 0 \\ 0 & 0 & 0 & 0 & 2(S_{11} - S_{12}) & 0 \\ 0 & 0 & 0 & 0 & 0 & 2(S_{11} - S_{12}) \end{bmatrix}. \quad (2.45)$$

We summarize the number of independent elastic constants for various types of materials:

- Anisotropic: 21
- Monoclinic: 13
- Orthotropic: 9
- Transversely isotropic: 5
- Isotropic: 2

Example 2.3

Show the reduction of anisotropic material stress–strain Equation (2.25) to those of a monoclinic material stress–strain Equation (2.35).

Solution

Assume direction 3 is perpendicular to the plane of symmetry. Now in the coordinate system 1–2–3, Equation (2.25) with $C_{ij} = C_{ji}$ from Equation (2.34) is

$$\begin{bmatrix} \sigma_1 \\ \sigma_2 \\ \sigma_3 \\ \tau_{23} \\ \tau_{31} \\ \tau_{12} \end{bmatrix} = \begin{bmatrix} C_{11} & C_{12} & C_{13} & C_{14} & C_{15} & C_{16} \\ C_{12} & C_{22} & C_{23} & C_{24} & C_{25} & C_{26} \\ C_{13} & C_{23} & C_{33} & C_{34} & C_{35} & C_{36} \\ C_{14} & C_{24} & C_{34} & C_{44} & C_{45} & C_{46} \\ C_{15} & C_{25} & C_{35} & C_{45} & C_{55} & C_{56} \\ C_{16} & C_{26} & C_{36} & C_{46} & C_{56} & C_{66} \end{bmatrix} \begin{bmatrix} \varepsilon_1 \\ \varepsilon_2 \\ \varepsilon_3 \\ \gamma_{23} \\ \gamma_{31} \\ \gamma_{12} \end{bmatrix}, \qquad (2.46)$$

Also, in the coordinate system 1′–2′–3′ (Figure 2.11),

$$\begin{bmatrix} \sigma_{1'} \\ \sigma_{2'} \\ \sigma_{3'} \\ \tau_{2'3'} \\ \tau_{3'1'} \\ \tau_{1'2'} \end{bmatrix} = \begin{bmatrix} C_{11} & C_{12} & C_{13} & C_{14} & C_{15} & C_{16} \\ C_{12} & C_{22} & C_{23} & C_{24} & C_{25} & C_{26} \\ C_{13} & C_{23} & C_{33} & C_{34} & C_{35} & C_{36} \\ C_{14} & C_{24} & C_{34} & C_{44} & C_{45} & C_{46} \\ C_{15} & C_{25} & C_{35} & C_{45} & C_{55} & C_{56} \\ C_{16} & C_{26} & C_{36} & C_{46} & C_{56} & C_{66} \end{bmatrix} \begin{bmatrix} \varepsilon_{1'} \\ \varepsilon_{2'} \\ \varepsilon_{3'} \\ \gamma_{2'3'} \\ \gamma_{3'1'} \\ \gamma_{1'2'} \end{bmatrix}, \qquad (2.47)$$

Because there is a plane of symmetry normal to direction 3, the stresses and strains in the 1–2–3 and 1′–2′–3′ coordinate systems are related by

$$\sigma_1 = \sigma_{1'}, \sigma_2 = \sigma_{2'}, \sigma_3 = \sigma_{3'}$$

$$\tau_{23} - \tau_{2'3'}, \tau_{31} = -\tau_{3'1'}, \tau_{12} = \tau_{1'2'}, \qquad (2.48a–f)$$

$$\varepsilon_1 = \varepsilon_{1'}, \varepsilon_2 = \varepsilon_{2'}, \varepsilon_3 = \varepsilon_{3'},$$

$$\gamma_{23} = -\gamma_{2'3'}, \gamma_{31} = -\gamma_{3'1'}, \gamma_{12} = \gamma_{1'2'}. \tag{2.49a–f}$$

The terms in the first equation of Equation (2.46) and Equation (2.47) can be written as

$$\sigma_1 = C_{11}\varepsilon_1 + C_{12}\varepsilon_2 + C_{13}\varepsilon_3 + C_{14}\gamma_{23} + C_{15}\gamma_{31} + C_{16}\gamma_{12},$$

$$\sigma_{1'} = C_{11}\varepsilon_{1'} + C_{12}\varepsilon_{2'} + C_{13}\varepsilon_{3'} + C_{1'4}\gamma_{2'3'} + C_{15}\gamma_{3'1'}, + C_{16}\gamma_{1'2'} \tag{2.50a–b}$$

Substituting Equation (2.48) and Equation (2.49) in Equation (2.50b),

$$\sigma_1 = C_{11}\varepsilon_1 + C_{12}\varepsilon_2 + C_{13}\varepsilon_3 - C_{14}\gamma_{23} - C_{15}\gamma_{31} + C_{16}\gamma_{12}\,. \tag{2.51}$$

Subtracting Equation (2.51) from Equation (2.50a) gives

$$0 = 2C_{14}\gamma_{23} + 2C_{15}\gamma_{31}\,. \tag{2.52}$$

Because γ_{23} and γ_{31} are arbitrary,

$$C_{14} = C_{15} = 0. \tag{2.53a}$$

Similarly, one can show that

$$C_{24} = C_{25} = 0,$$

$$C_{34} = C_{35} = 0,$$

$$C_{46} = C_{56} = 0. \tag{2.54b-d}$$

Thus, only 13 independent elastic constants are present in a monoclinic material.

Example 2.4

The stress–strain relation is given in terms of compliance matrix for an orthotropic material in Equation (2.26) and Equation (2.39). Rewrite the compliance matrix equations in terms of the nine engineering constants for

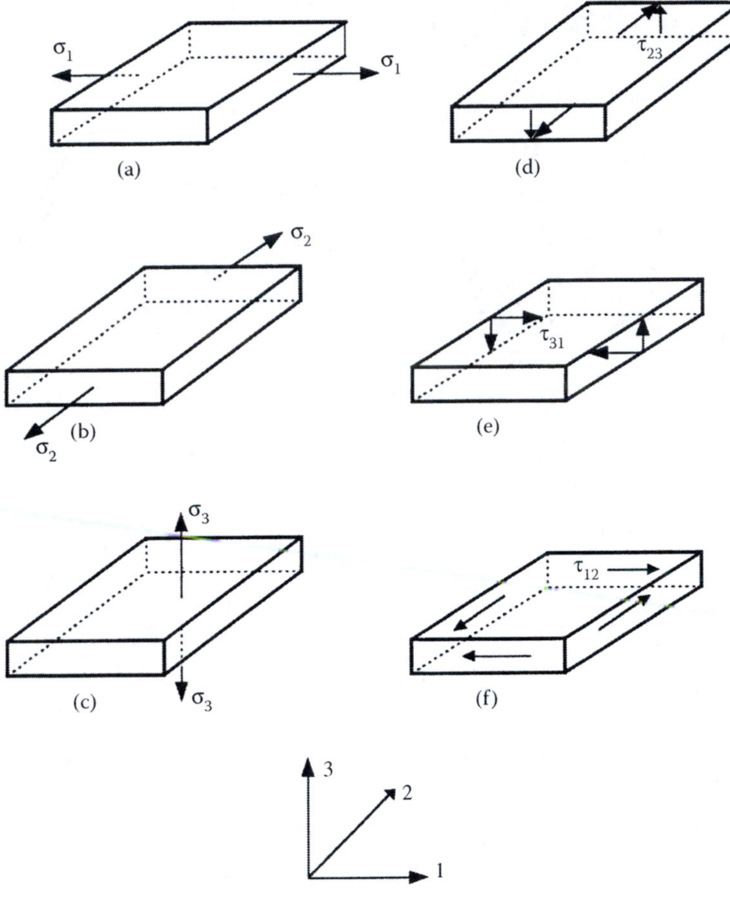

FIGURE 2.16
Application of stresses to find engineering constants of a three-dimensional orthotropic body.

an orthotropic material. What is the stiffness matrix in terms of the engineering constants?

Solution

Let us see how the compliance matrix and engineering constants of an orthotropic material are related. As shown in Figure 2.16a, apply $\sigma_1 \neq 0$, $\sigma_2 = 0$, $\sigma_3 = 0$, $\tau_{23} = 0$, $\tau_{31} = 0$, $\tau_{12} = 0$. Then, from Equation (2.26) and Equation (2.39):

$$\varepsilon_1 = S_{11}\sigma_1$$

$$\varepsilon_2 = S_{12}\sigma_1$$

$$\varepsilon_3 = S_{13}\sigma_1$$

$$\gamma_{23} = 0$$

$$\gamma_{31} = 0$$

$$\gamma_{12} = 0.$$

The Young's modulus in direction 1, E_1, is defined as

$$E_1 \equiv \frac{\sigma_1}{\varepsilon_1} = \frac{1}{S_{11}} . \tag{2.55}$$

The Poisson's ratio, ν_{12}, is defined as

$$\nu_{12} \equiv -\frac{\varepsilon_2}{\varepsilon_1} = -\frac{S_{12}}{S_{11}} . \tag{2.56}$$

In general terms, ν_{ij} is defined as the ratio of the negative of the normal strain in direction j to the normal strain in direction i, when the load is applied in the normal direction i.

The Poisson's ratio ν_{13} is defined as

$$\nu_{13} \equiv -\frac{\varepsilon_3}{\varepsilon_1} = -\frac{S_{13}}{S_{11}} . \tag{2.57}$$

Similarly, as shown in Figure 2.16b, apply $\sigma_1 = 0$, $\sigma_2 = 0$, $\sigma_3 \neq 0$, $\tau_{23} = 0$, $\tau_{31} = 0$, $\tau_{12} = 0$. Then, from Equation (2.26) and Equation (2.39),

$$E_2 = \frac{1}{S_{22}} \tag{2.58}$$

$$\nu_{21} = -\frac{S_{12}}{S_{22}} \tag{2.59}$$

$$\nu_{23} = -\frac{S_{23}}{S_{22}} . \tag{2.60}$$

Similarly, as shown in Figure 2.16c, apply $\sigma_1 = 0$, $\sigma_2 = 0$, $\sigma_3 \neq 0$, $\tau_{23} = 0$, $\tau_{31} = 0$, $\tau_{12} = 0$. From Equation (2.26) and Equation (2.39),

$$E_3 = \frac{1}{S_{33}} \tag{2.61}$$

$$\nu_{31} = -\frac{S_{13}}{S_{33}} \tag{2.62}$$

$$\nu_{32} = -\frac{S_{23}}{S_{33}}. \tag{2.63}$$

Apply, as shown in Figure 2.16d, $\sigma_1 = 0$, $\sigma_2 = 0$, $\sigma_3 = 0$, $\tau_{23} \neq 0$, $\tau_{31} = 0$, $\tau_{12} = 0$. Then, from Equation (2.26) and Equation (2.39),

$$\varepsilon_1 = 0$$

$$\varepsilon_2 = 0$$

$$\varepsilon_3 = 0$$

$$\gamma_{23} = S_{44}\tau_{23}$$

$$\gamma_{31} = 0$$

$$\gamma_{12} = 0$$

The shear modulus in plane 2–3 is defined as

$$G_{23} \equiv \frac{\tau_{23}}{\gamma_{23}} = \frac{1}{S_{44}}. \tag{2.64}$$

Similarly, as shown in Figure 2.16e, apply $\sigma_1 = 0$, $\sigma_2 = 0$, $\sigma_3 = 0$, $\tau_{23} = 0$, $\tau_{31} \neq 0$, $\tau_{12} = 0$. Then, from Equation (2.26) and Equation (2.39),

$$G_{31} = \frac{1}{S_{55}}. \tag{2.65}$$

Similarly, as shown in Figure 2.16f, apply $\sigma_1 = 0$, $\sigma_2 = 0$, $\sigma_3 = 0$, $\tau_{23} = 0$, $\tau_{31} = 0$, $\tau_{12} \neq 0$. Then, from Equation (2.26) and Equation (2.39),

$$G_{12} = \frac{1}{S_{66}}. \tag{2.66}$$

In Equation (2.55) through Equation (2.66), 12 engineering constants have been defined as follows:

Three Young's moduli, E_1, E_2, and E_3, one in each material axis

Six Poisson's ratios, ν_{12}, ν_{13}, ν_{21}, ν_{23}, ν_{31}, and ν_{32}, two for each plane
Three shear moduli, G_{23}, G_{31}, and G_{12}, one for each plane

However, the six Poisson's ratios are not independent of each other. For example, from Equation (2.55), Equation (2.56), Equation (2.58), and Equation (2.59),

$$\frac{\nu_{12}}{E_1} = \frac{\nu_{21}}{E_2} . \tag{2.67}$$

Similarly, from Equation (2.55), Equation (2.57), Equation (2.61), and Equation (2.62),

$$\frac{\nu_{13}}{E_1} = \frac{\nu_{31}}{E_3} , \tag{2.68}$$

and from Equation (2.58), Equation (2.60), Equation (2.61), and Equation (2.63),

$$\frac{\nu_{23}}{E_2} = \frac{\nu_{32}}{E_3} . \tag{2.69}$$

Equation (2.67), Equation (2.68), and Equation (2.69) are called reciprocal Poisson's ratio equations. These relations reduce the total independent engineering constants to nine. This is the same number as the number of independent constants in the stiffness or the compliance matrix.

Rewriting the compliance matrix in terms of the engineering constants gives

$$[S] = \begin{bmatrix} \frac{1}{E_1} & -\frac{\nu_{12}}{E_1} & -\frac{\nu_{13}}{E_1} & 0 & 0 & 0 \\ -\frac{\nu_{21}}{E_2} & \frac{1}{E_2} & -\frac{\nu_{23}}{E_2} & 0 & 0 & 0 \\ -\frac{\nu_{31}}{E_3} & -\frac{\nu_{32}}{E_3} & \frac{1}{E_3} & 0 & 0 & 0 \\ 0 & 0 & 0 & \frac{1}{G_{23}} & 0 & 0 \\ 0 & 0 & 0 & 0 & \frac{1}{G_{31}} & 0 \\ 0 & 0 & 0 & 0 & 0 & \frac{1}{G_{12}} \end{bmatrix} . \tag{2.70}$$

Inversion of Equation (2.70) would be the compliance matrix [C] and is given by

$$\left[C\right]=\begin{bmatrix} \frac{1-\nu_{23}\nu_{32}}{E_2E_3\Delta} & \frac{\nu_{21}+\nu_{23}\nu_{31}}{E_2E_3\Delta} & \frac{\nu_{31}+\nu_{21}\nu_{32}}{E_2E_3\Delta} & 0 & 0 & 0 \\ \frac{\nu_{21}+\nu_{23}\nu_{31}}{E_2E_3\Delta} & \frac{1-\nu_{13}\nu_{31}}{E_1E_3\Delta} & \frac{\nu_{32}+\nu_{12}\nu_{31}}{E_1E_3\Delta} & 0 & 0 & 0 \\ \frac{\nu_{31}+\nu_{21}\nu_{32}}{E_2E_3\Delta} & \frac{\nu_{32}+\nu_{12}\nu_{31}}{E_1E_3\Delta} & \frac{1-\nu_{12}\nu_{21}}{E_1E_2\Delta} & 0 & 0 & 0 \\ 0 & 0 & 0 & G_{23} & 0 & 0 \\ 0 & 0 & 0 & 0 & G_{31} & 0 \\ 0 & 0 & 0 & 0 & 0 & G_{12} \end{bmatrix}, \quad (2.71)$$

where

$$\Delta=\left(1-\nu_{12}\nu_{21}-\nu_{23}\nu_{32}-\nu_{13}\nu_{31}-2\nu_{21}\nu_{32}\nu_{13}\right)/\left(E_1E_2E_3\right). \quad (2.72)$$

Although nine independent elastic constants are in the compliance matrix [S] and, correspondingly, in the stiffness matrix [C] for orthotropic materials, constraints on the values of these constants exist. Based on the first law of thermodynamics, the stiffness and compliance matrices must be positive definite. Thus, the diagonal terms of [C] and [S] in Equation (2.71) and Equation (2.70), respectively, need to be positive. From the diagonal elements of the compliance matrix [S], this gives

$$E_1>0\ ,\ E_2>0\ ,\ E_3>0\ ,\ G_{12}>0\ ,\ G_{23}>0\ ,\ G_{31}>0 \quad (2.73)$$

and, from the diagonal elements of the stiffness matrix [C], gives

$$1-\nu_{23}\nu_{32}>0\ ,\ 1-\nu_{31}\nu_{13}>0\ ,\ 1-\nu_{12}\nu_{21}>0\ , \quad (2.74)$$

$$\Delta=1-\nu_{12}\nu_{21}-\nu_{23}\nu_{32}-\nu_{31}\nu_{13}-2\nu_{13}\nu_{21}\nu_{32}>0$$

Using the reciprocal relations given by Equation (2.67) through Equation (2.69),

$$\frac{\nu_{ij}}{E_i}=\frac{\nu_{ji}}{E_j} \text{ for } i\neq j \text{ and } i,j = 1,2,3,$$

we can rewrite the inequalities as follows.

For example, because

$$1 - \nu_{12}\nu_{21} > 0 \ ,$$

then

$$\nu_{12} < \frac{1}{\nu_{21}} = \frac{E_1}{E_2}\frac{1}{\nu_{12}}$$

$$|\nu_{12}| < \left|\frac{E_1}{E_2}\frac{1}{\nu_{12}}\right|$$

$$|\nu_{12}| < \sqrt{\frac{E_1}{E_2}} \ . \tag{2.75a}$$

Similarly, five other such relationships can be developed to give

$$|\nu_{21}| < \sqrt{\frac{E_2}{E_1}} \tag{2.75b}$$

$$|\nu_{32}| < \sqrt{\frac{E_3}{E_2}} \tag{2.75c}$$

$$|\nu_{23}| < \sqrt{\frac{E_2}{E_3}} \tag{2.75d}$$

$$|\nu_{31}| < \sqrt{\frac{E_3}{E_1}} \tag{2.75e}$$

$$|\nu_{13}| < \sqrt{\frac{E_1}{E_3}} \ . \tag{2.75f}$$

These restrictions on the elastic moduli are important in optimizing properties of a composite because they show that the nine independent properties cannot be varied without influencing the limits of the others.

Example 2.5

Find the compliance and stiffness matrix for a graphite/epoxy lamina. The material properties are given as

$$E_1 = 181 GPa \ , \ E_2 = 10.3 GPa \ , \ E_3 = 10.3 GPa$$

$$\nu_{12} = 0.28 \ , \ \nu_{23} = 0.60 \ , \ \nu_{13} = 0.27$$

$$G_{12} = 7.17 GPa \ , \ G_{23} = 3.0 GPa \ , \ G_{31} = 7.00 GPa \ .$$

Solution

$$S_{11} = \frac{1}{E_1} = \frac{1}{181 \times 10^9} = 5.525 \times 10^{-12} \, Pa^{-1}$$

$$S_{22} = \frac{1}{E_2} = \frac{1}{10.3 \times 10^9} = 9.709 \times 10^{-11} \, Pa^{-1}$$

$$S_{33} = \frac{1}{E_3} = \frac{1}{10.3 \times 10^9} = 9.709 \times 10^{-11} \, Pa^{-1}$$

$$S_{12} = -\frac{\nu_{12}}{E_1} = -\frac{0.28}{181 \times 10^9} = -1.547 \times 10^{-12} \, Pa^{-1}$$

$$S_{13} = -\frac{\nu_{13}}{E_1} = -\frac{0.27}{181 \times 10^9} = -1.492 \times 10^{-12} \, Pa^{-1}$$

$$S_{23} = -\frac{\nu_{23}}{E_2} = -\frac{0.6}{10.3 \times 10^9} = -5.825 \times 10^{-11} \, Pa^{-1}$$

$$S_{44} = \frac{1}{G_{23}} = \frac{1}{3 \times 10^9} = 3.333 \times 10^{-10} \, Pa^{-1}$$

$$S_{55} = \frac{1}{G_{31}} = \frac{1}{7 \times 10^9} = 1.429 \times 10^{-10} \, Pa^{-1}$$

$$S_{66} = \frac{1}{G_{12}} = \frac{1}{7.17 \times 10^{9}} = 1.395 \times 10^{-10}\,Pa^{-1}.$$

Thus, the compliance matrix for the orthotropic lamina is given by

$$[S] =$$

$$\begin{bmatrix} 5.525\times10^{-12} & -1.547\times10^{-12} & -1.492\times10^{-12} & 0 & 0 & 0 \\ -1.547\times10^{-12} & 9.709\times10^{-11} & -5.825\times10^{-11} & 0 & 0 & 0 \\ -1.492\times10^{-12} & -5.825\times10^{-11} & 9.709\times10^{-11} & 0 & 0 & 0 \\ 0 & 0 & 0 & 3.333\times10^{-10} & 0 & 0 \\ 0 & 0 & 0 & 0 & 1.429\times10^{-10} & 0 \\ 0 & 0 & 0 & 0 & 0 & 1.395\times10^{-10} \end{bmatrix} Pa^{-1}$$

The stiffness matrix can be found by inverting the compliance matrix and is given by

$$[C] = [S]^{-1}$$

$$[C] =$$

$$\begin{bmatrix} 0.1850\times10^{12} & 0.7269\times10^{10} & 0.7204\times10^{10} & 0 & 0 & 0 \\ 0.7269\times10^{10} & 0.1638\times10^{11} & 0.9938\times10^{10} & 0 & 0 & 0 \\ 0.7204\times10^{10} & 0.9938\times10^{10} & 0.1637\times10^{11} & 0 & 0 & 0 \\ 0 & 0 & 0 & 0.3000\times10^{10} & 0 & 0 \\ 0 & 0 & 0 & 0 & 0.6998\times10^{10} & 0 \\ 0 & 0 & 0 & 0 & 0 & 0.7168\times10^{10} \end{bmatrix} Pa$$

The preceding stiffness matrix [C] can also be found directly by using Equation (2.71).

2.4 Hooke's Law for a Two-Dimensional Unidirectional Lamina

2.4.1 Plane Stress Assumption

A thin plate is a prismatic member having a small thickness, and it is the case for a typical lamina. If a plate is thin and there are no out-of-plane loads, it can be considered to be under plane stress (Figure 2.17). If the upper and lower surfaces of the plate are free from external loads, then $\sigma_3 = 0$, $\tau_{31} = 0$, and $\tau_{23} = 0$. Because the plate is thin, these three stresses within the plate are

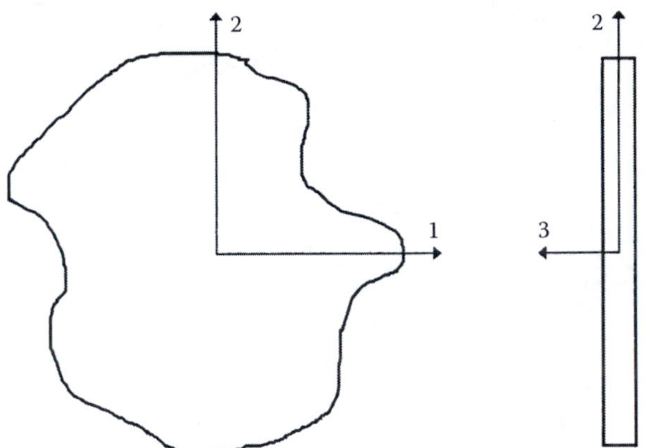

FIGURE 2.17
Plane stress conditions for a thin plate.

assumed to vary little from the magnitude of stresses at the top and the bottom surfaces. Thus, they can be assumed to be zero within the plate also. A lamina is thin and, if no out-of-plane loads are applied, one can assume that it is under plane stress. This assumption then reduces the three-dimensional stress–strain equations to two-dimensional stress–strain equations.

2.4.2 Reduction of Hooke's Law in Three Dimensions to Two Dimensions

A unidirectional lamina falls under the orthotropic material category. If the lamina is thin and does not carry any out-of-plane loads, one can assume plane stress conditions for the lamina. Therefore, taking Equation (2.26) and Equation (2.39) and assuming $\sigma_3 = 0$, $\tau_{23} = 0$, and $\tau_{31} = 0$, then

$$\varepsilon_3 = S_{13}\sigma_1 + S_{23}\sigma_2,$$

$$\gamma_{23} = \gamma_{31} = 0. \tag{2.76a,b}$$

The normal strain, ε_3, is not an independent strain because it is a function of the other two normal strains, ε_1 and ε_2. Therefore, the normal strain, ε_3, can be omitted from the stress–strain relationship (2.39). Also, the shearing strains, γ_{23} and γ_{31}, can be omitted because they are zero. Equation (2.39) for an orthotropic plane stress problem can then be written as

$$\begin{bmatrix} \varepsilon_1 \\ \varepsilon_2 \\ \gamma_{12} \end{bmatrix} = \begin{bmatrix} S_{11} & S_{12} & 0 \\ S_{12} & S_{22} & 0 \\ 0 & 0 & S_{66} \end{bmatrix} \begin{bmatrix} \sigma_1 \\ \sigma_2 \\ \tau_{12} \end{bmatrix}, \tag{2.77}$$

where S_{ij} are the elements of the compliance matrix. Note the four independent compliance elements in the matrix.

Inverting Equation (2.77) gives the stress–strain relationship as

$$\begin{bmatrix} \sigma_1 \\ \sigma_2 \\ \tau_{12} \end{bmatrix} = \begin{bmatrix} Q_{11} & Q_{12} & 0 \\ Q_{12} & Q_{22} & 0 \\ 0 & 0 & Q_{66} \end{bmatrix} \begin{bmatrix} \varepsilon_1 \\ \varepsilon_2 \\ \gamma_{12} \end{bmatrix}, \tag{2.78}$$

where Q_{ij} are the reduced stiffness coefficients, which are related to the compliance coefficients as

$$Q_{11} = \frac{S_{22}}{S_{11}S_{22} - S_{12}^2},$$

$$Q_{12} = -\frac{S_{12}}{S_{11}S_{22} - S_{12}^2}, \tag{2.79a–d}$$

$$Q_{22} = \frac{S_{11}}{S_{11}S_{22} - S_{12}^2},$$

$$Q_{66} = \frac{1}{S_{66}}.$$

Note that the elements of the reduced stiffness matrix, Q_{ij}, are not the same as the elements of the stiffness matrix, C_{ij} (see Exercise 2.13).

2.4.3 Relationship of Compliance and Stiffness Matrix to Engineering Elastic Constants of a Lamina

Equation (2.77) and Equation (2.78) show the relationship of stress and strain through the compliance [S] and reduced stiffness [Q] matrices. However, stress and strains are generally related through engineering elastic constants. For a unidirectional lamina, these engineering elastics constants are

E_1 = longitudinal Young's modulus (in direction 1)

E_2 = transverse Young's modulus (in direction 2)

ν_{12} = major Poisson's ratio, where the general Poisson's ratio, ν_{ij} is defined as the ratio of the negative of the normal strain in direction j to the normal strain in direction i, when the only normal load is applied in direction i

G_{12} = in-plane shear modulus (in plane 1–2)

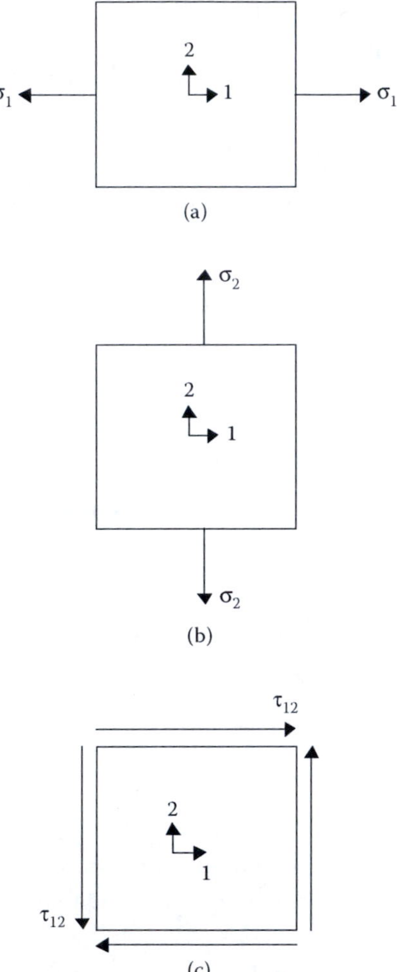

FIGURE 2.18
Application of stresses to find engineering constants of a unidirectional lamina.

Experimentally, the four independent engineering elastic constants are measured as follows and can be related to the four independent elements of the compliance matrix [S] of Equation (2.77).

- Apply a pure tensile load in direction 1 (Figure 2.18a), that is,

$$\sigma_1 \neq 0, \sigma_2 = 0, \tau_{12} = 0. \tag{2.80}$$

Then, from Equation (2.77),

$$\varepsilon_1 = S_{11}\sigma_1,$$
$$\varepsilon_2 = S_{12}\sigma_1, \quad (2.81a\text{–}c)$$
$$\gamma_{12} = 0.$$

By definition, if the only nonzero stress is σ_1, as is the case here, then

$$E_1 \equiv \frac{\sigma_1}{\varepsilon_1} = \frac{1}{S_{11}}, \quad (2.82)$$

$$\nu_{12} \equiv -\frac{\varepsilon_2}{\varepsilon_1} = -\frac{S_{12}}{S_{11}}. \quad (2.83)$$

- Apply a pure tensile load in direction 2 (Figure 2.18b), that is

$$\sigma_1 = 0,\ \sigma_2 \neq 0,\ \tau_{12} = 0. \quad (2.84)$$

Then, from Equation (2.77),

$$\varepsilon_1 = S_{12}\sigma_2,$$
$$\varepsilon_2 = S_{22}\sigma_2, \quad (2.85a\text{–}c)$$
$$\gamma_{12} = 0.$$

By definition, if the only nonzero stress is σ_2, as is the case here, then

$$E_2 \equiv \frac{\sigma_2}{\varepsilon_2} = \frac{1}{S_{22}}, \quad (2.86)$$

$$\nu_{21} \equiv -\frac{\varepsilon_1}{\varepsilon_2} = -\frac{S_{12}}{S_{22}}. \quad (2.87)$$

The ν_{21} term is called the minor Poisson's ratio. From Equation (2.82), Equation (2.83), Equation (2.86), and Equation (2.87), we have the reciprocal relationship

$$\frac{\nu_{12}}{E_1} = \frac{\nu_{21}}{E_2}. \quad (2.88)$$

- Apply a pure shear stress in the plane 1–2 (Figure 2.18c) — that is,

$$\sigma_1 = 0,\ \sigma_2 = 0 \text{ and } \tau_{12} \neq 0. \tag{2.89}$$

Then, from Equation (2.77),

$$\varepsilon_1 = 0,$$

$$\varepsilon_2 = 0,$$

$$\gamma_{12} = S_{66}\tau_{12}. \tag{2.90a–c}$$

By definition, if τ_{12} is the only nonzero stress, as is the case here, then

$$G_{12} \equiv \frac{\tau_{12}}{\gamma_{12}} = \frac{1}{S_{66}}. \tag{2.91}$$

Thus, we have proved that

$$S_{11} = \frac{1}{E_1},$$

$$S_{12} = -\frac{\nu_{12}}{E_1},$$

$$S_{22} = \frac{1}{E_2},$$

$$S_{66} = \frac{1}{G_{12}}. \tag{2.92a–d}$$

Also, the stiffness coefficients Q_{ij} are related to the engineering constants through Equation (2.98) and Equation (2.92) as

$$Q_{11} = \frac{E_1}{1 - \nu_{21}\nu_{12}},$$

$$Q_{12} = \frac{\nu_{12}E_2}{1-\nu_{21}\nu_{12}},$$

$$Q_{22} = \frac{E_2}{1-\nu_{21}\nu_{12}}, \text{ and}$$

$$Q_{66} = G_{12}. \tag{2.93a–d}$$

Equation (2.77), Equation (2.78), Equation (2.92), and Equation (2.93) relate stresses and strains through any of the following combinations of four constants.

Q_{11}, Q_{12}, Q_{22}, Q_{66}, or
S_{11}, S_{12}, S_{22}, S_{66}, or
E_1, E_2, ν_{12}, G_{12}

The unidirectional lamina is a *specially orthotropic* lamina because normal stresses applied in the 1–2 direction do not result in any shearing strains in the 1–2 plane because $Q_{16} = Q_{26} = 0 = S_{16} = S_{26}$. Also, the shearing stresses applied in the 1–2 plane do not result in any normal strains in the 1 and 2 directions because $Q_{16} = Q_{26} = 0 = S_{16} = S_{26}$.

A woven composite with its weaves perpendicular to each other and short fiber composites with fibers arranged perpendicularly to each other or aligned in one direction also are *specially orthotropic*. Thus, any discussion in this chapter or in Chapter 4 ("Macromechanics of a Laminate") is valid for such a lamina as well. Mechanical properties of some typical unidirectional lamina are given in Table 2.1 and Table 2.2.

Example 2.6

For a graphite/epoxy unidirectional lamina, find the following

1. Compliance matrix
2. Minor Poisson's ratio
3. Reduced stiffness matrix
4. Strains in the 1–2 coordinate system if the applied stresses (Figure 2.19) are

$$\sigma_1 = 2MPa,\ \sigma_2 = -3MPa,\ \tau_{12} = 4MPa.$$

Use the properties of unidirectional graphite/epoxy lamina from Table 2.1.

TABLE 2.1

Typical Mechanical Properties of a Unidirectional Lamina (SI System of Units)

Property	Symbol	Units	Glass/ epoxy	Boron/ epoxy	Graphite/ epoxy
Fiber volume fraction	V_f		0.45	0.50	0.70
Longitudinal elastic modulus	E_1	GPa	38.6	204	181
Transverse elastic modulus	E_2	GPa	8.27	18.50	10.30
Major Poisson's ratio	ν_{12}		0.26	0.23	0.28
Shear modulus	G_{12}	GPa	4.14	5.59	7.17
Ultimate longitudinal tensile strength	$(\sigma_1^T)_{ult}$	MPa	1062	1260	1500
Ultimate longitudinal compressive strength	$(\sigma_1^C)_{ult}$	MPa	610	2500	1500
Ultimate transverse tensile strength	$(\sigma_2^T)_{ult}$	MPa	31	61	40
Ultimate transverse compressive strength	$(\sigma_2^C)_{ult}$	MPa	118	202	246
Ultimate in-plane shear strength	$(\tau_{12})_{ult}$	MPa	72	67	68
Longitudinal coefficient of thermal expansion	α_1	μm/m/°C	8.6	6.1	0.02
Transverse coefficient of thermal expansion	α_2	μm/m/°C	22.1	30.3	22.5
Longitudinal coefficient of moisture expansion	β_1	m/m/kg/kg	0.00	0.00	0.00
Transverse coefficient of moisture expansion	β_2	m/m/kg/kg	0.60	0.60	0.60

Source: Tsai, S.W. and Hahn, H.T., *Introduction to Composite Materials*, CRC Press, Boca Raton, FL, Table 1.7, p. 19; Table 7.1, p. 292; Table 8.3, p. 342. Reprinted with permission.

Solution

From Table 2.1, the engineering elastic constants of the unidirectional graphite/epoxy lamina are

$$E_1 = 181\ GPa,\ E_2 = 10.3\ GPa,\ \nu_{12} = 0.28,\ G_{12} = 7.17\ GPa.$$

1. Using Equation (2.92), the compliance matrix elements are

$$S_{11} = \frac{1}{181 \times 10^9} = 0.5525 \times 10^{-11}\ Pa^{-1},$$

$$S_{12} = -\frac{0.28}{181 \times 10^9} = -0.1547 \times 10^{-11}\ Pa^{-1},$$

TABLE 2.2

Typical Mechanical Properties of a Unidirectional Lamina (USCS System of Units)

Property	Symbol	Units	Glass/ epoxy	Boron/ epoxy	Graphite/ epoxy
Fiber volume fraction	V_f	—	0.45	0.50	0.70
Longitudinal elastic modulus	E_1	Msi	5.60	29.59	26.25
Transverse elastic modulus	E_2	Msi	1.20	2.683	1.49
Major Poisson's ratio	υ_{12}		0.26	0.23	0.28
Shear modulus	G_{12}	Msi	0.60	0.811	1.040
Ultimate longitudinal tensile strength	$(\sigma_1^T)_{ult}$	ksi	154.03	182.75	217.56
Ultimate longitudinal compressive strength	$(\sigma_1^C)_{ult}$	ksi	88.47	362.6	217.56
Ultimate transverse tensile strength	$(\sigma_2^T)_{ult}$	ksi	4.496	8.847	5.802
Ultimate transverse compressive strength	$(\sigma_2^C)_{ult}$	ksi	17.12	29.30	35.68
Ultimate in-plane shear strength	$(\tau_{12})_{ult}$	ksi	10.44	9.718	9.863
Longitudinal coefficient of thermal expansion	α_1	μin./in./°F	4.778	3.389	0.0111
Transverse coefficient of thermal expansion	α_2	μin./in./°F	12.278	16.83	12.5
Longitudinal coefficient of moisture expansion	β_1	in./in./lb/lb	0.00	0.00	0.00
Transverse coefficient of moisture expansion	β_2	in./in./lb/lb	0.60	0.60	0.60

Source: Tsai, S.W. and Hahn, H.T., *Introduction to Composite Materials*, CRC Press, Boca Raton, FL, Table 1.7, p. 19; Table 7.1, p. 292; Table 8.3, p. 342. USCS system used for tables reprinted with permission.

$$S_{22} = \frac{1}{10.3 \times 10^9} = 0.9709 \times 10^{-10}\, Pa^{-1},$$

$$S_{66} = \frac{1}{7.17 \times 10^9} = 0.1395 \times 10^{-9}\, Pa^{-1}.$$

2. Using the reciprocal relationship (2.88), the minor Poisson's ratio is

$$\nu_{21} = \frac{0.28}{181 \times 10^9} \times (10.3 \times 10^9) = 0.01593.$$

3. Using Equation (2.93), the reduced stiffness matrix $[Q]$ elements are

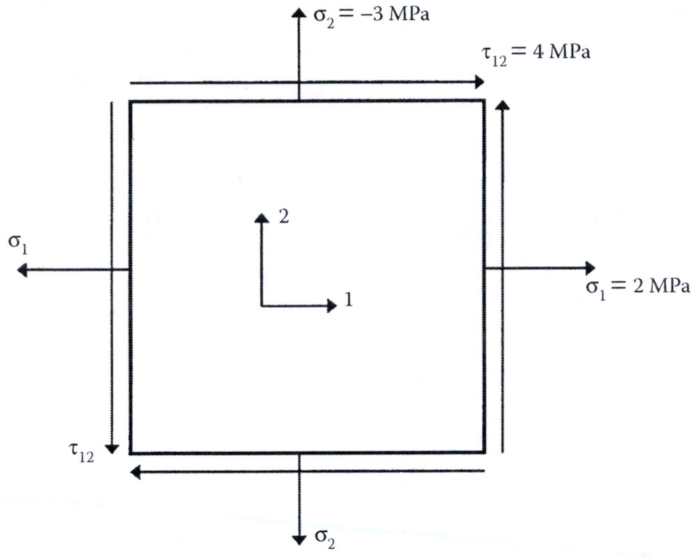

FIGURE 2.19
Applied stresses in a unidirectional lamina in Example 2.6.

$$Q_{11} = \frac{181 \times 10^9}{1-(0.28)(0.01593)} = 181.8 \times 10^9 \, Pa,$$

$$Q_{12} = \frac{(0.28)(10.3 \times 10^9)}{1-(0.28)(0.01593)} = 2.897 \times 10^9 \, Pa,$$

$$Q_{22} = \frac{10.3 \times 10^9}{1-(0.28)(0.01593)} = 10.35 \times 10^9 \, Pa,$$

$$Q_{66} = 7.17 \times 10^9 \, Pa .$$

The reduced stiffness matrix [Q] could also be obtained by inverting the compliance matrix [S] of part 1:

$$[Q] = [S]^{-1} = \begin{bmatrix} 0.5525 \times 10^{-11} & -0.1547 \times 10^{-11} & 0 \\ -0.1547 \times 10^{-11} & 0.9709 \times 10^{-10} & 0 \\ 0 & 0 & 0.1395 \times 10^{-9} \end{bmatrix}^{-1} .$$

$$= \begin{bmatrix} 181.8\times10^9 & 2.897\times10^9 & 0 \\ 2.897\times10^9 & 10.35\times10^9 & 0 \\ 0 & 0 & 7.17\times10^9 \end{bmatrix} Pa\,.$$

4. Using Equation (2.77), the strains in the 1–2 coordinate system are

$$\begin{bmatrix} \varepsilon_1 \\ \varepsilon_2 \\ \gamma_{12} \end{bmatrix} = \begin{bmatrix} 0.5525\times10^{-11} & -0.1547\times10^{-11} & 0 \\ -0.1547\times10^{-11} & 0.9709\times10^{-10} & 0 \\ 0 & 0 & 0.1395\times10^{-9} \end{bmatrix} \begin{bmatrix} 2\times10^6 \\ -3\times10^6 \\ 4\times10^6 \end{bmatrix}$$

$$= \begin{bmatrix} 15.69 \\ -294.4 \\ 557.9 \end{bmatrix} (10^{-6}).$$

Thus, the strains in the local axes are

$$\varepsilon_1 = 15.69\frac{\mu m}{m},$$

$$\varepsilon_2 = 294.4\frac{\mu m}{m},$$

$$\gamma_{12} = 557.9\frac{\mu m}{m}.$$

2.5 Hooke's Law for a Two-Dimensional Angle Lamina

Generally, a laminate does not consist only of unidirectional laminae because of their low stiffness and strength properties in the transverse direction. Therefore, in most laminates, some laminae are placed at an angle. It is thus necessary to develop the stress–strain relationship for an angle lamina.

The coordinate system used for showing an angle lamina is as given in Figure 2.20. The axes in the 1–2 coordinate system are called the local axes or the material axes. The direction 1 is parallel to the fibers and the direction 2 is perpendicular to the fibers. In some literature, direction 1 is also called

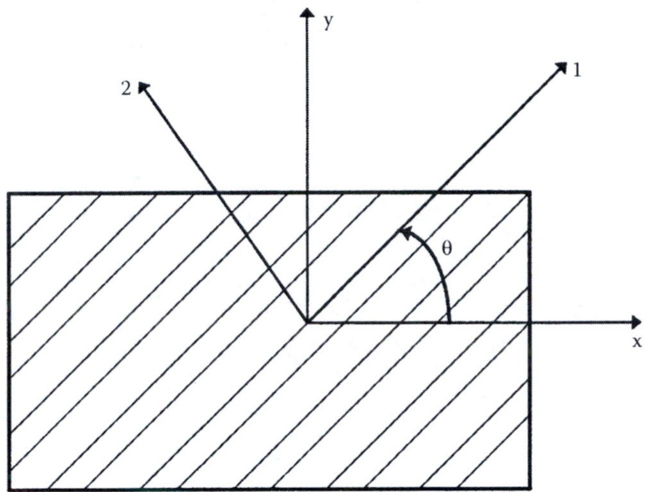

FIGURE 2.20
Local and global axes of an angle lamina.

the longitudinal direction L and the direction 2 is called the transverse direction T. The axes in the x–y coordinate system are called the global axes or the off-axes. The angle between the two axes is denoted by an angle θ. The stress–strain relationship in the 1–2 coordinate system has already been established in Section 2.4 and we are now going to develop the stress–strain equations for the x–y coordinate system.

The global and local stresses in an angle lamina are related to each other through the angle of the lamina, θ (Appendix B):

$$\begin{bmatrix} \sigma_x \\ \sigma_y \\ \tau_{xy} \end{bmatrix} = [T]^{-1} \begin{bmatrix} \sigma_1 \\ \sigma_2 \\ \tau_{12} \end{bmatrix}, \tag{2.94}$$

where $[T]$ is called the transformation matrix and is defined as

$$[T]^{-1} = \begin{bmatrix} c^2 & s^2 & -2sc \\ s^2 & c^2 & 2sc \\ sc & -sc & c^2 - s^2 \end{bmatrix}, \tag{2.95}$$

and

$$[T] = \begin{bmatrix} c^2 & s^2 & 2sc \\ s^2 & c^2 & -2sc \\ -sc & sc & c^2 - s^2 \end{bmatrix}, \tag{2.96}$$

$$c = \text{Cos}\,(\theta),$$

$$s = \text{Sin}\,(\theta). \tag{2.97a,b}$$

Using the stress–strain Equation (2.78) in the local axes, Equation (2.94) can be written as

$$\begin{bmatrix} \sigma_x \\ \sigma_y \\ \tau_{xy} \end{bmatrix} = [T]^{-1}[Q] \begin{bmatrix} \varepsilon_1 \\ \varepsilon_2 \\ \gamma_{12} \end{bmatrix}. \tag{2.98}$$

The global and local strains are also related through the transformation matrix (Appendix B):

$$\begin{bmatrix} \varepsilon_1 \\ \varepsilon_2 \\ \gamma_{12}/2 \end{bmatrix} = [T] \begin{bmatrix} \varepsilon_x \\ \varepsilon_y \\ \gamma_{xy}/2 \end{bmatrix}, \tag{2.99}$$

which can be rewritten as

$$\begin{bmatrix} \varepsilon_1 \\ \varepsilon_2 \\ \gamma_{12} \end{bmatrix} = [R][T][R]^{-1} \begin{bmatrix} \varepsilon_x \\ \varepsilon_y \\ \gamma_{xy} \end{bmatrix}, \tag{2.100}$$

where $[R]$ is the Reuter matrix[3] and is defined as

$$[R] = \begin{bmatrix} 1 & 0 & 0 \\ 0 & 1 & 0 \\ 0 & 0 & 2 \end{bmatrix}. \tag{2.101}$$

Then, substituting Equation (2.100) in Equation (2.98) gives

$$\begin{bmatrix} \sigma_x \\ \sigma_y \\ \tau_{xy} \end{bmatrix} = [T]^{-1}[Q][R][T][R]^{-1} \begin{bmatrix} \varepsilon_x \\ \varepsilon_y \\ \gamma_{xy} \end{bmatrix}. \quad (2.102)$$

On carrying the multiplication of the first five matrices on the right-hand side of Equation (2.102),

$$\begin{bmatrix} \sigma_x \\ \sigma_y \\ \tau_{xy} \end{bmatrix} = \begin{bmatrix} \bar{Q}_{11} & \bar{Q}_{12} & \bar{Q}_{16} \\ \bar{Q}_{12} & \bar{Q}_{22} & \bar{Q}_{26} \\ \bar{Q}_{16} & \bar{Q}_{26} & \bar{Q}_{66} \end{bmatrix} \begin{bmatrix} \varepsilon_x \\ \varepsilon_y \\ \gamma_{xy} \end{bmatrix}, \quad (2.103)$$

where $\bar{Q}_{ij}$ are called the elements of the transformed reduced stiffness matrix $[\bar{Q}]$ and are given by

$$\bar{Q}_{11} = Q_{11}c^4 + Q_{22}s^4 + 2(Q_{12} + 2Q_{66})s^2c^2,$$

$$\bar{Q}_{12} = (Q_{11} + Q_{22} - 4Q_{66})s^2c^2 + Q_{12}(c^4 + s^4),$$

$$\bar{Q}_{22} = Q_{11}s^4 + Q_{22}c^4 + 2(Q_{12} + 2Q_{66})s^2c^2,$$

$$\bar{Q}_{16} = (Q_{11} - Q_{12} - 2Q_{66})c^3s - (Q_{22} - Q_{12} - 2Q_{66})s^3c,$$

$$\bar{Q}_{26} = (Q_{11} - Q_{12} - 2Q_{66})cs^3 - (Q_{22} - Q_{12} - 2Q_{66})c^3s,$$

$$\bar{Q}_{66} = (Q_{11} + Q_{22} - 2Q_{12} - 2Q_{66})s^2c^2 + Q_{66}(s^4 + c^4). \quad (2.104a–f)$$

Note that six elements are in the $[\bar{Q}]$ matrix. However, by looking at Equation (2.104), it can be seen that they are just functions of the four stiffness elements, Q_{11}, Q_{12}, Q_{22}, and Q_{66}, and the angle of the lamina, θ.

Inverting Equation (2.103) gives

$$\begin{bmatrix} \varepsilon_x \\ \varepsilon_y \\ \gamma_{xy} \end{bmatrix} = \begin{bmatrix} \bar{S}_{11} & \bar{S}_{12} & \bar{S}_{16} \\ \bar{S}_{12} & \bar{S}_{22} & \bar{S}_{26} \\ \bar{S}_{16} & \bar{S}_{26} & \bar{S}_{66} \end{bmatrix} \begin{bmatrix} \sigma_x \\ \sigma_y \\ \sigma_{xy} \end{bmatrix}, \quad (2.105)$$

where $\bar{S}_{ij}$ are the elements of the transformed reduced compliance matrix and are given by

$$\bar{S}_{11} = S_{11}c^4 + (2S_{12} + S_{66})s^2c^2 + S_{22}s^4,$$

$$\bar{S}_{12} = S_{12}(s^4 + c^4) + (S_{11} + S_{22} - S_{66})s^2c^2,$$

$$\bar{S}_{22} = S_{11}s^4 + (2S_{12} + S_{66})s^2c^2 + S_{22}c^4,$$

$$\bar{S}_{16} = (2S_{11} - 2S_{12} - S_{66})sc^3 - (2S_{22} - 2S_{12} - S_{66})s^3c,$$

$$\bar{S}_{26} = (2S_{11} - 2S_{12} - S_{66})s^3c - (2S_{22} - 2S_{12} - S_{66})sc^3,$$

$$\bar{S}_{66} = 2(2S_{11} + 2S_{22} - 4S_{12} - S_{66})s^2c^2 + S_{66}(s^4 + c^4). \qquad (2.106a\text{–}f)$$

From Equation (2.77) and Equation (2.78), for a unidirectional lamina loaded in the material axes directions, no coupling occurs between the normal and shearing terms of strains and stresses. However, for an angle lamina, from Equation (2.103) and Equation (2.105), coupling takes place between the normal and shearing terms of strains and stresses. If only normal stresses are applied to an angle lamina, the shear strains are nonzero; if only shearing stresses are applied to an angle lamina, the normal strains are nonzero. Therefore, Equation (2.103) and Equation (2.105) are stress–strain equations for what is called a *generally orthotropic* lamina.

Example 2.7

Find the following for a 60° angle lamina (Figure 2.21) of graphite/epoxy. Use the properties of unidirectional graphite/epoxy lamina from Table 2.1.

1. Transformed compliance matrix
2. Transformed reduced stiffness matrix

If the applied stress is $\sigma_x = 2$ MPa, $\sigma_y = -3$ MPa, and $\tau_{xy} = 4$ MPa, also find

3. Global strains
4. Local strains
5. Local stresses
6. Principal stresses
7. Maximum shear stress

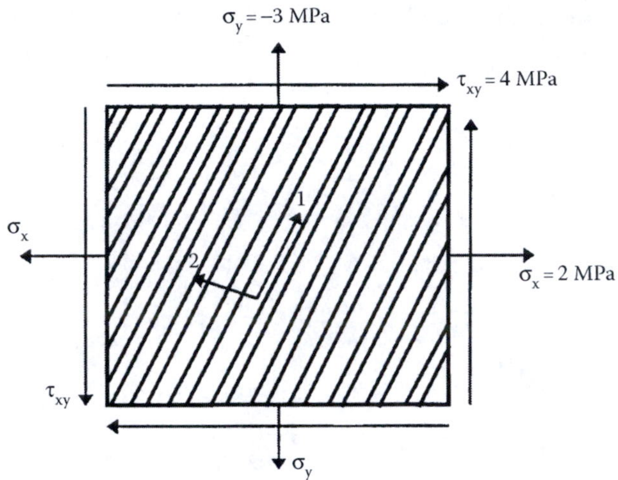

FIGURE 2.21
Applied stresses to an angle lamina in Example 2.7.

8. Principal strains
9. Maximum shear strain

Solution

$c = \text{Cos}(60°) = 0.500$
$s = \text{Sin}(60°) = 0.866$

1. From Example 2.6,

$$S_{11} = 0.5525 \times 10^{-11} \frac{1}{Pa},$$

$$S_{22} = 0.9709 \times 10^{-10} \frac{1}{Pa},$$

$$S_{12} = -0.1547 \times 10^{-11} \frac{1}{Pa},$$

$$S_{66} = 0.1395 \times 10^{-9} \frac{1}{Pa}.$$

Now, using Equation (2.106a),

$$\bar{S}_{11} = 0.5525 \times 10^{-11}(0.500)^4 + [2(-0.1547 \times 10^{-11})$$

$$+0.1395 \times 10^{-9}](0.866)^2(0.5)^2 + 0.9709 \times 10^{-10}(0.866)^4$$

$$= 0.8053 \times 10^{-10} \frac{1}{Pa}$$

Similarly, using Equation (2.106b–f), one can evaluate

$$\bar{S}_{12} = -0.7878 \times 10^{-11} \frac{1}{Pa},$$

$$\bar{S}_{16} = -0.3234 \times 10^{-10} \frac{1}{Pa},$$

$$\bar{S}_{22} = 0.3475 \times 10^{-10} \frac{1}{Pa},$$

$$\bar{S}_{26} = -0.4696 \times 10^{-10} \frac{1}{Pa},$$

$$\bar{S}_{66} = 0.1141 \times 10^{-9} \frac{1}{Pa}.$$

2. Invert the transformed compliance matrix [$\bar{S}$] to obtain the transformed reduced stiffness matrix [$\bar{Q}$]:

$$[\bar{Q}] = \begin{bmatrix} 0.8053 \times 10^{-10} & -0.7878 \times 10^{-11} & -0.3234 \times 10^{-10} \\ -0.7878 \times 10^{-11} & 0.3475 \times 10^{-10} & -0.4696 \times 10^{-10} \\ -0.3234 \times 10^{-10} & -0.4696 \times 10^{-10} & 0.1141 \times 10^{-9} \end{bmatrix}^{-1}$$

$$= \begin{bmatrix} 0.2365 \times 10^{11} & 0.3246 \times 10^{11} & 0.2005 \times 10^{11} \\ 0.3246 \times 10^{11} & 0.1094 \times 10^{12} & 0.5419 \times 10^{11} \\ 0.2005 \times 10^{11} & 0.5419 \times 10^{11} & 0.3674 \times 10^{11} \end{bmatrix} Pa.$$

3. The global strains in the x–y plane are given by Equation (2.105) as

$$\begin{bmatrix} \varepsilon_x \\ \varepsilon_y \\ \gamma_{xy} \end{bmatrix} = \begin{bmatrix} 0.8053\times10^{-10} & -0.7878\times10^{-11} & -0.3234\times10^{-10} \\ -0.7878\times10^{-11} & 0.3475\times10^{-10} & -0.4696\times10^{-10} \\ -0.3234\times10^{-10} & -0.4696\times10^{-10} & 0.1141\times10^{-9} \end{bmatrix} \begin{bmatrix} 2\times10^{6} \\ -3\times10^{6} \\ 4\times10^{6} \end{bmatrix}$$

$$= \begin{bmatrix} 0.5534\times10^{-4} \\ -0.3078\times10^{-3} \\ 0.5328\times10^{-3} \end{bmatrix}.$$

4. Using transformation Equation (2.99), the local strains in the lamina are

$$\begin{bmatrix} \varepsilon_1 \\ \varepsilon_2 \\ \gamma_{12}/2 \end{bmatrix} = \begin{bmatrix} 0.2500 & 0.7500 & 0.8660 \\ 0.7500 & 0.2500 & -0.8660 \\ -0.4330 & 0.4330 & -0.500 \end{bmatrix} \begin{bmatrix} 0.5534\times10^{-4} \\ -0.3078\times10^{-3} \\ 0.5328\times10^{-3}/2 \end{bmatrix}$$

$$\begin{bmatrix} \varepsilon_1 \\ \varepsilon_2 \\ \gamma_{12} \end{bmatrix} = \begin{bmatrix} 0.1367\times10^{-4} \\ -0.2662\times10^{-3} \\ -0.5809\times10^{-3} \end{bmatrix}.$$

5. Using transformation Equation (2.94), the local stresses in the lamina are

$$\begin{bmatrix} \sigma_1 \\ \sigma_2 \\ \tau_{12} \end{bmatrix} = \begin{bmatrix} 0.2500 & 0.7500 & 0.8660 \\ 0.7500 & 0.2500 & -0.8660 \\ -0.4330 & 0.4330 & -0.500 \end{bmatrix} \begin{bmatrix} 2\times10^{6} \\ -3\times10^{6} \\ 4\times10^{6} \end{bmatrix}$$

$$= \begin{bmatrix} 0.1714\times10^{7} \\ -0.2714\times10^{7} \\ -0.4165\times10^{7} \end{bmatrix} Pa.$$

6. The principal normal stresses are given by[4]

$$\sigma_{\max,\min} = \frac{\sigma_x + \sigma_y}{2} \pm \sqrt{\left(\frac{\sigma_x - \sigma_y}{2}\right)^2 + \tau_{xy}^2} \tag{2.107}$$

$$= \frac{2\times10^6 - 3\times10^6}{2} \pm \sqrt{\left(\frac{2\times10^6 + 3\times10^6}{2}\right)^2 + (4\times10^6)^2}$$

$$= 4.217, -5.217\ MPa.$$

The value of the angle at which the maximum normal stresses occur is[4]

$$\theta_p = \frac{1}{2}\tan^{-1}\left(\frac{2\tau_{xy}}{\sigma_x - \sigma_y}\right) \tag{2.108}$$

$$= \frac{1}{2}\tan^{-1}\left(\frac{2(4\times10^6)}{2\times10^6 + 3\times10^6}\right)$$

$$= 29.00^0.$$

Note that the principal normal stresses do not occur along the material axes. This should be also evident from the nonzero shear stresses in the local axes.

7. The maximum shear stress is given by[4]

$$\tau_{max} = \sqrt{\left(\frac{\sigma_x - \sigma_y}{2}\right)^2 + \tau_{xy}^2}$$

$$= \sqrt{\left(\frac{2\times10^6 - 3\times10^6}{2}\right)^2 + (4\times10^6)^2} \tag{2.109}$$

$$= 4.717\ MPa.$$

The angle at which the maximum shear stress occurs is[4]

$$\theta_s = \frac{1}{2}\tan^{-1}\left(-\frac{\sigma_x - \sigma_y}{2\tau_{xy}}\right) \tag{2.110}$$

$$= \frac{1}{2}\tan^{-1}\left(-\frac{2\times10^6+3\times10^6}{2(4\times10^6)}\right)$$

$$= 16.00^0$$

8. The principal strains are given by[4]

$$\varepsilon_{\max,\min} = \frac{\varepsilon_x+\varepsilon_y}{2} \pm \sqrt{\left(\frac{\varepsilon_x-\varepsilon_y}{2}\right)^2+\left(\frac{\gamma_{xy}}{2}\right)^2}$$

$$= \frac{0.5534\times10^{-4}+0.3078\times10^{-3}}{2} \pm \sqrt{\left(\frac{0.5534\times10^{-4}+0.3078\times10^{-3}}{2}\right)^2+\left(\frac{0.5328\times10^{-3}}{2}\right)^2} \quad (2.111)$$

$$= 1.962\times10^{-4},\ -4.486\times10^{-4}\ .$$

The value of the angle at which the maximum normal strains occur is[4]

$$\theta_p = \frac{1}{2}\tan^{-1}\left(\frac{\gamma_{xy}}{\varepsilon_x-\varepsilon_y}\right)$$

$$= \frac{1}{2}\tan^{-1}\left(\frac{0.5328\times10^{-3}}{0.5534\times10^{-4}+0.3078\times10^{-3}}\right) \quad (2.112)$$

$$= 27.86^0.$$

Note that the principal normal strains do not occur along the material axes. This should also be clear from the nonzero shear strain in the local axes. In addition, the axes of principal normal stresses and principal normal strains do not match, unlike in isotropic materials.

9. The maximum shearing strain is given by[4]

$$\gamma_{\max} = \sqrt{(\varepsilon_x - \varepsilon_y)^2 + \gamma_{xy}^2}$$

$$= \sqrt{(0.5534 \times 10^{-4} + 0.3078 \times 10^{-3})^2 + (0.532 \times 10^{-3})^2} \tag{2.113}$$

$$= 6.448 \times 10^{-4}.$$

The value of the angle at which the maximum shearing strain occurs is[4]

$$\theta_s = \frac{1}{2}\tan^{-1}\left(-\frac{\varepsilon_x - \varepsilon_y}{\gamma_{xy}}\right)$$

$$= \frac{1}{2}\tan^{-1}\left(-\frac{0.5534 \times 10^{-4} + 0.3078 \times 10^{-3}}{0.5328 \times 10^{-3}}\right) \tag{2.114}$$

$$= -17.14^0.$$

Example 2.8

As shown in Figure 2.22, a 60° angle graphite/epoxy lamina is subjected only to a shear stress τ_{xy} = 2 MPa in the global axes. What would be the value of the strains measured by the strain gage rosette — that is, what

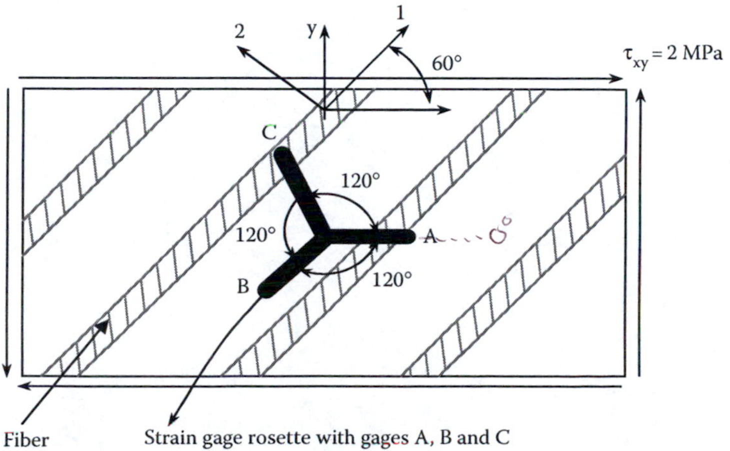

FIGURE 2.22
Strain gage rosette on an angle lamina.

would be the normal strains measured by strain gages A, B, and C? Use the properties of unidirectional graphite/epoxy lamina from Table 2.1.

Solution

Per Example 2.7, the reduced compliance matrix $[\bar{S}]$ is

$$\begin{bmatrix} 0.8053\times10^{-10} & -0.7878\times10^{-11} & -0.3234\times10^{-10} \\ -0.7878\times10^{-11} & 0.3475\times10^{-10} & -0.4696\times10^{-10} \\ -0.3234\times10^{-10} & -0.4696\times10^{-10} & 0.1141\times10^{-9} \end{bmatrix}\frac{1}{Pa}.$$

The global strains in the x–y plane are given by Equation (2.105) as

$$\begin{bmatrix} \varepsilon_x \\ \varepsilon_y \\ \gamma_{xy} \end{bmatrix} = \begin{bmatrix} 0.8053\times10^{-10} & -0.7878\times10^{-11} & -0.3234\times10^{-10} \\ -0.7878\times10^{-11} & 0.3475\times10^{-10} & -0.4696\times10^{-10} \\ -0.3234\times10^{-10} & -0.4696\times10^{-10} & 0.1141\times10^{-9} \end{bmatrix}\begin{bmatrix} 0 \\ 0 \\ 2\times10^{6} \end{bmatrix}$$

$$= \begin{bmatrix} -6.468\times10^{-5} \\ -9.392\times10^{-5} \\ 2.283\times10^{-4} \end{bmatrix}.$$

For a strain gage placed at an angle, ϕ, to the x-axis, the normal strain recorded by the strain gage is given by Equation (B.15) in Appendix B.

$$\varepsilon_\phi = \varepsilon_x \mathrm{Cos}^2\phi + \varepsilon_y\ \mathrm{Sin}^2\phi + \gamma_{xy}\,\mathrm{Sin}\,\phi\,\mathrm{Cos}\,\phi\ .$$

For strain gage A, $\phi = 0°$:

$$\varepsilon_A = -6.468\times10^{-5}\,Cos^2\,0° + (-9.392\times10^{-5})\mathrm{Sin}^2\,0° + 2.283\times10^{-4}\,\mathrm{Sin}\,0°\,\mathrm{Cos}\,0°$$

$$= -6.468\times10^{-5}\ .$$

For strain gage B, $\phi = 240°$:

$$\varepsilon_B = -6.468\times10^{-5}\,\mathrm{Cos}^2\,240° + (-9.392\times10^{-5})\mathrm{Sin}^2\,240°$$
$$+2.283\times10^{-4}\,\mathrm{Sin}\,240°\,\mathrm{Cos}\,240°$$

$$= 1.724 \times 10^{-4} .$$

For strain gage C, $\phi = 120°$:

$$\varepsilon_C = -6.468 \times 10^{-5} \text{Cos}^2 120° + (-9.392 \times 10^{-5}) \text{Sin}^2 120°$$
$$+ 2.283 \times 10^{-4} \text{Sin} 120° \text{Cos} 120°$$

$$= 1.083 \times 10^{-5} .$$

2.6 Engineering Constants of an Angle Lamina

The engineering constants for a unidirectional lamina were related to the compliance and stiffness matrices in Section 2.4.3. In this section, similar techniques are applied to relate the engineering constants of an angle ply to its transformed stiffness and compliance matrices.

1. For finding the engineering elastic moduli in direction x (Figure 2.23a), apply

$$\sigma_x \neq 0, \sigma_y = 0, \tau_{xy} = 0. \quad (2.115)$$

Then, from Equation (2.105),

$$\varepsilon_x = \bar{S}_{11} \sigma_x ,$$

$$\varepsilon_y = \bar{S}_{12} \sigma_x ,$$

$$\gamma_{xy} = \bar{S}_{16} \sigma_x . \quad (2.116a\text{–}c)$$

The elastic moduli in direction x is defined as

$$E_x \equiv \frac{\sigma_x}{\varepsilon_x} = \frac{1}{\bar{S}_{11}} . \quad (2.117)$$

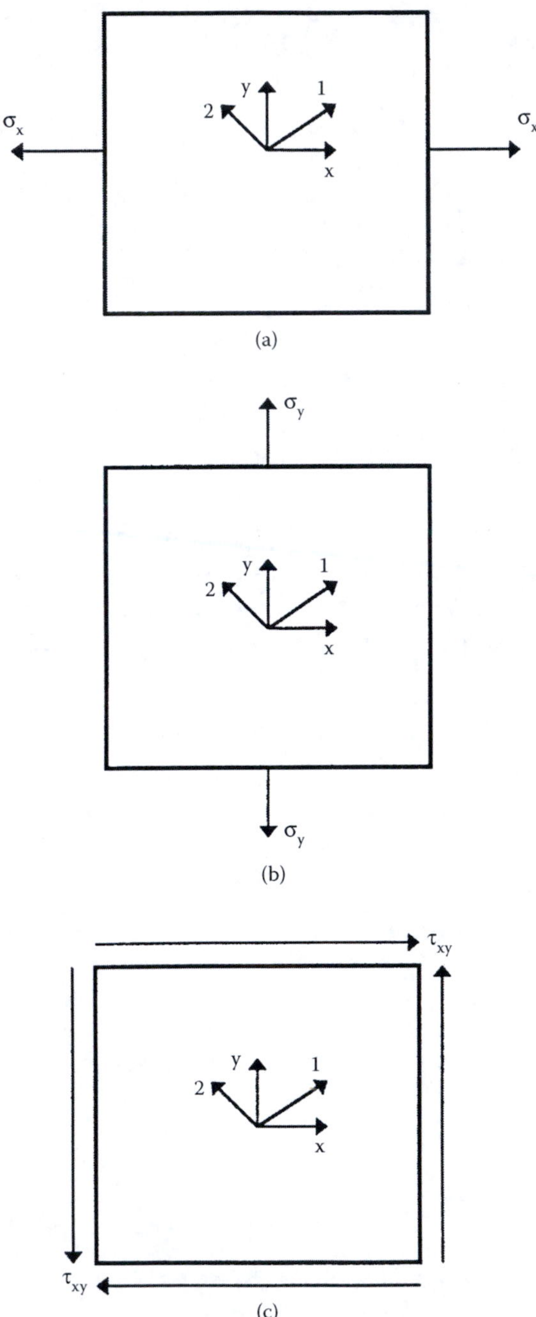

FIGURE 2.23
Application of stresses to find engineering constants of an angle lamina.

Also, the Poisson's ratio, ν_{xy}, is defined as

$$\nu_{xy} \equiv -\frac{\varepsilon_y}{\varepsilon_x} = -\frac{\bar{S}_{12}}{\bar{S}_{11}}. \tag{2.118}$$

In an angle lamina, unlike in a unidirectional lamina, interaction also occurs between the shear strain and the normal stresses. This is called shear coupling. The shear coupling term that relates the normal stress in the x-direction to the shear strain is denoted by m_x and is defined as

$$\frac{1}{m_x} \equiv -\frac{\sigma_x}{\gamma_{xy} E_1} = -\frac{1}{\bar{S}_{16} E_1}. \tag{2.119}$$

Note that m_x is a nondimensional parameter like the Poisson's ratio. Later, note that the same parameter, m_x, relates the shearing stress in the x–y plane to the normal strain in direction-x.

The shear coupling term is particularly important in tensile testing of angle plies. For example, if an angle lamina is clamped at the two ends, it will not allow shearing strain to occur. This will result in bending moments and shear forces at the clamped ends.[5]

2. Similarly, by applying stresses

$$\sigma_x = 0,\ \sigma_y \neq 0,\ \tau_{xy} = 0\ , \tag{2.120}$$

as shown in Figure 2.23b, it can be found

$$E_y = \frac{1}{\bar{S}_{22}}, \tag{2.121}$$

$$\nu_{yx} = -\frac{\bar{S}_{12}}{\bar{S}_{22}},\ \text{and} \tag{2.122}$$

$$\frac{1}{m_y} = -\frac{1}{\bar{S}_{26} E_1}. \tag{2.123}$$

The shear coupling term m_y relates the normal stress σ_y to the shear strain γ_{xy}. In the following section (3), note that the same parameter m_y relates the shear stress τ_{xy} in the x–y plane to the normal strain ε_y.

From Equation (2.117), Equation (2.118), Equation (2.121), and Equation (2.122), the reciprocal relationship is given by

$$\frac{\nu_{yx}}{E_y} = \frac{\nu_{xy}}{E_x} . \tag{2.124}$$

3. Also, by applying the stresses

$$\sigma_x = 0,\ \sigma_y = 0,\ \tau_{xy} \neq 0 , \tag{2.125}$$

as shown in Figure 2.23c, it is found that

$$\frac{1}{m_x} = -\frac{1}{\bar{S}_{16} E_1}, \tag{2.126}$$

$$\frac{1}{m_y} = -\frac{1}{\bar{S}_{26} E_1}, \text{ and} \tag{2.127}$$

$$G_{xy} = \frac{1}{\bar{S}_{66}}. \tag{2.128}$$

Thus, the strain–stress Equation (2.105) of an angle lamina can also be written in terms of the engineering constants of an angle lamina in matrix form as

$$\begin{bmatrix} \varepsilon_x \\ \varepsilon_y \\ \gamma_{xy} \end{bmatrix} = \begin{bmatrix} \frac{1}{E_x} & -\frac{\nu_{xy}}{E_x} & -\frac{m_x}{E_1} \\ -\frac{\nu_{xy}}{E_x} & \frac{1}{E_y} & -\frac{m_y}{E_1} \\ -\frac{m_x}{E_1} & -\frac{m_y}{E_1} & \frac{1}{G_{xy}} \end{bmatrix} \begin{bmatrix} \sigma_x \\ \sigma_y \\ \tau_{xy} \end{bmatrix}. \tag{2.129}$$

The preceding six engineering constants of an angle ply can also be written in terms of the engineering constants of a unidirectional ply using Equation (2.92) and Equation (2.106) in Equation (2.117) through Equation (2.119), Equation (2.121), Equation (2.123), and Equation (2.128):

$$\frac{1}{E_x} = \bar{S}_{11}$$

$$= S_{11}c^4 + (2S_{12} + S_{66})s^2c^2 + S_{22}s^4.$$

$$= \frac{1}{E_1}c^4 + \left(\frac{1}{G_{12}} - \frac{2\nu_{12}}{E_1}\right)s^2c^2 + \frac{1}{E_2}s^4, \tag{2.130}$$

$$\nu_{xy} = -E_x\bar{S}_{12}$$

$$= -E_x[S_{12}(s^4 + c^4) + (S_{11} + S_{22} - S_{66})s^2c^2]$$

$$= E_x\left[\frac{\nu_{12}}{E_1}\left(s^4 + c^4\right) - \left(\frac{1}{E_1} + \frac{1}{E_2} - \frac{1}{G_{12}}\right)s^2c^2\right], \tag{2.131}$$

$$\frac{1}{E_y} = \bar{S}_{22}$$

$$= S_{11}s^4 + (2S_{12} + S_{66})c^2s^2 + S_{22}c^4$$

$$= \frac{1}{E_1}s^4 + \left(-\frac{2\nu_{12}}{E_1} + \frac{1}{G_{12}}\right)c^2s^2 + \frac{1}{E_2}c^4, \tag{2.132}$$

$$\frac{1}{G_{xy}} = \bar{S}_{66}$$

$$= 2(2S_{11} + 2S_{22} - 4S_{12} - S_{66})s^2c^2 + S_{66}(s^4 + c^4)$$

$$= 2\left(\frac{2}{E_1} + \frac{2}{E_2} + \frac{4\nu_{12}}{E_1} - \frac{1}{G_{12}}\right)s^2c^2 + \frac{1}{G_{12}}(s^4 + c^4), \tag{2.133}$$

$$m_x = -\bar{S}_{16}E_1$$

$$= -E_1[(S_{11} - 2S_{12} - S_{66})sc^3 - (2S_{22} - 2S_{12} - S_{66})s^3c]$$

$$= E_1\left[\left(-\frac{2}{E_1} - \frac{2\nu_{12}}{E_1} + \frac{1}{G_{12}}\right)sc^3 + \left(\frac{2}{E_2} + \frac{2\nu_{12}}{E_1} - \frac{1}{G_{12}}\right)s^3c\right], \tag{2.134}$$

$$m_y = -\bar{S}_{26} E_1$$

$$= -E_1[(2S_{11} - 2S_{12} - S_{66})s^3c - (2S_{22} - 2S_{12} - S_{66})sc^3]$$

$$= E_1\left[\left(-\frac{2}{E_1} - \frac{2\nu_{12}}{E_1} + \frac{1}{G_{12}}\right)s^3c + \left(\frac{2}{E_2} + \frac{2\nu_{12}}{E_1} - \frac{1}{G_{12}}\right)sc^3\right]. \tag{2.135}$$

Example 2.9

Find the engineering constants of a 60° graphite/epoxy lamina. Use the properties of a unidirectional graphite/epoxy lamina from Table 2.1.

Solution

From Example 2.7, we have

$$\bar{S}_{11} = 0.8053 \times 10^{-10} \frac{1}{Pa},$$

$$\bar{S}_{12} = -0.7878 \times 10^{-11} \frac{1}{Pa},$$

$$\bar{S}_{16} = -0.3234 \times 10^{-10} \frac{1}{Pa},$$

$$\bar{S}_{22} = 0.3475 \times 10^{-10} \frac{1}{Pa},$$

$$\bar{S}_{26} = -0.4696 \times 10^{-10} \frac{1}{Pa}, \text{ and}$$

$$\bar{S}_{66} = 0.1141 \times 10^{-9} \frac{1}{Pa}.$$

From Equation (2.117),

$$E_x = \frac{1}{0.8053 \times 10^{-10}}$$

$$= 12.42 GPa.$$

From Equation (2.118),

$$\nu_{xy} = -\frac{-0.7878 \times 10^{-11}}{0.8053 \times 10^{-10}}$$

$$= 0.09783.$$

From Equation (2.119),

$$\frac{1}{m_x} = -\frac{1}{(-0.3234 \times 10^{-10})(181 \times 10^9)}$$

$$m_x = 5.854 \, .$$

From Equation (2.121),

$$E_y = \frac{1}{0.3475 \times 10^{-10}}$$

$$= 28.78 \; GPa.$$

From Equation (2.123),

$$\frac{1}{m_y} = -\frac{1}{(-0.4696 \times 10^{-10})(181 \times 10^9)}$$

$$m_y = 8.499.$$

From Equation (2.128),

$$G_{xy} = \frac{1}{0.1141 \times 10^{-9}}$$

$$= 8.761 \; GPa.$$

The variations of the six engineering elastic constants are shown as a function of the angle for the preceding graphite/epoxy lamina in Figure 2.24 through Figure 2.29.

The variations of the Young's modulus, E_x and E_y are inverses of each other. As the fiber orientation (angle of ply) varies from 0° to 90°, the value of E_x

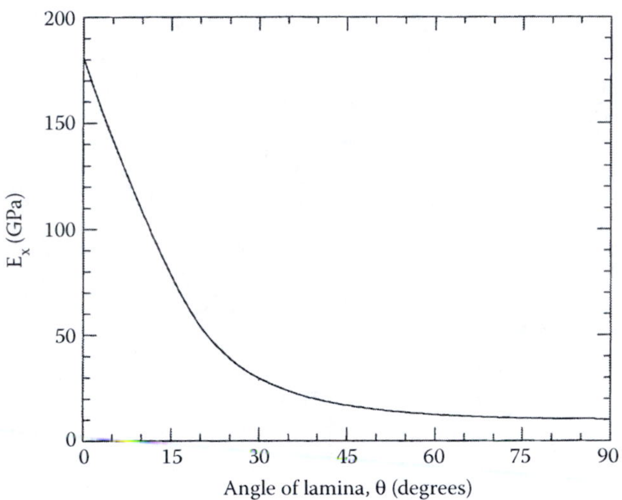

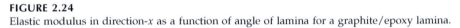

FIGURE 2.24
Elastic modulus in direction-x as a function of angle of lamina for a graphite/epoxy lamina.

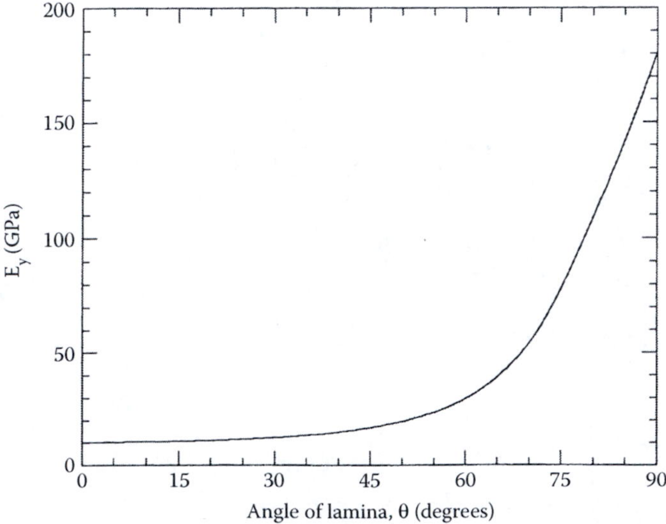

FIGURE 2.25
Elastic modulus in direction-y as a function of angle of lamina for a graphite/epoxy lamina.

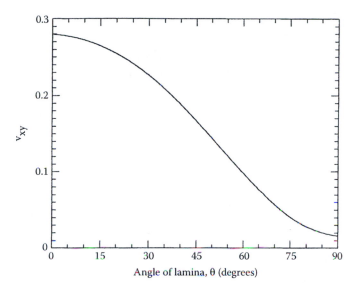

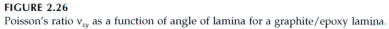

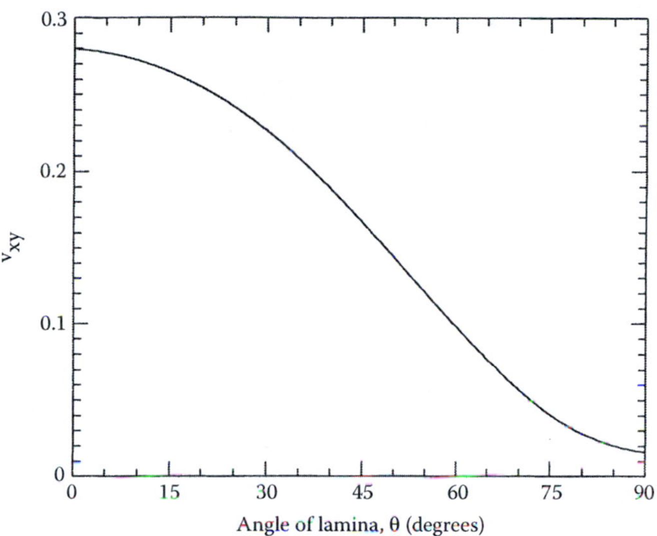

FIGURE 2.26
Poisson's ratio ν_{xy} as a function of angle of lamina for a graphite/epoxy lamina.

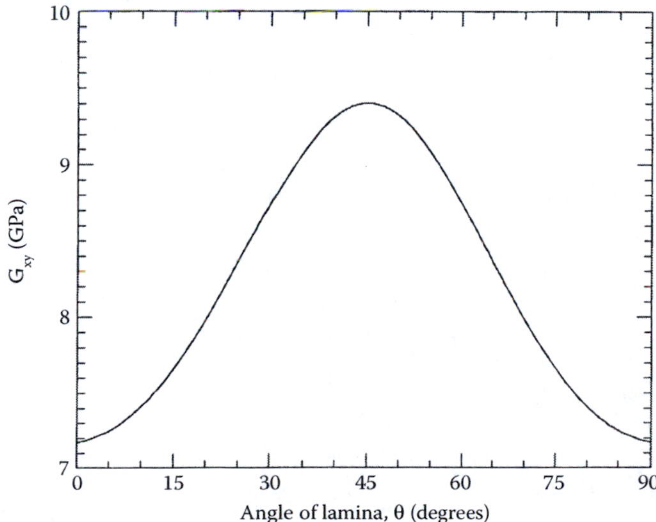

FIGURE 2.27
In-plane shear modulus in *xy*-plane as a function of angle of lamina for a graphite/epoxy lamina.

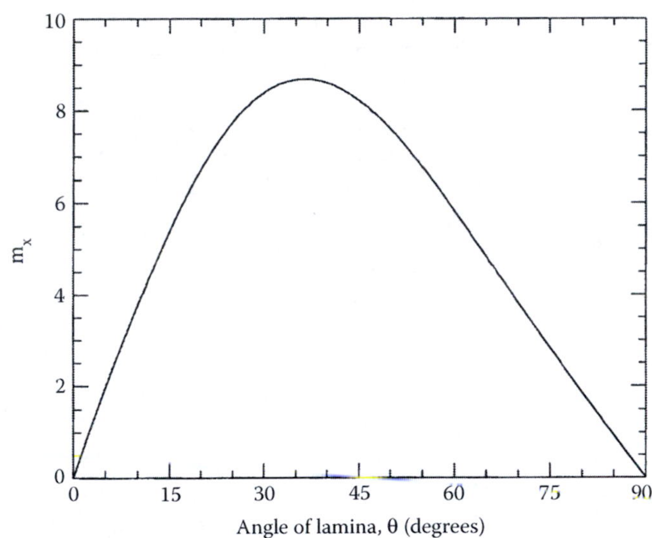

FIGURE 2.28
Shear coupling coefficient m_x as a function of angle of lamina for a graphite/epoxy lamina.

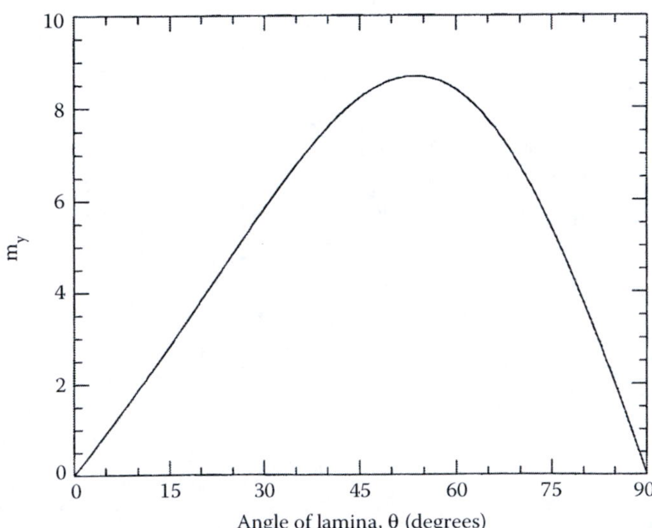

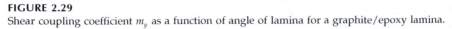

FIGURE 2.29
Shear coupling coefficient m_y as a function of angle of lamina for a graphite/epoxy lamina.

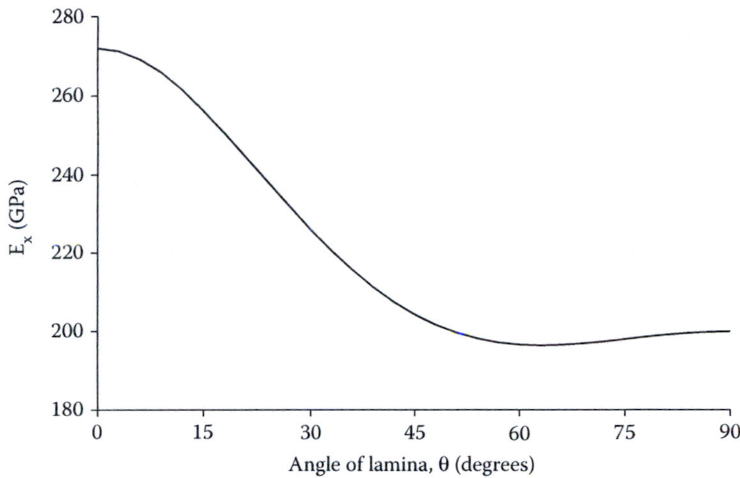

FIGURE 2.30
Variation of elastic modulus in direction-x as a function of angle of lamina for a typical $SCS - 6/Ti6 - Al - 4V$ lamina.

varies from the value of the longitudinal (E_1) to the transverse Young's modulus E_2. However, the maximum and minimum values of E_x do not necessarily exist for $\theta = 0°$ and $\theta = 90°$, respectively, for every lamina.

Consider the case of a metal matrix composite such as a typical $SCS - 6/Ti6 - Al - 4V$ composite. The elastic moduli of such a lamina with a 55% fiber volume fraction is

$$E_1 = 272 \text{ GPa}$$

$$E_2 = 200 \text{ GPa}$$

$$\nu_{12} = 0.2770$$

$$G_{12} = 77.33 \text{ GPa}$$

In Figure 2.30, the lowest modulus value of E_x is found for $\theta = 63°$. In fact, the angle of 63° at which E_x is minimum is independent of the fiber volume fraction, if one uses the "mechanics of materials approach" (Section 3.3.1) to evaluate the preceding four elastic moduli of a unidirectional lamina. See Exercise 3.13.

In Figure 2.27, the shear modulus G_{xy} is maximum for $\theta = 45°$ and is minimum for 0 and 90° plies. The shear modulus G_{xy} becomes maximum for 45° because the principal stresses for pure shear load on a 45° ply are along the material axis.

From Equation (2.133), the expression for G_{xy} for a 45° ply is

$$G_{xy/45^\circ} = \frac{E_1}{\left(1 + 2\nu_{12} + \frac{E_1}{E_2}\right)}. \tag{2.136}$$

In Figure 2.28 and Figure 2.29, the shear coupling coefficients m_x and m_y are maximum at $\theta = 36.2°$ and $\theta = 53.78°$, respectively. The values of these coefficients are quite extreme, showing that the normal-shear coupling terms have a stronger effect than the Poisson's effect. This phenomenon of shear coupling terms is missing in isotropic materials and unidirectional plies, but cannot be ignored in angle plies.

2.7 Invariant Form of Stiffness and Compliance Matrices for an Angle Lamina

Equation (2.104) and Equation (2.106) for the $[\bar{Q}]$ and $[\bar{S}]$ matrices are not analytically convenient because they do not allow a direct study of the effect of the angle of the lamina on the $[\bar{Q}]$ and $[\bar{S}]$ matrices. The stiffness elements can be written in invariant form as[6]

$$\bar{Q}_{11} = U_1 + U_2 \operatorname{Cos} 2\theta + U_3 \operatorname{Cos} 4\theta,$$

$$\bar{Q}_{12} = U_4 - U_3 \operatorname{Cos} 4\theta,$$

$$\bar{Q}_{22} = U_1 - U_2 \operatorname{Cos} 2\theta + U_3 \operatorname{Cos} 4\theta$$

$$\bar{Q}_{16} = \frac{U_2}{2} \operatorname{Sin} 2\theta + U_3 \operatorname{Sin} 4\theta,$$

$$\bar{Q}_{26} = \frac{U_2}{2} \operatorname{Sin} 2\theta - U_3 \operatorname{Sin} 4\theta,$$

$$\bar{Q}_{66} = \frac{1}{2}(U_1 - U_4) - U_3 \operatorname{Cos} 4\theta, \tag{2.137a–f}$$

where

$$U_1 = \frac{1}{8}(3Q_{11} + 3Q_{22} + 2Q_{12} + 4Q_{66})$$

$$U_2 = \frac{1}{2}(Q_{11} - Q_{22}),$$

$$U_3 = \frac{1}{8}(Q_{11} + Q_{22} - 2Q_{12} - 4Q_{66}),$$

$$U_4 = \frac{1}{8}(Q_{11} + Q_{22} + 6Q_{12} - 4Q_{66}). \quad (2.138a\text{–}d)$$

The terms U_1, U_2, U_3, and U_4 are the four invariants and are combinations of the Q_{ij}, which are invariants as well.

The transformed reduced compliance [$\bar{S}$] matrix can similarly be written as

$$\bar{S}_{11} = V_1 + V_2 \operatorname{Cos} 2\theta + V_3 \operatorname{Cos} 4\theta,$$

$$\bar{S}_{12} = V_4 - V_3 \operatorname{Cos} 4\theta,$$

$$\bar{S}_{22} = V_1 - V_2 \operatorname{Cos} 2\theta + V_3 \operatorname{Cos} 4\theta,$$

$$\bar{S}_{16} = V_2 \operatorname{Sin} 2\theta + 2V_3 \operatorname{Sin} 4\theta,$$

$$\bar{S}_{26} = V_2 \operatorname{Sin} 2\theta - 2V_3 \operatorname{Sin} 4\theta, \text{ and}$$

$$\bar{S}_{66} = 2(V_1 - V_4) - 4V_3 \operatorname{Cos} 4\theta, \quad (2.139a\text{–}f)$$

where

$$V_1 = \frac{1}{8}(3S_{11} + 3S_{22} + 2S_{12} + S_{66}),$$

$$V_2 = \frac{1}{2}(S_{11} - S_{22}),$$

$$V_3 = \frac{1}{8}(S_{11} + S_{22} - 2S_{12} - S_{66}),$$

$$V_4 = \frac{1}{8}(S_{11} + S_{22} + 6S_{12} - S_{66}). \quad (2.140a–d)$$

The terms V_1, V_2, V_3, and V_4 are invariants and are combinations of S_{ij}, which are also invariants.

The main advantage of writing the equations in this form is that one can easily examine the effect of the lamina angle on the reduced stiffness matrix elements. Also, formulas given by Equation (2.137) and Equation (2.139) are easier to manipulate for integration, differentiation, etc. The concept is mainly important in deriving the laminate stiffness properties in Chapter 4.

The elastic moduli of quasi-isotropic laminates that behave like isotropic material are directly given in terms of these invariants. Because quasi-isotropic laminates have the minimum stiffness of any laminate, these can be used as a comparative measure of the stiffness of other types of laminates.[7]

Example 2.10

Starting with the expression for $\bar{Q}_{11}$ from Equation (2.104a), $\bar{Q}_{11} = Q_{11}\,\text{Cos}^4\,\theta,$ $+ Q_{22}\,\text{Sin}^4\,\theta + 2(Q_{12} + 2Q_{66})\text{Sin}^2\,\theta\,\text{Cos}^2\,\theta$, reduce it to the expression for $\bar{Q}_{11}$ of Equation (2.137a) — that is,

$$\bar{Q}_{11} = U_1 + U_2\,\text{Cos}\,2\theta + U_3\,\text{Cos}\,4\theta$$

Solution

Given

$$\bar{Q}_{11} = Q_{11}\,\text{Cos}^4\,\theta + Q_{22}\,\text{Sin}^4\,\theta + 2(Q_{12} + 2Q_{66})\text{Sin}^2\,\theta\,\text{Cos}^2\,\theta\,,$$

and substituting

$$\text{Cos}^2\,\theta = \frac{1 + \text{Cos}\,2\theta}{2},$$

$$\text{Sin}^2\,\theta = \frac{1 - \text{Cos}\,2\theta}{2},$$

$$\mathrm{Cos}^2 2\theta = \frac{1+\mathrm{Cos}\,4\theta}{2}, \text{ and}$$

$$2\,\mathrm{Sin}\,\theta\,\mathrm{Cos}\,\theta = \mathrm{Sin}\,2\theta,$$

$$\mathrm{Sin}^2 2\theta = \frac{1-\mathrm{Cos}\,4\theta}{2},$$

we get

$$\bar{Q}_{11} = U_1 + U_2\,\mathrm{Cos}\,2\theta + U_3\,\mathrm{Cos}\,4\theta\,,$$

where

$$U_1 = \frac{1}{8}(3Q_{11} + 3Q_{22} + 2Q_{12} + 4Q_{66}),$$

$$U_2 = \frac{1}{2}(Q_{11} - Q_{22})$$

$$U_3 = \frac{1}{8}(Q_{11} + Q_{22} - 2Q_{12} - 4Q_{66})\,.$$

Example 2.11

Evaluate the four compliance and four stiffness invariants for a graphite/epoxy angle lamina. Use the properties for a unidirectional graphite/epoxy lamina from Table 2.1.

Solution

From Example 2.6, the compliance matrix [S] elements are

$$S_{11} = 0.5525 \times 10^{-11}\,\frac{1}{Pa},$$

$$S_{12} = -0.1547 \times 10^{-11}\,\frac{1}{Pa},$$

$$S_{22} = 0.9709 \times 10^{-10} \frac{1}{Pa},$$

$$S_{66} = 0.1395 \times 10^{-9} \frac{1}{Pa}.$$

The stiffness matrix $[Q]$ elements are

$$[Q] = [S]^{-1},$$

$$Q_{11} = 0.1818 \times 10^{12} \; Pa,$$

$$Q_{12} = 0.2897 \times 10^{10} \; Pa,$$

$$Q_{22} = 0.1035 \times 10^{11} \; Pa,$$

$$Q_{66} = 0.7170 \times 10^{10} \; Pa.$$

Using Equation (2.138),

$$U_1 = \frac{1}{8}[3(0.1818 \times 10^{12}) + 3(0.1035 \times 10^{11}) + 2(0.2897 \times 10^{10}) + 4(0.7171 \times 10^{10})]$$

$$= 0.7637 \times 10^{11} \; Pa,$$

$$U_2 = \frac{1}{2}(0.1818 \times 10^{12} - 0.1035 \times 10^{11})$$

$$= 0.8573 \times 10^{11} \; Pa,$$

$$U_3 = \frac{1}{8}[0.1818 \times 10^{12} + 0.1035 \times 10^{11} - 2(0.2897 \times 10^{10}) - 4(0.7171 \times 10^{10})]$$

$$= 0.1971 \times 10^{11} \; Pa,$$

$$U_4 = \frac{1}{8}[0.1818 \times 10^{12} + 0.1035 \times 10^{11} + 6(0.2897 \times 10^{10}) - 4(0.7171 \times 10^{10})]$$

$$= 0.2261 \times 10^{11} \; Pa.$$

Using Equation (2.140),

$$V_1 = \frac{1}{8}[3(0.5525\times10^{-11})+3(-0.1547\times10^{-11})+2(0.9709\times10^{-10})+0.1395\times10^{-9}]$$

$$= 0.5553\times10^{-10}\frac{1}{Pa},$$

$$V_2 = \frac{1}{2}[(0.5525\times10^{-11}-(-0.1547\times10^{-11})]$$

$$= -0.4578\times10^{-10}\frac{1}{Pa},$$

$$V_3 = \frac{1}{8}[0.5525\times10^{-11}+0.9709\times10^{-10}-2(0.1547\times10^{-11})-0.1395\times10^{-9}]$$

$$= -0.4220\times10^{-11}\frac{1}{Pa},$$

$$V_4 = \frac{1}{8}[0.5525\times10^{-11}+0.9709\times10^{-10}+6(0.1547\times10^{-11})-0.1395\times10^{-9}]$$

$$= -0.5767\times10^{-11}\frac{1}{Pa}.$$

2.8 Strength Failure Theories of an Angle Lamina

A successful design of a structure requires efficient and safe use of materials. Theories need to be developed to compare the state of stress in a material to failure criteria. It should be noted that failure theories are only stated and their application is validated by experiments.

For a laminate, the strength is related to the strength of each individual lamina. This allows for a simple and economical method for finding the strength of a laminate. Various theories have been developed for studying the failure of an angle lamina. The theories are generally based on the normal and shear strengths of a unidirectional lamina.

An isotropic material, such as steel, generally has two strength parameters: normal strength and shear strength. In some cases, such as concrete or gray cast iron, the normal strengths are different in the tension and compression. A simple failure theory for an isotropic material is based on finding the principal normal stresses and the maximum shear stresses. These maximum stresses, if greater than any of the corresponding ultimate strengths, indicate failure in the material.

Example 2.12

A cylindrical rod made of gray cast iron is subjected to a uniaxial tensile load, *P*. Given:

Cross-sectional area of rod = 2 in.^{2}
Ultimate tensile strength = 25 ksi
Ultimate compressive strength = 95 ksi
Ultimate shear strength = 35 ksi
Modulus of elasticity = 10 Msi

Find the maximum load, *P*, that can be applied using maximum stress failure theory.

Solution

At any location, the stress state in the rod is $\sigma = P/2$. From a typical Mohr's circle analysis, the maximum principal normal stress is $P/2$. The maximum shear stress is $P/4$ and acts at a cross-section 45° to the plane of maximum normal stress. Comparing these maximum stresses to the corresponding ultimate strengths, we have

$$\frac{P}{2} < 25 \times 10^3 \text{ } or \text{ } P < 50{,}000 \text{ lb},$$

and

$$\frac{P}{4} < 35 \times 10^3 \text{ } or \text{ } P < 140{,}000 \text{ lb}.$$

Thus, the maximum load is 50,000 lb.

However, in a lamina, the failure theories are not based on principal normal stresses and maximum shear stresses. Rather, they are based on the stresses in the material or local axes because a lamina is orthotropic and its properties are different at different angles, unlike an isotropic material.

In the case of a unidirectional lamina, there are two material axes: one parallel to the fibers and one perpendicular to the fibers. Thus, there are four normal strength parameters for a unidirectional lamina, one for tension and one for compression, in each of the two material axes directions. The fifth strength parameter is the shear strength of a unidirectional lamina. The shear stress, whether positive or negative, does not have an effect on the reported shear strengths of a unidirectional lamina. However, we will find later that the sign of the shear stress does affect the strength of an angle lamina. The five strength parameters of a unidirectional lamina are therefore

$(\sigma_1^T)_{ult}$ = Ultimate longitudinal tensile strength (in direction 1),
$(\sigma_1^C)_{ult}$ = Ultimate longitudinal compressive strength (in direction 1),
$(\sigma_2^T)_{ult}$ = Ultimate transverse tensile strength (in direction 2),
$(\sigma_2^C)_{ult}$ = Ultimate transverse compressive strength (in direction 2), and
$(\tau_{12})_{ult}$ = Ultimate in-plane shear strength (in plane 12).

Unlike the stiffness parameters, these strength parameters cannot be transformed directly for an angle lamina. Thus, the failure theories are based on first finding the stresses in the local axes and then using these five strength parameters of a unidirectional lamina to find whether a lamina has failed. Four common failure theories are discussed here. Related concepts of strength ratio and the development of failure envelopes are also discussed.

2.8.1 Maximum Stress Failure Theory

Related to the maximum normal stress theory by Rankine and the maximum shearing stress theory by Tresca, this theory is similar to those applied to isotropic materials. The stresses acting on a lamina are resolved into the normal and shear stresses in the local axes. Failure is predicted in a lamina, if any of the normal or shear stresses in the local axes of a lamina is equal to or exceeds the corresponding ultimate strengths of the unidirectional lamina.

Given the stresses or strains in the global axes of a lamina, one can find the stresses in the material axes by using Equation (2.94). The lamina is considered to be failed if

$$-(\sigma_1^C)_{ult} < \sigma_1 < (\sigma_1^T)_{ult}, \text{ or}$$

$$-(\sigma_2^C)_{ult} < \sigma_2 < (\sigma_2^T)_{ult}, \text{ or}$$

$$-(\tau_{12})_{ult} < \tau_{12} < (\tau_{12})_{ult} \tag{2.141a–c}$$

is violated. Note that all five strength parameters are treated as positive numbers, and the normal stresses are positive if tensile and negative if compressive.

Each component of stress is compared with the corresponding strength; thus, each component of stress does not interact with the others.

Example 2.13

Find the maximum value of $S > 0$ if a stress of $\sigma_x = 2S$, $\sigma_y = -3S$, and $\tau_{xy} = 4S$ is applied to the 60° lamina of graphite/epoxy. Use maximum stress failure theory and the properties of a unidirectional graphite/epoxy lamina given in Table 2.1.

Solution

Using Equation (2.94), the stresses in the local axes are

$$\begin{bmatrix} \sigma_1 \\ \sigma_2 \\ \tau_{12} \end{bmatrix} = \begin{bmatrix} 0.2500 & 0.7500 & 0.8660 \\ 0.7500 & 0.2500 & -0.8660 \\ -0.4330 & 0.4330 & -0.5000 \end{bmatrix} \begin{bmatrix} 2S \\ -3S \\ 4S \end{bmatrix}$$

$$= \begin{bmatrix} 0.1714 \times 10^1 \\ -0.2714 \times 10^1 \\ -0.4165 \times 10^1 \end{bmatrix} S.$$

From Table 2.1, the ultimate strengths of a unidirectional graphite/epoxy lamina are

$$(\sigma_1^T)_{ult} = 1500 \text{ MPa}$$

$$(\sigma_1^C)_{ult} = 1500 \text{ MPa}$$

$$(\sigma_2^T)_{ult} = 40 \text{ MPa}$$

$$(\sigma_2^C)_{ult} = 246 \text{ MPa}$$

$$(\tau_{12})_{ult} = 68 \text{ MPa}$$

Then, using the inequalities (2.141) of the maximum stress failure theory,

$$-1500 \times 10^6 < 0.1714 \times 10^1 S < 1500 \times 10^6$$

$$-246 \times 10^6 < -0.2714 \times 10^1 S < 40 \times 10^6$$

$$-68 \times 10^6 < -0.4165 \times 10^1 S < 68 \times 10^6$$

or

$$-875.1 \times 10^6 < S < 875.1 \times 10^6$$

$$-14.73 \times 10^6 < S < 90.64 \times 10^6$$

$$-16.33 \times 10^6 < S < 16.33 \times 10^6.$$

All the inequality conditions (and $S > 0$) are satisfied if $0 < S < 16.33$ MPa. The preceding inequalities also show that the angle lamina will fail in shear. The maximum stress that can be applied before failure is

$$\sigma_x = 32.66\,MPa, \sigma_y = -48.99\,MPa, \tau_{xy} = 65.32\,MPa.$$

Example 2.14

Find the off-axis shear strength of a 60° graphite/epoxy lamina. Use the properties of unidirectional graphite/epoxy from Table 2.1 and apply the maximum stress failure theory.

Solution

The off-axis shear strength of a lamina is defined as the minimum of the magnitude of positive and negative shear stress (Figure 2.31) that can be applied to an angle lamina before failure.

Assume the following stress state

$$\sigma_x = 0,\ \sigma_y = 0,\ \tau_{xy} = \tau.$$

Then, using the transformation Equation (2.94),

$$\begin{bmatrix} \sigma_1 \\ \sigma_2 \\ \tau_{12} \end{bmatrix} = \begin{bmatrix} 0.2500 & 0.7500 & 0.8660 \\ 0.7500 & 0.2500 & -0.8660 \\ -0.4330 & 0.4330 & -0.5000 \end{bmatrix} \begin{bmatrix} 0 \\ 0 \\ \tau \end{bmatrix}$$

$$\sigma_1 = 0.866\tau$$

$$\sigma_2 - 0.866\tau$$

$$\tau_{12} = -0.500\tau\ .$$

Using the inequalities (2.141) of the maximum stress failure theory, we have

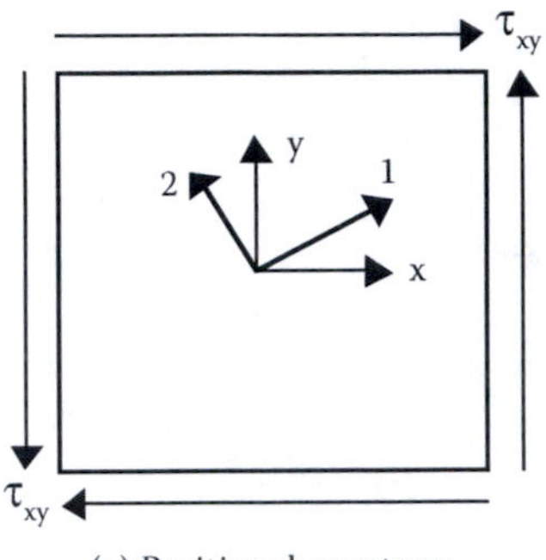

(a) Positive shear stress

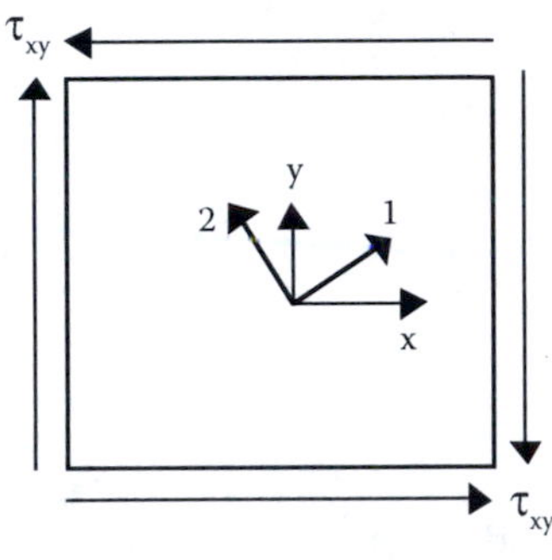

(b) Negative shear stress

FIGURE 2.31
Positive and negative shear stresses applied to an angle lamina.

$$-1500 < 0.866\tau < 1500 \text{ or } -1732 < \tau < 1732$$

$$-246 < -0.866\tau < 40 \text{ or } -46.19 < \tau < 284.1$$

$$-68 < -0.500\tau < 68 \text{ or } -136.0 < \tau < 136.0,$$

which shows that τ_{xy} = 46.19 MPa is the largest magnitude of shear stress that can be applied to the 60° graphite/epoxy lamina. However, the largest positive shear stress that could be applied is τ_{xy} = 136.0 MPa, and the largest negative shear stress is τ_{xy} = –46.19 MPa.

This shows that the maximum magnitude of allowable shear stress in other than the material axes' direction depends on the sign of the shear stress. This is mainly because the local axes' stresses in the direction perpendicular to the fibers are opposite in sign to each other for opposite signs of shear stress ($\sigma_2 = -0.866\tau$ for positive τ_{xy} and $\sigma_2 = 0.866\tau$ for negative τ_{xy}). Because the tensile strength perpendicular to the fiber direction is much lower than the compressive strength perpendicular to the fiber direction, the two limiting values of τ_{xy} are different.

TABLE 2.3

Effect of Sign of Shear Stress as a Function of Angle of Lamina

Angle, Degrees	Positive τ_{xy} MPa	Negative τ_{xy} MPa	Shear strength MPa
0	68.00 (S)	68.00 (S)	68.00
15	78.52 (S)	78.52 (S)	78.52
30	136.0 (S)	46.19 (2T)	46.19
45	246.0 (2C)	40.00 (2T)	40.00
60	136.0 (S)	46.19 (2T)	46.19
75	78.52 (S)	78.52 (S)	78.52
90	68.00 (S)	68.00 (S)	68.00

Note: The notation in the parentheses denotes the mode of failure of the angle lamina as follows:
(1T) — longitudinal tensile failure;
(1C) — longitudinal compressive failure;
(2T) — transverse tensile failure;
(2C) — transverse compressive failure;
(S) — shear failure.

Table 2.3 shows the maximum negative and positive values of shear stress that can be applied to different angle plies of graphite/epoxy of Table 2.1. The minimum magnitude of the two stresses is the shear strength of the angle lamina.

2.8.2 Strength Ratio

In a failure theory such as the maximum stress failure theory of Section 2.8.1, it can be determined whether a lamina has failed if any of the inequalities of Equation (2.141) are violated. However, this does not give the information about how much the load can be increased if the lamina is safe or how much the load should be decreased if the lamina has failed. The definition of strength ratio (SR) is helpful here. The strength ratio is defined as

$$SR = \frac{\textit{Maximum Load Which Can Be Applied}}{\textit{Load Applied}}. \tag{2.142}$$

The concept of strength ratio is applicable to any failure theory. If SR > 1, then the lamina is safe and the applied stress can be increased by a factor of SR. If SR < 1, the lamina is unsafe and the applied stress needs to be reduced by a factor of SR. A value of SR = 1 implies the failure load.

Example 2.15

Assume that one is applying a load of

$$\sigma_x = 2\,MPa, \sigma_y = -3\,MPa, \tau_{xy} = 4\,MPa$$

to a 60° angle lamina of graphite/epoxy. Find the strength ratio using the maximum stress failure theory.

Solution

If the strength ratio is R, then the maximum stress that can be applied is

$$\sigma_x = 2R,\ \sigma_y = -3R,\ \tau_{xy} = 4R\ .$$

Following Example 2.13 for finding the local stresses gives

$$\sigma_1 = 0.1714 \times 10^1\, R$$

$$\sigma_2 = -0.2714 \times 10^1\, R$$

$$\tau_{12} = -0.4165 \times 10^1\, R\ .$$

Using the maximum stress failure theory as given by Equation (2.141) yields

$$R = 16.33.$$

Thus, the load that can be applied just before failure is

$$\sigma_x = 16.33 \times 2\ MPa,\ \sigma_y = 16.33 \times (-3)\ MPa,\ \tau_{xy} = 16.33 \times 4\ Mpa,$$

$$\sigma_x = 32.66\ MPa,\ \sigma_y = -48.99\ MPa,\ \tau_{xy} = 65.32\ MPa.$$

Note that all the components of the stress vector must be multiplied by the strength ratio.

2.8.3 Failure Envelopes

A failure envelope is a three-dimensional plot of the combinations of the normal and shear stresses that can be applied to an angle lamina just before failure. Because drawing three dimensional graphs can be time consuming, one may develop failure envelopes for constant shear stress τ_{xy} and then use the two normal stresses σ_x and σ_y as the two axes. Then, if the applied stress is within the failure envelope, the lamina is safe; otherwise, it has failed.

Example 2.16

Develop a failure envelope for the 60° lamina of graphite/epoxy for a constant shear stress of τ_{xy} = 24 MPa. Use the properties for the unidirectional graphite/epoxy lamina from Table 2.1.

Solution

From Equation (2.94), the stresses in the local axes for a 60° lamina are given by

$$\sigma_1 = 0.2500\,\sigma_x + 0.7500\,\sigma_y + 20.78\,MPa,$$

$$\sigma_2 = 0.7500\,\sigma_x + 0.2500\,\sigma_y - 20.78\,MPa,$$

$$\tau_{12} = -0.4330\,\sigma_x + 0.4330\,\sigma_y - 12.00\,MPa,$$

where σ_x and σ_y are also in units of MPa.

Using the preceding inequalities,

$$-1500 < 0.2500\,\sigma_x + 0.7500\,\sigma_y + 20.78 < 1500$$

$$-246 < 0.7500\,\sigma_x + 0.2500\,\sigma_y - 20.78 < 40$$

$$-68 < -0.4330\,\sigma_x + 0.4330\,\sigma_y - 12.00 < 68\ .$$

Various combinations of (σ_x, σ_y) can be found to satisfy the preceding inequalities. However, the objective is to find the points on the failure envelope. These are combinations of σ_x and σ_y, where one of the three inequalities is just violated and the other two are satisfied. Some of the values of (σ_x, σ_y) obtained on the failure envelope are given in Table 2.4.

Several methods can be used to obtain the points on the failure envelope for a constant shear stress. One way is to fix the value of σ_x and find the maximum value of σ_y that can be applied without violating any of the conditions. For example, for σ_x = 100 MPa, from the inequalities we have

$$-2061 < \sigma_y < 1939,$$

$$-1201 < \sigma_y < -56.88,$$

$$-29.33 < \sigma_y < 284.80.$$

TABLE 2.4

Typical Values of (σ_x, σ_y) on the Failure Envelope for Example 2.16

σ_x (MPa)	σ_y (MPa)
50.0	93.1
50.0	–79.3
–50.0	179
–50.0	–135
25.0	168
25.0	–104
–25.0	160
–25.0	–154

The preceding three inequalities show no allowable value of σ_y for this value of σ_x = 100 MPa.

As another example, for σ_x = 50 MPa, we have from inequalities,

$$-2044 < \sigma_y < 1956,$$

$$-1051 < \sigma_y < 93.12,$$

$$-79.33 < \sigma_y < 234.80.$$

The preceding three inequalities show two maximum allowable values of the normal stress, σ_y. These are σ_y = 93.12 MPa and σ_y = –79.33 MPa. The failure envelope for τ_{xy} = 24 MPa is shown in Figure 2.32.

2.8.4 Maximum Strain Failure Theory

This theory is based on the maximum normal strain theory by St. Venant and the maximum shear stress theory by Tresca as applied to isotropic materials. The strains applied to a lamina are resolved to strains in the local axes. Failure is predicted in a lamina, if any of the normal or shearing strains in the local axes of a lamina equal or exceed the corresponding ultimate strains of the unidirectional lamina. Given the strains/stresses in an angle lamina, one can find the strains in the local axes. A lamina is considered to be failed if

$$-(\varepsilon_1^C)_{ult} < \varepsilon_1 < (\varepsilon_1^T)_{ult}, \text{ or}$$

$$-(\varepsilon_2^C)_{ult} < \varepsilon_2 < (\varepsilon_2^T)_{ult}, \text{ or}$$

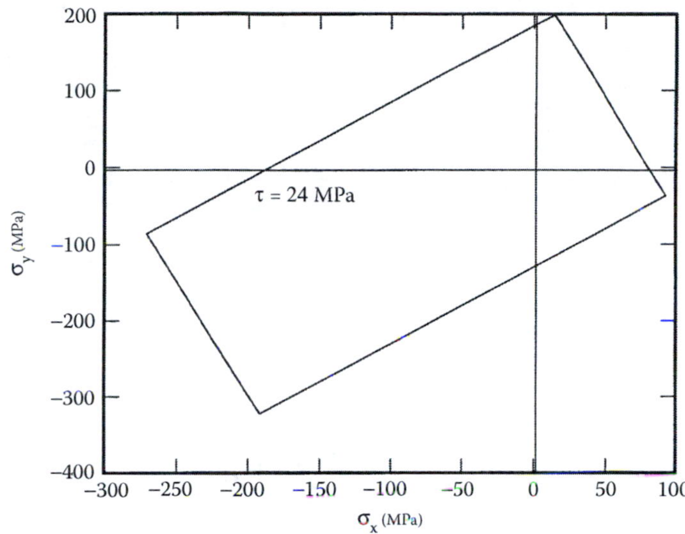

FIGURE 2.32
Failure envelopes for constant shear stress using maximum stress failure theory.

$$-(\gamma_{12})_{ult} < \gamma_{12} < (\gamma_{12})_{ult} \tag{2.143a–c}$$

is violated, where

$(\varepsilon_1^T)_{ult}$ = ultimate longitudinal tensile strain (in direction 1)
$(\varepsilon_1^C)_{ult}$ = ultimate longitudinal compressive strain (in direction 1)
$(\varepsilon_2^T)_{ult}$ = ultimate transverse tensile strain (in direction 2)
$(\varepsilon_2^C)_{ult}$ = ultimate transverse compressive strain (in direction 2)
$(\gamma_{12})_{ult}$ = ultimate in-plane shear strain (in plane 1–2)

The ultimate strains can be found directly from the ultimate strength parameters and the elastic moduli, assuming the stress–strain response is linear until failure. The maximum strain failure theory is similar to the maximum stress failure theory in that no interaction occurs between various components of strain. However, the two failure theories give different results because the local strains in a lamina include the Poisson's ratio effect. In fact, if the Poisson's ratio is zero in the unidirectional lamina, the two failure theories will give identical results.

Example 2.17

Find the maximum value of $S > 0$ if a stress, $\sigma_x = 2S$, $\sigma_y = -3S$, and $\tau_{xy} = 4S$, is applied to a 60° graphite/epoxy lamina. Use maximum strain failure theory. Use the properties of the graphite/epoxy unidirectional lamina given in Table 2.1.

Solution

In Example 2.6, the compliance matrix [S] was obtained and, in Example 2.13, the local stresses for this problem were obtained. Then, from Equation (2.77),

$$\begin{bmatrix} \varepsilon_1 \\ \varepsilon_2 \\ \gamma_{12} \end{bmatrix} = [S] \begin{bmatrix} \sigma_1 \\ \sigma_2 \\ \tau_{12} \end{bmatrix}$$

$$= \begin{bmatrix} 0.5525 \times 10^{-11} & -0.1547 \times 10^{-11} & 0 \\ -0.1547 \times 10^{-11} & 0.9709 \times 10^{-10} & 0 \\ 0 & 0 & 0.1395 \times 10^{-9} \end{bmatrix} \begin{bmatrix} 0.1714 \times 10^{1} \\ -0.2714 \times 10^{1} \\ -0.4165 \times 10^{1} \end{bmatrix} S$$

$$= \begin{bmatrix} 0.1367 \times 10^{-10} \\ -0.2662 \times 10^{-9} \\ -0.5809 \times 10^{-9} \end{bmatrix} S.$$

Assume a linear relationship between all the stresses and strains until failure; then the ultimate failure strains are

$$(\varepsilon_1^T)_{ult} = \frac{(\sigma_1^T)_{ult}}{E_1} = \frac{1500 \times 10^6}{181 \times 10^9} = 8.287 \times 10^{-3},$$

$$(\varepsilon_1^C)_{ult} = \frac{(\sigma_1^C)_{ult}}{E_1} = \frac{1500 \times 10^6}{181 \times 10^9} = 8.287 \times 10^{-3},$$

$$(\varepsilon_2^T)_{ult} = \frac{(\sigma_2^T)_{ult}}{E_2} = \frac{40 \times 10^6}{10.3 \times 10^9} = 3.883 \times 10^{-3},$$

$$(\varepsilon_2^C)_{ult} = \frac{(\sigma_2^C)_{ult}}{E_2} = \frac{246 \times 10^6}{10.3 \times 10^9} = 2.388 \times 10^{-2},$$

$$(\gamma_{12})_{ult} = \frac{(\tau_{12})_{ult}}{G_{12}} = \frac{68 \times 10^6}{7.17 \times 10^6} = 9.483 \times 10^{-3}.$$

The preceding values for the ultimate strains also assume that the compressive and tensile stiffnesses are identical. Using the inequalities (2.143) and recognizing that $S > 0$,

$$-8.287 \times 10^{-3} < 0.1367 \times 10^{-10} S < 8.287 \times 10^{-3},$$

$$-2.388 \times 10^{-2} < -0.2662 \times 10^{-9} S < 3.883 \times 10^{-3},$$

$$-9.483 \times 10^{-3} < -0.5809 \times 10^{-9} S < 9.483 \times 10^{-3},$$

or

$$-606.2 \times 10^6 < S < 606.2 \times 10^6,$$

$$-14.58 \times 10^6 < S < 89.71 \times 10^6$$

$$-16.33 \times 10^6 < S < 16.33 \times 10^6,$$

which give

$$0 < S < 16.33\, MPa.$$

The maximum value of S before failure is 16.33 MPa. The same maximum value of S = 16.33 MPa is also found using maximum stress failure theory. There is no difference between the two values because the mode of failure is shear. However, if the mode of failure were other than shear, a difference in the prediction of failure loads would have been present due to the Poisson's ratio effect, which couples the normal strains and stresses in the local axes.

Neither the maximum stress failure theory nor the maximum strain failure theory has any coupling among the five possible modes of failure. The following theories are based on the interaction failure theory.

2.8.5 Tsai–Hill Failure Theory

This theory is based on the distortion energy failure theory of Von-Mises' distortional energy yield criterion for isotropic materials as applied to anisotropic materials. Distortion energy is actually a part of the total strain energy in a body. The strain energy in a body consists of two parts; one due to a change in volume and is called the dilation energy and the second is due to a change in shape and is called the distortion energy. It is assumed that failure in the material takes place only when the distortion energy is greater than the failure distortion energy of the material. Hill[8] adopted the Von-Mises' distortional energy yield criterion to anisotropic materials. Then, Tsai[7] adapted it to a unidirectional lamina. Based on the distortion energy theory, he proposed that a lamina has failed if

$$(G_2+G_3)\sigma_1^2+(G_1+G_3)\sigma_2^2+(G_1+G_2)\sigma_3^2-2G_3\sigma_1\sigma_2-2G_2\sigma_1\sigma_3 \tag{2.144}$$

$$-\,2G_1\sigma_2\sigma_3+2G_4\tau_{23}^2+2G_5\tau_{13}^2+2G_6\tau_{12}^2<1$$

is violated. The components G_1, G_2, G_3, G_4, G_5, and G_6 of the strength criterion depend on the failure strengths and are found as follows.

1. Apply $\sigma_1=(\sigma_1^T)_{ult}$ to a unidirectional lamina; then, the lamina will fail. Thus, Equation (2.144) reduces to

$$(G_2+G_3)(\sigma_1^T)_{ult}^2=1. \tag{2.145}$$

2. Apply $\sigma_2=(\sigma_2^T)_{ult}$ to a unidirectional lamina; then, the lamina will fail. Thus, Equation (2.144) reduces to

$$(G_1+G_3)(\sigma_2^T)_{ult}^2=1. \tag{2.146}$$

3. Apply $\sigma_3=(\sigma_2^T)_{ult}$ to a unidirectional lamina and, assuming that the normal tensile failure strength is same in directions (2) and (3), the lamina will fail. Thus, Equation (2.144) reduces to

$$(G_1+G_2)(\sigma_2^T)_{ult}^2=1. \tag{2.147}$$

4. Apply $\tau_{12}=(\tau_{12})_{ult}$ to a unidirectional lamina; then, the lamina will fail. Thus, Equation (2.144) reduces to

$$2G_6(\tau_{12})_{ult}^2=1. \tag{2.148}$$

From Equation (2.145) to Equation (2.148),

$$G_1 = \frac{1}{2}\left(\frac{2}{[(\sigma_2^T)_{ult}]^2} - \frac{1}{[(\sigma_1^T)_{ult}]^2}\right),$$

$$G_2 = \frac{1}{2}\left(\frac{1}{[(\sigma_1^T)_{ult}]^2}\right),$$

$$G_3 = \frac{1}{2}\left(\frac{1}{[(\sigma_1^T)_{ult}]^2}\right),$$

$$G_6 = \frac{1}{2}\left(\frac{1}{[(\tau_{12})_{ult}]^2}\right). \qquad (2.149\text{a–d})$$

Because the unidirectional lamina is assumed to be under plane stress — that is, $\sigma_3 = \tau_{31} = \tau_{23} = 0$, then Equation (2.144) reduces through Equation (2.149) to

$$\left[\frac{\sigma_1}{(\sigma_1^T)_{ult}}\right]^2 - \left[\frac{\sigma_1\sigma_2}{(\sigma_1^T)_{ult}^2}\right] + \left[\frac{\sigma_2}{(\sigma_2^T)_{ult}}\right]^2 + \left[\frac{\tau_{12}}{(\tau_{12})_{ult}}\right]^2 < 1. \qquad (2.150)$$

Given the global stresses in a lamina, one can find the local stresses in a lamina and apply the preceding failure theory to determine whether the lamina has failed.

Example 2.18

Find the maximum value of $S > 0$ if a stress of $\sigma_x = 2S$, $\sigma_y = -3S$, and $\tau_{xy} = 4S$ is applied to a 60° graphite/epoxy lamina. Use Tsai–Hill failure theory. Use the unidirectional graphite/epoxy lamina properties given in Table 2.1.

Solution

From Example 2.13,

$$\sigma_1 = 1.714\ S,$$

$$\sigma_2 = -2.714\ S,$$

$$\tau_{12} = -4.165\ S.$$

Using the Tsai–Hill failure theory from Equation (2.150),

$$\left(\frac{1.714S}{1500\times10^{6}}\right)^{2}-\left(\frac{1.714S}{1500\times10^{6}}\right)\left(\frac{-2.714S}{1500\times10^{6}}\right)+\left(\frac{-2.714S}{40\times10^{6}}\right)^{2}+\left(\frac{-4.165S}{68\times10^{6}}\right)^{2}<1$$

$$S < 10.94\ MPa$$

1. Unlike the maximum strain and maximum stress failure theories, the Tsai–Hill failure theory considers the interaction among the three unidirectional lamina strength parameters.
2. The Tsai–Hill failure theory does not distinguish between the compressive and tensile strengths in its equations. This can result in underestimation of the maximum loads that can be applied when compared to other failure theories. For the load of σ_x = 2 MPa, σ_y = –3 MPa, and τ_{xy} = 4 MPa, as found in Example 2.15, Example 2.17, and Example 2.18, the strength ratios are given by

   SR = 10.94 (Tsai–Hill failure theory)

   SR = 16.33 (maximum stress failure theory)

   SR = 16.33 (maximum strain failure theory)

Tsai–Hill failure theory underestimates the failure stress because the transverse tensile strength of a unidirectional lamina is generally much less than its transverse compressive strength. The compressive strengths are not used in the Tsai–Hill failure theory, but it can be modified to use corresponding tensile or compressive strengths in the failure theory as follows

$$\left[\frac{\sigma_1}{X_1}\right]^2-\left[\left(\frac{\sigma_1}{X_2}\right)\left(\frac{\sigma_2}{X_2}\right)\right]+\left[\frac{\sigma_2}{Y}\right]^2+\left[\frac{\tau_{12}}{S}\right]^2<1, \qquad (2.151)$$

where

$$X_1 = (\sigma_1^T)_{ult} \text{ if } \sigma_1 > 0$$

$$= (\sigma_1^C)_{ult} \text{ if } \sigma_1 < 0;$$

$$X_2 = (\sigma_1^T)_{ult} \text{ if } \sigma_2 > 0$$

$$= (\sigma_1^C)_{ult} \text{ if } \sigma_2 < 0;$$

$$Y = (\sigma_2^T)_{ult} \text{ if } \sigma_2 > 0$$

$$= (\sigma_2^C)_{ult} \text{ if } \sigma_2 < 0$$

$$S = (\tau_{12})_{ult}.$$

For Example 2.18, the modified Tsai–Hill failure theory given by Equation (2.151) now gives

$$\left(\frac{1.714\sigma}{1500\times10^6}\right)^2 - \left(\frac{1.714\sigma}{1500\times10^6}\right)\left(\frac{-2.714\sigma}{1500\times10^6}\right) + \left(\frac{-2.714\sigma}{246\times10^6}\right)^2 + \left(\frac{-4.165\sigma}{68\times10^6}\right)^2 < 1$$

$$\sigma < 16.06\ MPa,$$

which implies that the strength ratio is $SR = 16.06$ (modified Tsai–Hill failure theory). This value is closer to the values obtained using maximum stress and maximum strain failure theories.

3. The Tsai–Hill failure theory is a unified theory and thus does not give the mode of failure like the maximum stress and maximum strain failure theories do. However, one can make a reasonable guess of the failure mode by calculating $\left|\sigma_1/(\sigma_1^T)_{ult}\right|$, $\left|\sigma_2/(\sigma_2^T)_{ult}\right|$ and $|\tau_{12}/(\tau_{12})_{ult}|$. The maximum of these three values gives the associated mode of failure. In the modified Tsai–Hill failure theory, calculate the maximum of $|\sigma_1/X_1|$, $|\sigma_2/Y|$, and $|\tau_{12}/S|$ for the associated mode of failure.

2.8.6 Tsai–Wu Failure Theory

This failure theory is based on the total strain energy failure theory of Beltrami. Tsai-Wu[9] applied the failure theory to a lamina in plane stress. A lamina is considered to be failed if

$$H_1\sigma_1 + H_2\sigma_2 + H_6\tau_{12} + H_{11}\sigma_1^2 + H_{22}\sigma_2^2 + H_{66}\tau_{12}^2 + 2H_{12}\sigma_1\sigma_2 < 1 \quad (2.152)$$

is violated. This failure theory is more general than the Tsai–Hill failure theory because it distinguishes between the compressive and tensile strengths of a lamina.

The components H_1, H_2, H_6, H_{11}, H_{22}, and H_{66} of the failure theory are found using the five strength parameters of a unidirectional lamina as follows:

1. Apply $\sigma_1 = (\sigma_1^T)_{ult}$, $\sigma_2 = 0$, $\tau_{12} = 0$ to a unidirectional lamina; the lamina will fail. Equation (2.152) reduces to

$$H_1(\sigma_1^T)_{ult} + H_{11}(\sigma_1^T)_{ult}^2 = 1. \tag{2.153}$$

2. Apply $\sigma_1 = -(\sigma_1^C)_{ult}$, $\sigma_2 = 0$, $\tau_{12} = 0$ to a unidirectional lamina; the lamina will fail. Equation (2.152) reduces to

$$-H_1(\sigma_1^C)_{ult} + H_{11}(\sigma_1^C)_{ult}^2 = 1. \tag{2.154}$$

From Equation (2.153) and Equation (2.154),

$$H_1 = \frac{1}{(\sigma_1^T)_{ult}} - \frac{1}{(\sigma_1^C)_{ult}}, \tag{2.155}$$

$$H_{11} = \frac{1}{(\sigma_1^T)_{ult}(\sigma_1^C)_{ult}}. \tag{2.156}$$

3. Apply $\sigma_1 = 0$, $\sigma_2 = (\sigma_2^T)_{ult}$, $\tau_{12} = 0$ to a unidirectional lamina; the lamina will fail. Equation (2.152) reduces to

$$H_2(\sigma_2^T)_{ult} + H_{22}(\sigma_2^T)_{ult}^2 = 1. \tag{2.157}$$

4. Apply $\sigma_1 = 0$, $\sigma_2 = -(\sigma_2^C)_{ult}$, $\tau_{12} = 0$ to a unidirectional lamina; the lamina will fail. Equation (2.152) reduces to

$$-H_2(\sigma_2^C)_{ult} + H_{22}(\sigma_2^C)_{ult}^2 = 1. \tag{2.158}$$

From Equation (2.157) and Equation (2.158),

$$H_2 = \frac{1}{(\sigma_2^T)_{ult}} - \frac{1}{(\sigma_2^C)_{ult}}, \tag{2.159}$$

$$H_{22} = \frac{1}{(\sigma_2^T)_{ult}(\sigma_2^C)_{ult}}. \tag{2.160}$$

5. Apply $\sigma_1 = 0$, $\sigma_2 = 0$, and $\tau_{12} = (\tau_{12})_{ult}$ to a unidirectional lamina; it will fail. Equation (2.152) reduces to

$$H_6(\tau_{12})_{ult} + H_{66}(\tau_{12})^2_{ult} = 1. \tag{2.161}$$

6. Apply $\sigma_1 = 0$, $\sigma_2 = 0$, and $\tau_{12} = -(\tau_{12})_{ult}$ to a unidirectional lamina; the lamina will fail. Equation (2.152) reduces to

$$-H_6(\tau_{12})_{ult} + H_{66}(\tau_{12})^2_{ult} = 1. \tag{2.162}$$

From Equation (2.161) and Equation (2.162),

$$H_6 = 0, \tag{2.163}$$

$$H_{66} = \frac{1}{(\tau_{12})^2_{ult}}. \tag{2.164}$$

The only component of the failure theory that cannot be found directly from the five strength parameters of the unidirectional lamina is H_{12}. This can be found experimentally by knowing a biaxial stress at which the lamina fails and then substituting the values of σ_1, σ_2, and τ_{12} in the Equation (2.152). Note that σ_1 and σ_2 need to be nonzero to find H_{12}. Experimental methods to find H_{12} include the following.

1. Apply equal tensile loads along the two material axes in a unidirectional composite. If $\sigma_x = \sigma_y = \sigma$, $\tau_{xy} = 0$ is the load at which the lamina fails, then

$$(H_1 + H_2)\sigma + (H_{11} + H_{22} + 2H_{12})\sigma^2 = 1. \tag{2.165}$$

The solution of Equation (2.165) gives

$$H_{12} = \frac{1}{2\sigma^2}[1 - (H_1 + H_2)\sigma - (H_{11} + H_{22})\sigma^2]. \tag{2.166}$$

It is not necessary to pick tensile loads in the preceding biaxial test, but one may apply any combination of

$$\sigma_1 = \sigma, \ \sigma_2 = \sigma,$$

$$\sigma_1 = -\sigma, \ \sigma_2 = -\sigma,$$

$$\sigma_1 = \sigma, \ \sigma_2 = -\sigma,$$

$$\sigma_1 = -\sigma, \ \sigma_2 = \sigma. \tag{2.167}$$

This will give four different values of H_{12}, each corresponding to the four tests.

2. Take a 45° lamina under uniaxial tension σ_x. The stress σ_x at failure is noted. If this stress is $\sigma_x = \sigma$, then, using Equation (2.94), the local stresses at failure are

$$\sigma_1 = \frac{\sigma}{2},$$

$$\sigma_2 = \frac{\sigma}{2}, \tag{2.168a–c}$$

$$\tau_{12} = -\frac{\sigma}{2}.$$

Substituting the preceding local stresses in Equation (2.152),

$$(H_1 + H_2)\frac{\sigma}{2} + \frac{\sigma^2}{4}(H_{11} + H_{22} + H_{66} + 2H_{12}) = 1. \tag{2.169}$$

$$H_{12} = \frac{2}{\sigma^2} - \frac{(H_1 + H_2)}{\sigma} - \frac{1}{2}(H_{11} + H_{22} + H_{66}). \tag{2.170}$$

Some empirical suggestions for finding the value of H_{12} include

$$H_{12} = -\frac{1}{2(\sigma_1^T)_{ult}^2}, \text{ per Tsai–Hill failure theory}^{8} \tag{2.171a–c}$$

$$H_{12} = -\frac{1}{2(\sigma_1^T)_{ult}(\sigma_1^C)_{ult}}, \text{ per Hoffman criterion}^{10}$$

$$H_{12} = -\frac{1}{2}\sqrt{\frac{1}{(\sigma_1^T)_{ult}(\sigma_1^C)_{ult}(\sigma_2^T)_{ult}(\sigma_2^C)_{ult}}}, \text{ per Mises–Hencky criterion.}^{11}$$

Example 2.19

Find the maximum value of $S > 0$ if a stress $\sigma_x = 2S$, $\sigma_y = -3S$, and $\tau_{xy} = 4S$ are applied to a 60° lamina of graphite/epoxy. Use Tsai–Wu failure theory. Use the properties of a unidirectional graphite/epoxy lamina from Table 2.1.

Solution

From Example 2.13,

$$\sigma_1 = 1.714S,$$

$$\sigma_2 = -2.714S,$$

$$\tau_{12} = -4.165S.$$

From Equations (2.155), (2.156), (2.159), (2.160), (2.163), and (2.164),

$$H_1 = \frac{1}{1500 \times 10^6} - \frac{1}{1500 \times 10^6} = 0 \ Pa^{-1},$$

$$H_2 = \frac{1}{40 \times 10^6} - \frac{1}{246 \times 10^6} = 2.093 \times 10^{-8} \ Pa^{-1},$$

$$H_6 = 0 \ Pa^{-1},$$

$$H_{11} = \frac{1}{(1500 \times 10^6)(1500 \times 10^6)} = 4.4444 \times 10^{-19} \ Pa^{-2},$$

$$H_{22} = \frac{1}{(40 \times 10^6)(246 \times 10^6)} = 1.0162 \times 10^{-16} \ Pa^{-2},$$

$$H_{66} = \frac{1}{(68 \times 10^6)^2} = 2.1626 \times 10^{-16} \, Pa^{-2}.$$

Using the Mises–Hencky criterion for evaluation of H_{12}, (Equation 2.165c),

$$H_{12} = -\frac{1}{2}\sqrt{\frac{1}{(1500 \times 10^6)(1500 \times 10^6)(40 \times 10^6)(246 \times 10^6)}} = 3.360 \times 10^{-18} \, Pa^{-2}.$$

Substituting these values in Equation (2.152), we obtain

$$(0)(1.714S) + (2.093 \times 10^{-8})(-2.714S)$$

$$+ (0)(-4.165S) + (4.444 \times 10^{-19})(1.714S)^2$$

$$+ (1.0162 \times 10^{-16})(-2.714S)^2 + (2.1626 \times 10^{-16})(-4.165S)^2$$

$$+ 2(-3.360 \times 10^{-18})(1.714S)(-2.714S) < 1,$$

or

$$S < 22.39 \; MPa \; .$$

If one uses the other two empirical criteria for H_{12}, per Equation (2.171), this yields

$$S < 22.49 \; MPa \text{ for } H_{12} = -\frac{1}{2(\sigma_1^T)_{ult}^2},$$

$$S < 22.49 \; MPa \; for \; H_{12} = -\frac{1}{2}\frac{1}{(\sigma_1^T)_{ult}(\sigma_1^C)_{ult}}.$$

Summarizing the four failure theories for the same stress state, the value of S obtained is

S = 16.33 (maximum stress failure theory)
S = 16.33 (maximum strain failure theory)
S = 10.94 (Tsai–Hill failure theory)
S = 16.06 (modified Tsai–Hill failure theory)
S = 22.39 (Tsai–Wu failure theory)

2.8.7 Comparison of Experimental Results with Failure Theories

Tsai[7] compared the results from various failure theories to some experimental results. He considered an angle lamina subjected to a uniaxial load in the x-direction, σ_x, as shown in Figure 2.33. The failure stresses were obtained experimentally for tensile and compressive stresses for various angles of the lamina.

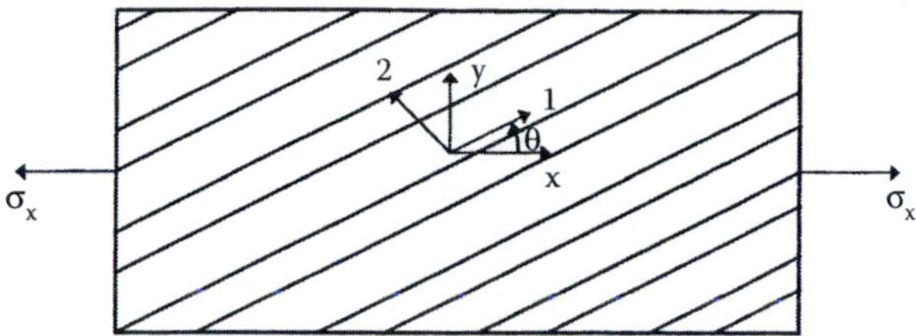

FIGURE 2.33
Off-axis loading in the x-direction in Figure 2.34 to Figure 2.37.

The experimental results can be compared with the four failure theories by finding the stresses in the material axes for an arbitrary stress, σ_x, for an angle lamina with an angle, θ, between the fiber and loading direction as

$$\sigma_1 = \sigma_x \operatorname{Cos}^2\theta,$$

$$\sigma_2 = \sigma_x \operatorname{Sin}^2\theta, \tag{2.172}$$

$$\tau_{12} = -\sigma_x \operatorname{Sin}\theta \operatorname{Cos}\theta,$$

per Equation (2.94).

The corresponding strains in the material axes are

$$\varepsilon_1 = \frac{1}{E_1}(\operatorname{Cos}^2\theta - \nu_{12}\operatorname{Sin}^2\theta)\sigma_x,$$

$$\varepsilon_2 = \frac{1}{E_2}(\operatorname{Sin}^2\theta - \nu_{21}\operatorname{Cos}^2\theta)\sigma_x, \tag{2.173}$$

$$\gamma_{12} = -\frac{1}{G_{12}}(\operatorname{Sin}\theta \operatorname{Cos}\theta)\sigma_x,$$

per Equation (2.99).

Using the preceding local strains and stresses in the four failure theories given by Equation (2.141), Equation (2.143), Equation (2.150), and Equation (2.152), one can find the ultimate off-axis load, σ_x, that can be applied as a function of the angle of the lamina.

The following values were used in the failure theories for the unidirectional lamina stiffnesses and strengths:

$$E_1 = 7.8\ Msi,$$

$$E_2 = 2.6\ Msi,$$

$$\nu_{12} = 0.25,$$

$$G_{12} = 1.3\ Msi,$$

$$(\sigma_1^T)_{ult} = 150\ Ksi,$$

$$(\sigma_1^C)_{ult} = 150\ Ksi,$$

$$(\sigma_2^T)_{ult} = 4\ Ksi,$$

$$(\sigma_2^C)_{ult} = 20\ Ksi,$$

$$(\tau_{12})_{ult} = 6\ Ksi.$$

The comparison for the four failure theories is shown in Figure 2.34 through Figure 2.37. Observations from the figures are:

- The difference between the maximum stress and maximum strain failure theories and the experimental results is quite pronounced.
- Tsai–Hill and Tsai–Wu failure theories' results are in good agreement with experimentally obtained results.
- The variation of the strength of the angle lamina as a function of angle is smooth in the Tsai–Hill and Tsai–Wu failure theories, but has cusps in the maximum stress and maximum strain failure theories. The cusps correspond to the change in failure modes in the maximum stress and maximum strain failure theories.

2.9 Hygrothermal Stresses and Strains in a Lamina

Composite materials are generally processed at high temperatures and then cooled down to room temperatures. For polymeric matrix composites, this temperature difference is in the range of 200 to 300°C; for ceramic matrix composites, it may be as high as 1000°C. Due to mismatch of the coefficients of thermal expansion of the fiber and matrix, residual stresses result in a lamina when it is cooled down. Also, the cooling down induces expansional

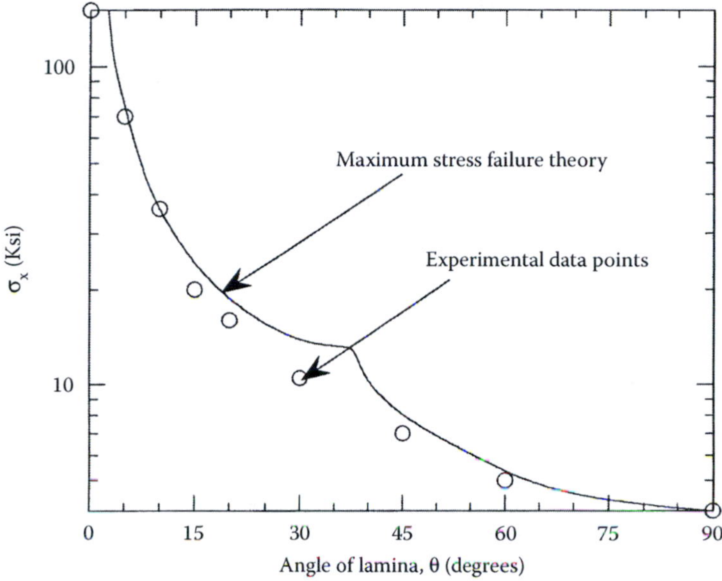

FIGURE 2.34
Maximum normal tensile stress in the *x*-direction as a function of angle of lamina using maximum stress failure theory. (Experimental data reprinted with permission from *Introduction to Composite Materials*, Tsai, S.W. and Hahn, H.T., 1980, CRC Press, Boca Raton, FL, 301.)

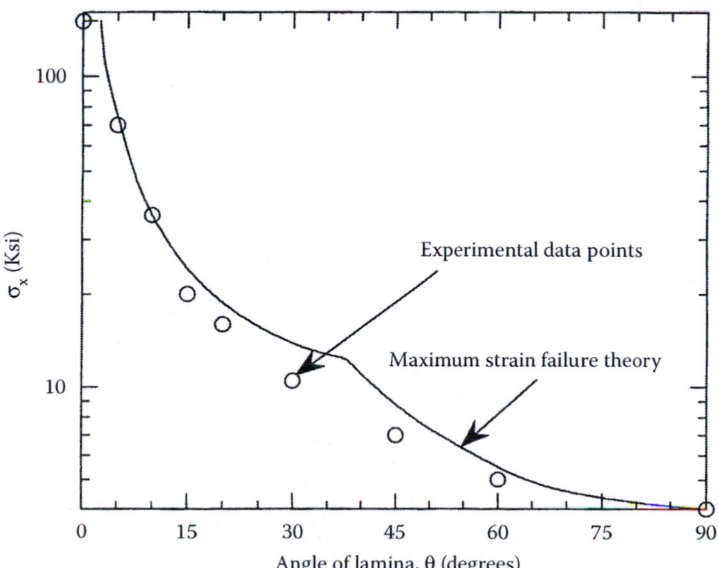

FIGURE 2.35
Maximum normal tensile stress in the *x*-direction as a function of angle of lamina using maximum strain failure theory. (Experimental data reprinted with permission from *Introduction to Composite Materials*, Tsai, S.W. and Hahn, H.T., 1980, CRC Press, Boca Raton, FL, 301.)

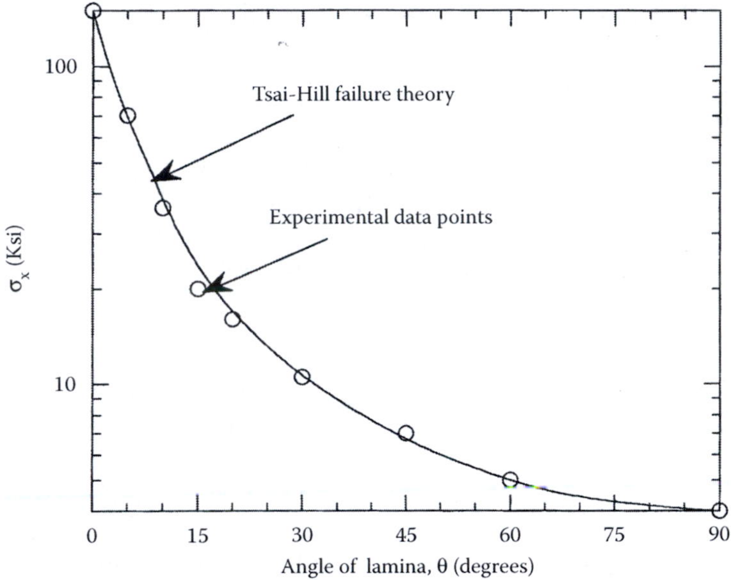

FIGURE 2.36
Maximum normal tensile stress in the x-direction as a function of angle of lamina using Tsai–Hill failure theory. (Experimental data reprinted with permission from *Introduction to Composite Materials*, Tsai, S.W. and Hahn, H.T., 1980, CRC Press, Boca Raton, FL, 301.)

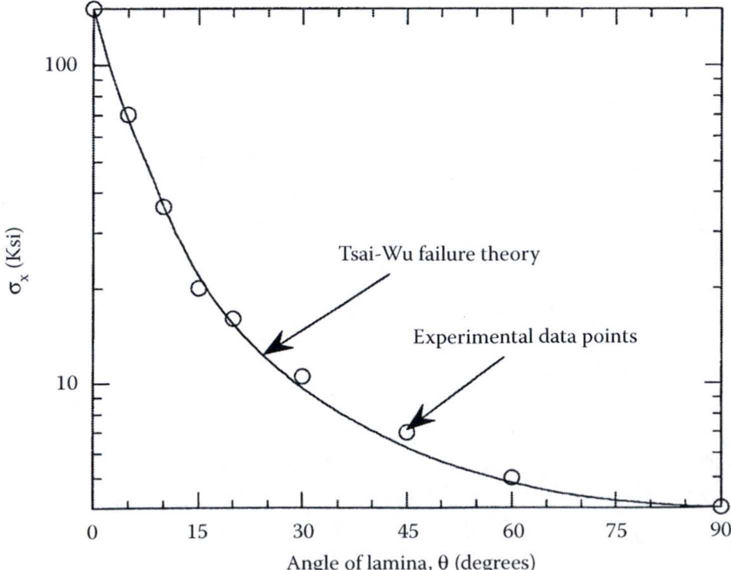

FIGURE 2.37
Maximum normal tensile stress in the x-direction as a function of angle of lamina using Tsai–Wu failure theory. (Experimental data reprinted with permission from *Introduction to Composite Materials*, Tsai, S.W. and Hahn, H.T., 1980, CRC Press, Boca Raton, FL, 301.)

strains in the lamina. In addition, most polymeric matrix composites can absorb or deabsorb moisture. This moisture change leads to swelling strains and stresses similar to those due to thermal expansion. Laminates in which laminae are placed at different angles have residual stresses in each lamina due to differing hygrothermal expansion of each lamina. The hygrothermal strains are not equal in a lamina in the longitudinal and transverse directions because the elastic constants and the thermal and moisture expansion coefficients of the fiber and matrix are different. In the following sections, stress–strain relationships are developed for unidirectional and angle laminae subjected to hygrothermal loads.

2.9.1 Hygrothermal Stress–Strain Relationships for a Unidirectional Lamina

For a unidirectional lamina, the stress–strain relationship with temperature and moisture difference gives

$$\begin{bmatrix} \varepsilon_1 \\ \varepsilon_2 \\ \gamma_{12} \end{bmatrix} = \begin{bmatrix} S_{11} & S_{12} & 0 \\ S_{12} & S_{22} & 0 \\ 0 & 0 & S_{66} \end{bmatrix} \begin{bmatrix} \sigma_1 \\ \sigma_2 \\ \tau_{12} \end{bmatrix} + \begin{bmatrix} \varepsilon_1^T \\ \varepsilon_2^T \\ 0 \end{bmatrix} + \begin{bmatrix} \varepsilon_1^C \\ \varepsilon_2^C \\ 0 \end{bmatrix}, \tag{2.174}$$

where the subscripts T and C are used to denote temperature and moisture, respectively. Note that the temperature and moisture change do not have any shearing strain terms because no shearing strains are induced in the material axes. The thermally induced strains are given by

$$\begin{bmatrix} \varepsilon_1^T \\ \varepsilon_2^T \\ 0 \end{bmatrix} = \Delta T \begin{bmatrix} \alpha_1 \\ \alpha_2 \\ 0 \end{bmatrix}, \tag{2.175}$$

where α_1 and α_2 are the longitudinal and transverse coefficients of thermal expansion, respectively, and ΔT is the temperature change. The moisture-induced strains are given by

$$\begin{bmatrix} \varepsilon_1^C \\ \varepsilon_2^C \\ 0 \end{bmatrix} = \Delta C \begin{bmatrix} \beta_1 \\ \beta_2 \\ 0 \end{bmatrix}, \tag{2.176}$$

where β_1 and β_2 are the longitudinal and transverse coefficients of moisture, respectively, and ΔC is the weight of moisture absorption per unit weight of the lamina.

Equation (2.174) can be inverted to give

$$\begin{bmatrix} \sigma_1 \\ \sigma_2 \\ \tau_{12} \end{bmatrix} = \begin{bmatrix} Q_{11} & Q_{12} & 0 \\ Q_{12} & Q_{22} & 0 \\ 0 & 0 & Q_{66} \end{bmatrix} \begin{bmatrix} \varepsilon_1 - \varepsilon_1^T - \varepsilon_1^C \\ \varepsilon_2 - \varepsilon_2^T - \varepsilon_2^C \\ \gamma_{12} \end{bmatrix}. \tag{2.177}$$

2.9.2 Hygrothermal Stress–Strain Relationships for an Angle Lamina

The stress–strain relationship for an angle lamina takes the following form:

$$\begin{bmatrix} \varepsilon_x \\ \varepsilon_y \\ \gamma_{xy} \end{bmatrix} = \begin{bmatrix} \bar{S}_{11} & \bar{S}_{12} & \bar{S}_{16} \\ \bar{S}_{12} & \bar{S}_{22} & \bar{S}_{26} \\ \bar{S}_{16} & \bar{S}_{26} & \bar{S}_{66} \end{bmatrix} \begin{bmatrix} \sigma_x \\ \sigma_y \\ \tau_{xy} \end{bmatrix} + \begin{bmatrix} \varepsilon_x^T \\ \varepsilon_y^T \\ \gamma_{xy}^T \end{bmatrix} + \begin{bmatrix} \varepsilon_x^C \\ \varepsilon_y^C \\ \gamma_{xy}^C \end{bmatrix}, \tag{2.178}$$

where

$$\begin{bmatrix} \varepsilon_x^T \\ \varepsilon_y^T \\ \gamma_{xy}^T \end{bmatrix} = \Delta T \begin{bmatrix} \alpha_x \\ \alpha_y \\ \alpha_{xy} \end{bmatrix}, \tag{2.179}$$

and

$$\begin{bmatrix} \varepsilon_x^C \\ \varepsilon_y^C \\ \gamma_{xy}^C \end{bmatrix} = \Delta C \begin{bmatrix} \beta_x \\ \beta_y \\ \beta_{xy} \end{bmatrix}. \tag{2.180}$$

The terms α_x, α_y, and α_{xy} are the coefficients of thermal expansion for an angle lamina and are given in terms of the coefficients of thermal expansion for a unidirectional lamina as

$$\begin{bmatrix} \alpha_x \\ \alpha_y \\ \alpha_{xy}/2 \end{bmatrix} = [T]^{-1} \begin{bmatrix} \alpha_1 \\ \alpha_2 \\ 0 \end{bmatrix}. \tag{2.181}$$

Similarly, β_x, β_y, and β_{xy} are the coefficients of moisture expansion for an angle lamina and are given in terms of the coefficients of moisture expansion for a unidirectional lamina as

$$\begin{bmatrix} \beta_x \\ \beta_y \\ \beta_{xy}/2 \end{bmatrix} = [T]^{-1} \begin{bmatrix} \beta_1 \\ \beta_2 \\ 0 \end{bmatrix}. \tag{2.182}$$

From Equation (2.174), if no constraints are placed on a lamina, no mechanical strains will be induced in it. This also implies then that no mechanical stresses are induced. However, in a laminate, even if the laminate has no constraints, the difference in the thermal/moisture expansion coefficients of the various layers induces different thermal/moisture expansions in each layer. This difference results in residual stresses and will be explained fully in Chapter 4.

Example 2.20

Find the following for a 60° angle lamina of glass/epoxy:

1. Coefficients of thermal expansion
2. Coefficients of moisture expansion
3. Strains under a temperature change of –100°C and a moisture absorption of 0.02 kg/kg.

Use properties of unidirectional glass/epoxy lamina from Table 2.1.

Solution

1. From Table 2.1,

$$\alpha_1 = 8.6 \times 10^{-6} \text{ m/m/°C},$$

$$\alpha_2 = 22.1 \times 10^{-6} \text{ m/m/°C}.$$

Using Equation (2.181) gives

$$\begin{bmatrix} \alpha_x \\ \alpha_y \\ \alpha_{xy}/2 \end{bmatrix} = \begin{bmatrix} 0.2500 & 0.7500 & -0.8660 \\ 0.7500 & 0.2500 & 0.8660 \\ 0.4330 & -0.4330 & -0.5000 \end{bmatrix} \begin{bmatrix} 8.6 \times 10^{-6} \\ 22.1 \times 10^{-6} \\ 0 \end{bmatrix},$$

$$\begin{bmatrix} \alpha_x \\ \alpha_y \\ \alpha_{xy} \end{bmatrix} = \begin{bmatrix} 18.73 \times 10^{-6} \\ 11.98 \times 10^{-6} \\ -11.69 \times 10^{-6} \end{bmatrix} \text{m/m/°C.}$$

2. From Table 2.1,

$$\beta_1 = 0 \text{ m/m/kg/kg,}$$

$$\beta_2 = 0.6 \text{ m/m/kg/kg.}$$

Using Equation (2.182) gives

$$\begin{bmatrix} \beta_x \\ \beta_y \\ \beta_{xy}/2 \end{bmatrix} = \begin{bmatrix} 0.2500 & 0.7500 & -0.8660 \\ 0.7500 & 0.2500 & 0.8660 \\ 0.4330 & -0.4330 & -0.5000 \end{bmatrix} \begin{bmatrix} 0.0 \\ 0.6 \\ 0 \end{bmatrix}$$

$$\begin{bmatrix} \beta_x \\ \beta_y \\ \beta_{xy} \end{bmatrix} = \begin{bmatrix} 0.4500 \\ 0.1500 \\ -0.5196 \end{bmatrix} \text{m/m/kg/kg .}$$

3. Now, use Equation (2.179) and Equation (2.180) to calculate the strains as

$$\begin{bmatrix} \varepsilon_x \\ \varepsilon_y \\ \gamma_{xy} \end{bmatrix} = \begin{bmatrix} 18.73 \times 10^{-6} \\ 11.98 \times 10^{-6} \\ -11.69 \times 10^{-6} \end{bmatrix} (-100) + \begin{bmatrix} 0.4500 \\ 0.1500 \\ -0.5196 \end{bmatrix} (0.02)$$

$$= \begin{bmatrix} 0.7127 \times 10^{-2} \\ 0.1802 \times 10^{-2} \\ -0.9223 \times 10^{-2} \end{bmatrix} \text{m/m.}$$

2.10 Summary

After reviewing the definitions of stress, strain, elastic moduli, and strain energy, we developed the three-dimensional stress–strain relationships for different materials. These materials range from anisotropic to isotropic. The number of independent constants ranges from 21 for anisotropic to 2 for isotropic materials, respectively. Using plane stress assumptions, we reduced the three-dimensional problem to a two-dimensional problem and developed a stress–strain relationship for a unidirectional/bidirectional lamina. These relationships were then found for an angle lamina, using transformation of strains and stresses. We introduced failure theories of an angle lamina in terms of strengths of unidirectional lamina. Finally, we developed stress–strain equations for an angle lamina under thermal and moisture loads. In the appendix of this chapter, we review matrix algebra and the transformation of stresses and strains.

Key Terms

Mechanical characterization
Stress
Strain
Elastic moduli
Strain energy
Anisotropic material
Monoclinic material
Orthotropic material
Transversely isotropic material
Isotropic material
Plane stress
Compliance matrix
Stiffness matrix
Angle ply
Engineering constants
Invariant stiffness and compliance
Failure theories
Maximum stress failure theory
Maximum strain failure theory
Tsai–Hill theory
Tsai–Wu theory
Failure envelopes
Hygrothermal stresses
Hygrothermal loads

Exercise Set

2.1 Write the number of independent elastic constants for three-dimensional anisotropic, monoclinic, orthotropic, transversely isotropic, and isotropic materials.

2.2 The engineering constants for an orthotropic material are found to be

$$E_1 = 4\,Msi,\; E_3 = 3\,Msi,\; E_3 = 3.1\,Msi,$$

$$\nu_{12} = 0.2,\; \nu_{23} = 0.4,\; \nu_{31} = 0.6,$$

$$G_{12} = 6\,Msi,\; G_{23} = 7\,Msi,\; G_{31} = 2\,Msi$$

Find the stiffness matrix [C] and the compliance matrix [S] for the preceding orthotropic material.

2.3 Consider an orthotropic material with the stiffness matrix given by

$$[C] = \begin{bmatrix} -0.67308 & -1.8269 & -1.0577 & 0 & 0 & 0 \\ -1.8269 & -0.67308 & -1.4423 & 0 & 0 & 0 \\ -1.0577 & -1.4423 & 0.48077 & 0 & 0 & 0 \\ 0 & 0 & 0 & 4 & 0 & 0 \\ 0 & 0 & 0 & 0 & 2 & 0 \\ 0 & 0 & 0 & 0 & 0 & 1.5 \end{bmatrix} GPa$$

Find:

1. The stresses in the principal directions of symmetry if the strains in the principal directions of symmetry at a point in the material are $\varepsilon_1 = 1\ \mu m/m$, $\varepsilon_2 = 3\ \mu m/m$, $\varepsilon_3 = 2\ \mu m/m$; $\gamma_{23} = 0$, $\gamma_{31} = 5\ \mu m/m$, $\gamma_{12} = 6\ \mu m/m$
2. The compliance matrix [S]
3. The engineering constants E_1, E_2, E_3, ν_{12}, ν_{23}, ν_{31}, G_{12}, G_{23}, G_{31}
4. The strain energy per unit volume at the point where strains are given in part (1.)

2.4 Reduce the monoclinic stress–strain relationships to those of an orthotropic material.

2.5 Show the difference between monoclinic and orthotropic materials by applying normal stress in principal directions and shear stress in principal planes, one at a time and studying the resulting nonzero and zero strains.

2.6 Write down the compliance matrix of a transversely isotropic material (where 2–3 is the plane of isotropy) in terms of the following engineering constants:

E is the Young's modulus in the plane of isotropy 2–3

E' is the Young's modulus in direction 1 that is perpendicular to plane of isotropy 2–3

ν is the Poisson's ratio in the plane of isotropy 2–3

ν' is the Poisson's ratio in the 1–2 plane

G' is the shear modulus in the 1–2 plane

2.7 Find the relationship between the engineering constants of a three-dimensional orthotropic material and its compliance matrix.

2.8 What are the values of stiffness matrix elements C_{11} and C_{12} in terms of the Young's modulus and Poisson's ratio for an isotropic material?

2.9 Are ν_{12} and ν_{21} independent of each other for a unidirectional orthotropic lamina?

2.10 Find the reduced stiffness $[Q]$ and the compliance $[S]$ matrices for a unidirectional lamina of boron/epoxy. Use the properties of a unidirectional boron/epoxy lamina from Table 2.1.

2.11 Find the strains in the 1–2 coordinate system (local axes) in a unidirectional boron/epoxy lamina, if the stresses in the 1–2 coordinate system applied to are $\sigma_1 = 4$ MPa, $\sigma_2 = 2$ MPa, and $\tau_{12} = -3$ MPa. Use the properties of a unidirectional boron/epoxy lamina from Table 2.1.

2.12. Write the reduced stiffness and the compliance matrix for an isotropic lamina.

2.13 Show that for an orthotropic material $Q_{11} \neq C_{11}$. Explain why. Also, show $Q_{66} = C_{66}$. Explain why.

2.14 Consider a unidirectional continuous fiber composite. Start from $[\sigma] = [Q]\,[\varepsilon]$ and follow the procedure in Section 2.4.3 to get

$$E_1 = Q_{11} - \frac{Q_{12}^2}{Q_{22}} \qquad \nu_{12} = \frac{Q_{12}}{Q_{22}}$$

$$E_2 = Q_{22} - \frac{Q_{12}^2}{Q_{11}} \qquad \nu_{21} = \frac{Q_{12}}{Q_{11}} \qquad G_{12} = Q_{66}.$$

2.15 The reduced stiffness matrix $[Q]$ is given for a unidirectional lamina is given as follows:

$$[Q] = \begin{bmatrix} 5.681 & 0.3164 & 0 \\ 0.3164 & 1.217 & 0 \\ 0 & 0 & 0.6006 \end{bmatrix} Msi \, .$$

What are the four engineering constants, E_1, E_2, ν_{12}, and G_{12}, of the lamina?

2.16 The stresses in the global axes of a 30° ply are given as $\sigma_x = 4$ MPa, $\sigma_y = 2$ MPa, and $\tau_{xy} = -3$ MPa. Find the stresses in the local axes. Are the stresses in the local axes independent of elastic moduli? Why or why not?

2.17 The strains in the global axes of a 30° ply are given as $\varepsilon_x = 4$ µin./in., $\varepsilon_y = 2$ µin./in., and $\gamma_{xy} = -3$ µin./in. Find the strains in the local axes. Are the strains independent of material properties? Why or why not?

2.18 Find the transformed reduced stiffness matrix $[\bar{Q}]$ and transformed compliance matrix $[\bar{S}]$ for a 60° angle lamina of a boron/epoxy lamina. Use the properties of a unidirectional boron/epoxy lamina from Table 2.1.

2.19 What is the relationship between the elements of the transformed compliance matrix $[\bar{S}]$ for a 0 and 90° lamina?

2.20 For a 60° angle lamina of boron/epoxy under stresses in global axes as $\sigma_x = 4$ MPa, $\sigma_y = 2$ MPa, and $\tau_{xy} = -3$ MPa, and using the properties of a unidirectional boron/epoxy lamina from Table 2.1, find the following

1. Global strains
2. Local stresses and strains
3. Principal normal stresses and principal normal strains
4. Maximum shear stress and maximum shear strain

2.21 An angle glass/epoxy lamina is subjected to a shear stress $\tau_{xy} = 0.4$ ksi in the global axes resulting in a shear strain $\gamma_{xy} = 468.3$ µin./in. in the global axes. What is the angle of the ply? Use the properties of unidirectional glass/epoxy lamina from Table 2.2.

2.22 Find the six engineering constants for a 60° boron/epoxy lamina. Use the properties of unidirectional boron/epoxy lamina from Table 2.2.

2.23 A bidirectional woven composite ply may yield equal longitudinal and transverse Young's modulus but is still orthotropic. Determine the angles of the ply for which the shear modulus (G_{xy}) are maximum and minimum. Also find these maximum and minimum values. Given: $E_1 = 69$ GPa, $E_2 = 69$ GPa, $\nu_{12} = 0.3$, $G_{12} = 20$ GPa.

2.24 A strain gage measures normal strain in a component. Experiments[12] suggest that errors due to strain gage misalignment are more appreciable for angle plies of composite materials than isotropic materials.

1. Take a graphite/epoxy angle ply of 8° under a uniaxial stress, σ_x = 4 Msi. Estimate the strain, ε_x, as measured by a strain gage aligned in the x-direction. Now, if the strain gage is misaligned by +3° to the x-axis, estimate the measured strain. Find the percentage of error due to misalignment. Use properties of unidirectional graphite/epoxy lamina from Table 2.2.
2. Take an aluminum layer under a uniaxial stress, σ_x = 4 Msi. Estimate the strain, ε_x, as measured by a strain gage in the x-direction. Now, if the strain gage is misaligned by +3° to the x-axis, estimate the measured strain. Find the percentage of error due to misalignment. Assume E = 10 Msi, v = 0.3 for aluminum.

2.25 A uniaxial load is applied to a 10° ply. The linear stress–strain curve along the line of load is related as $\sigma_x = 123\varepsilon_x$, where the stress is measured in GPa and strain in m/m. Given E_1 = 180 GPa, E_2 = 10 GPa and ν_{12} = 0.25, find the value of (1) shear modulus, G_{12};and (2) modulus E_x for a 60° ply.

2.26 The tensile modulus of a 0°, 90°, and 45° graphite/epoxy ply is measured as follows to give E_1 = 26.25 Msi, E_2 = 1.494 Msi, E_x = 2.427 Msi for the 45° ply, respectively.

1. What is the value E_x for a 30° ply?
2. Can you calculate the values of ν_{12} and G_{12} from the previous three measured values of elastic moduli?

2.27 Can the value of the modulus, E_x, of an angle lamina be less than both the longitudinal and transverse Young's modulus of a unidirectional lamina?

2.28 Can the value of the modulus, E_x, of an angle lamina be greater than both the longitudinal and transverse Young's modulus of a unidirectional lamina?

2.29 Is the ν_{xy} for a lamina maximum for a 45° boron/epoxy ply? Use properties of unidirectional boron/epoxy lamina from Table 2.2.

2.30 In finding the value of the Young's modulus, E_x, for an angle ply, length-to-width (L/W) ratio of the specimen affects the measured value of E_x. The Young's modulus E_x^1 for a finite length-to-width ratio specimen is related to the Young's modulus, E_x, for an infinite length-to-width ratio specimen by[5]

$$E_x^1 = \frac{E_x}{1-\zeta},$$

where

$$\zeta = \frac{1}{\bar{S}_{11}}\left[\frac{3\bar{S}_{16}^2}{3\bar{S}_{66} + 2\bar{S}_{11}(L/W)^2}\right].$$

Tabulate the values of ζ for $L/W = 2$, 8, 16, and 64 for a 30° glass/epoxy. Use properties of unidirectional glass/epoxy lamina from Table 2.2.

2.31 Starting from the expression for the reduced stiffness element

$$\bar{Q}_{66} = (Q_{11} + Q_{22} - 2Q_{12} - 2Q_{66})s^2c^2 + Q_{66}(s^4 + c^4),$$

derive the expression

$$\bar{Q}_{66} = \frac{1}{2}(U_1 - U_4) - U_3 \cos 4\theta .$$

2.32 Initial stress–strain data are given for a uniaxial tensile test of a 45° angle ply. Find the in-plane shear modulus of the unidirectional lamina, G_{12}. Use linear regression analysis for finding slopes of curves.

σ_x (KPa)	ε_x (%)	$-\varepsilon_y$ (%)
210	0.1	0.08
413	0.2	0.16
644	0.3	0.25
847	0.4	0.33
1092	0.5	0.42

If similar data were given for a 35° angle ply, would it be sufficient to find the in-plane shear modulus of the unidirectional lamina, G_{12}?

2.33 Calculate the four stiffness invariants, U_1, U_2, U_3, and U_4, and the four compliance invariants V_1, V_2, V_3, and V_4, for a boron/epoxy lamina. Use the properties of a unidirectional boron/epoxy lamina from Table 2.2.

2.34 Show that $\bar{Q}_{11} + \bar{Q}_{22} + \bar{Q}_{12} + \bar{Q}_{66}$ is not a function of the angle of ply.

2.35 Find the off-axis shear strength and mode of failure of a 60° boron/epoxy lamina. Use the properties of a unidirectional boron/epoxy lamina from Table 2.1. Apply the maximum stress failure, maximum strain, Tsai–Hill, and Tsai–Wu failure theories.

2.36 Give one advantage of the maximum stress failure theory over the Tsai–Wu failure theory.

2.37 Give one advantage of the Tsai–Wu failure theory over the maximum stress failure theory.

2.38 Find the maximum biaxial stress, $\sigma_x = -\sigma$, $\sigma_y = -\sigma$, $\sigma > 0$, that one can apply to a 60° lamina of graphite/epoxy. Use the properties of a unidirectional graphite/epoxy lamina from Table 2.1. Use maximum strain and Tsai–Wu failure theories.

2.39 Using Mohr's circle, show why the maximum shear stress that can be applied to angle laminae differs with the shear stress sign. Take a 45° graphite/epoxy lamina as an example. Use the properties of a unidirectional graphite/epoxy lamina from Table 2.1.

2.40 Reduce the Tsai–Wu failure theory for an isotropic material with equal ultimate tensile and compressive strengths and a shear strength that is 40% of the ultimate tensile strength.

2.41 An off-axis test is used to find the value of H_{12} for use in the Tsai–Wu failure theory for a boron/epoxy system. The five lamina strengths of a unidirectional boron/epoxy system are given as follows:

$$(\sigma_1^T)_{ult} = 188 \text{ ksi}, \ (\sigma_1^C)_{ult} = 361 \text{ ksi}, \ (\sigma_2^T)_{ult} = 9 \text{ ksi}, \ (\sigma_2^C)_{ult} = 45 \text{ ksi}, \text{ and } (\tau_{12})_{ult} = 10 \text{ ksi}.$$

A 15° specimen fails at a uniaxial load of 33.546 ksi. Find the value of H_{12}. Does it satisfy the inequality $H_{12}^2 < H_{11}H_{22}$, which is a stability criterion for Tsai–Wu failure theory that says failure surfaces intercept all stress axes and form a closed geometric surface[13]?

2.42 Give the units for the coefficient of thermal expansion in the USCS and SI systems.

2.43 Find the free-expansional strains of a glass/epoxy unidirectional lamina under a temperature change of –100°C and a moisture absorption of 0.002 kg/kg. Also find the temperature change for which the transverse expansional strains vanish for a moisture absorption of 0.002 kg/kg. Use the properties of a unidirectional glass/epoxy lamina from Table 2.1.

2.44 Find the coefficients of thermal expansion of a 60° glass/epoxy lamina. Use the properties of unidirectional glass/epoxy lamina from Table 2.2.

2.45 Give the units for coefficient of moisture expansion in the USCS and SI systems.

2.46 Find the coefficients of moisture expansion of a 60° glass/epoxy lamina. Use the properties of unidirectional glass/epoxy lamina from Table 2.1.

References

1. Timoshenko, S.P. and Goodier, J.N., *Theory of Elasticity,* McGraw–Hill, New York, 1970.
2. Lekhnitski, S.G., *Anisotropic Plates*, Gordon and Breach Science Publishers, New York, 1968.
3. Reuter, R.C. Jr., Concise property transformation relations for an anisotropic lamina, *J. Composite Mater.*, 6, 270, 1971.
4. Buchanan, G.R., *Mechanics of Materials*, HRW, Inc., New York, 1988.
5. Halphin, J.C. and Pagano, N.J., Influence of end constraint in the testing of anisotropic bodies, *J. Composite Mater.*, 2, 18, 1968.
6. Tsai, S.W. and Pagano, N.J., Composite materials workshop, in *Progress in Materials Science Series*, Tsai, S.W., Halftone, J.C., and Pagano, N.J., Eds., Technomic, Stamford, CT, 233, 1968.
7. Tsai, S.W., Strength theories of filamentary structures in *Fundamental Aspects of Fiber Reinforced Plastic Composites*, Schwartz, R.T. and Schwartz, H.S., Eds., Wiley Interscience, New York, 3, 1968.
8. Hill, R., *The Mathematical Theory of Plasticity,* Oxford University Press, London, 1950.
9. Tsai, S.W. and Wu, E.M., A general theory of strength for anisotropic materials, *J. Composite Mater.*, 5, 58, 1971.
10. Hoffman, O., The brittle strength of orthotropic materials, *J. Composite Mater.*, 1, 296, 1967.
11. Tsai, S.W. and Hahn, H.T., *Introduction to Composite Materials*, Technomic, Lancaster, PA, 1980.
12. Tuttle, M.E. and Brinson, H.F., Resistance-foil strain gage technology as applied to composite materials, *Exp. Mech.*, 24, 54, 1985.
13. Pipes, R.B. and Cole, B.W., On the off-axis strength test for anisotropic materials, in *Boron Reinforced Epoxy Systems*, Hilado, C.J, Ed., Technomic, Westport, CT, 74, 1974.
14. Chapra, S.C. and Canale, R.C., *Numerical Methods for Engineers*, 2nd ed., McGraw–Hill, New York, 1988.

Appendix A: Matrix Algebra*

What is a matrix?

A matrix is a rectangular array of elements. The elements can be symbolic expressions and/or numbers. Matrix [*A*] is denoted by

$$[A] = \begin{bmatrix} a_{11} & a_{12} & \ldots\ldots & a_{1n} \\ a_{21} & a_{22} & \ldots\ldots & a_{2n} \\ \vdots & & & \vdots \\ a_{m1} & a_{m2} & \ldots\ldots & a_{mn} \end{bmatrix}.$$

Look at the following matrix about the sale of tires — given by quarter and make of tires — in a Blowoutr'us store:

	Quarter 1	Quarter 2	Quarter 3	Quarter 4
Tirestone	25	20	3	2
Michigan	5	10	15	25
Copper	6	16	7	27

To determine how many Copper tires were sold in quarter 4, we go along the row *Copper* and column *quarter 4* and find that it is 27.

Row *i* of [*A*] has *n* elements and is $\begin{bmatrix} a_{i1} & a_{i2} \ldots . a_{in} \end{bmatrix}$ and

Column *j* of [*A*] has *m* elements and is $\begin{bmatrix} a_{1j} \\ a_{2j} \\ \vdots \\ a_{mj} \end{bmatrix}$.

Each matrix has rows and columns that define the size of the matrix. If a matrix [*A*] has *m* rows and *n* columns, the size of the matrix is denoted by $m \times n$. The matrix [*A*] may also be denoted by $[A]_{mxn}$ to show that [*A*] is a matrix with *m* rows and *n* columns.

Each entry in the matrix is called the *entry* or *element* of the matrix and is denoted by a_{ij}, where *i* is the row number ($i = 1, 2, \ldots m$) and *j* is the column number ($j = 1, 2, \ldots n$) of the element.

The matrix for the tire sales example given earlier could be denoted by the matrix [*A*] as

* This section on matrix algebra is adapted, with permission, from A.K. Kaw, *Introduction to Matrix Algebra*, E-book, http://numericalmethods.eng.usf.edu/, 2004. At the time of printing, the complete E-book can be downloaded free of charge from the given website.

$$[A] = \begin{bmatrix} 25 & 20 & 3 & 2 \\ 5 & 10 & 15 & 25 \\ 6 & 16 & 7 & 27 \end{bmatrix}.$$

The size of the matrix is 3×4 because there are three rows and four columns. In the preceding $[A]$ matrix, $a_{34} = 27$.

What are the special types of matrices?

Vector: A vector is a matrix that has only one row or one column. The two types of vectors are row vectors and column vectors.

Row vector: If a matrix has one row, it is called a row vector — $[B] = [b_1, b_2, \ldots b_m]$ and m is the dimension of the row vector.

Column vector: If a matrix has one column, it is called a column vector

$$[C] = \begin{bmatrix} c_1 \\ \vdots \\ \vdots \\ c_n \end{bmatrix}$$

and n is the dimension of the column vector.

Example A.1

Give an example of a row vector.

Solution

$[B] = [25\ 20\ 3\ 2\ 0]$ is an example of a row vector of dimension 5.

Example A.2

Give an example of a column vector.

Solution

An example of a column vector of dimension 3 is

$$[C] = \begin{bmatrix} 25 \\ 5 \\ 6 \end{bmatrix}.$$

Submatrix: If some row(s) or/and column(s) of a matrix [*A*] are deleted, the remaining matrix is called a submatrix of [*A*].

Example A.3

Find some of the submatrices of the matrix

$$[A]=\begin{bmatrix} 4 & 6 & 2 \\ 3 & -1 & 2 \end{bmatrix}.$$

Solution

Some submatrices of [*A*] are

$$\begin{bmatrix} 4 & 6 & 2 \\ 3 & -1 & 2 \end{bmatrix}, \begin{bmatrix} 4 & 6 \\ 3 & -1 \end{bmatrix}, \begin{bmatrix} 4 & 6 & 2 \end{bmatrix}, \begin{bmatrix} 4 \end{bmatrix}, \begin{bmatrix} 2 \\ 2 \end{bmatrix}$$

Can you find other submatrices of [*A*]?

Square matrix: If the number of rows, *m*, of a matrix is equal to the number of columns, *n*, of the matrix, ($m = n$), it is called a square matrix. The entries a_{11}, a_{22},...a_{nn} are called the *diagonal elements* of a square matrix. Sometimes the diagonal of the matrix is also called the *principal* or *main* of the matrix.

Example A.4

Give an example of a square matrix.

Solution

Because it has the same number of rows and columns (that is, three),

$$[A]=\begin{bmatrix} 25 & 20 & 3 \\ 5 & 10 & 15 \\ 6 & 15 & 7 \end{bmatrix}$$

is a square matrix.

The diagonal elements of [*A*] are $a_{11} = 25$, $a_{22} = 10$, and $a_{33} = 7$.

Diagonal matrix: A square matrix with all nondiagonal elements equal to zero is called a diagonal matrix — that is, only the diagonal entries of the square matrix can be nonzero ($a_{ij} = 0$, $i \neq j$).

Example A.5

Give examples of a diagonal matrix.

Solution

An example of a diagonal matrix is

$$\begin{bmatrix} 3 & 0 & 0 \\ 0 & 2.1 & 0 \\ 0 & 0 & 5 \end{bmatrix}.$$

Any or all the diagonal entries of a diagonal matrix can be zero. For example, the following is also a diagonal matrix:

$$[A] = \begin{bmatrix} 3 & 0 & 0 \\ 0 & 2.1 & 0 \\ 0 & 0 & 0 \end{bmatrix}.$$

Identity matrix: A diagonal matrix with all diagonal elements equal to one is called an identity matrix ($a_{ij} = 0$, $i \neq j$; and $a_{ii} = 1$ for all i).

Example A.6

Give an example of an identity matrix.

Solution

An identity matrix is

$$[A] = \begin{bmatrix} 1 & 0 & 0 & 0 \\ 0 & 1 & 0 & 0 \\ 0 & 0 & 1 & 0 \\ 0 & 0 & 0 & 1 \end{bmatrix}.$$

Zero matrix: A matrix whose entries are all zero is called a zero matrix ($a_{ij} = 0$ for all i and j).

Example A.7

Give examples of a zero matrix.

Solution

Examples of a zero matrix include:

$$[A] = \begin{bmatrix} 0 & 0 & 0 \\ 0 & 0 & 0 \\ 0 & 0 & 0 \end{bmatrix},$$

$$[B] = \begin{bmatrix} 0 & 0 & 0 \\ 0 & 0 & 0 \end{bmatrix},$$

$$[C] = \begin{bmatrix} 0 & 0 & 0 & 0 \\ 0 & 0 & 0 & 0 \\ 0 & 0 & 0 & 0 \end{bmatrix},$$

$$[D] = [0 \quad 0 \quad 0]\,.$$

When are two matrices considered equal?

Two matrices $[A]$ and $[B]$ are equal if

The size of $[A]$ and $[B]$ is the same (number of rows of $[A]$ is same as the number of rows of $[B]$ and the number of columns of $[A]$ is same as number of columns of $[B]$) and

$a_{ij} = b_{ij}$ for all i and j.

Example A.8

What would make

$$[A] = \begin{bmatrix} 2 & 3 \\ 6 & 7 \end{bmatrix}$$

equal to

$$[B] = \begin{bmatrix} b_{11} & 3 \\ 6 & b_{22} \end{bmatrix}?$$

Solution

The two matrices $[A]$ and $[B]$ would be equal if $b_{11} = 2$, $b_{22} = 7$.

How are two matrices added?

Two matrices $[A]$ and $[B]$ can be added only if they are the same size (number of rows of $[A]$ is same as the number of rows of $[B]$ and the number of columns of $[A]$ is same as number of columns of $[B]$). Then, the addition is shown as $[C] = [A] + [B]$, where $c_{ij} = a_{ij} + b_{ij}$ for all i and j.

Example A.9

Add the two matrices

$$[A] = \begin{bmatrix} 5 & 2 & 3 \\ 1 & 2 & 7 \end{bmatrix}$$

$$[B] = \begin{bmatrix} 6 & 7 & -2 \\ 3 & 5 & 19 \end{bmatrix}.$$

Solution

$$[C] = [A] + [B]$$

$$= \begin{bmatrix} 5 & 2 & 3 \\ 1 & 2 & 7 \end{bmatrix} + \begin{bmatrix} 6 & 7 & -2 \\ 3 & 5 & 19 \end{bmatrix}$$

$$= \begin{bmatrix} 5+6 & 2+7 & 3-2 \\ 1+3 & 2+5 & 7+19 \end{bmatrix}$$

$$= \begin{bmatrix} 11 & 9 & 1 \\ 4 & 7 & 26 \end{bmatrix}.$$

How are two matrices subtracted?

Two matrices $[A]$ and $[B]$ can be subtracted only if they are the same size (number of rows of $[A]$ is same as the number of rows of $[B]$ and the number of columns of $[A]$ is same as number of columns of $[B]$). The subtraction is given by $[D] = [A] - [B]$, where $d_{ij} = a_{ij} - b_{ij}$ for all i and j.

Example A.10

Subtract matrix [B] from matrix [A] — that is, find [A] – [B].

$$[A] = \begin{bmatrix} 5 & 2 & 3 \\ 1 & 2 & 7 \end{bmatrix}$$

$$[B] = \begin{bmatrix} 6 & 7 & -2 \\ 3 & 5 & 19 \end{bmatrix}.$$

Solution

$$[C] = [A] - [B]$$

$$= \begin{bmatrix} 5 & 2 & 3 \\ 1 & 2 & 7 \end{bmatrix} - \begin{bmatrix} 6 & 7 & -2 \\ 3 & 5 & 19 \end{bmatrix}$$

$$= \begin{bmatrix} 5-6 & 2-7 & 3-(-2) \\ 1-3 & 2-5 & 7-19 \end{bmatrix}$$

$$= \begin{bmatrix} -1 & -5 & 5 \\ -2 & -3 & -12 \end{bmatrix}.$$

How are two matrices multiplied?

A matrix [A] can be multiplied by another matrix [B] only if the number of columns of [A] is equal to the number of rows of [B] to give $[C]_{mxn} = [A]_{mxp}[B]_{pxn}$. If [$A$] is an $m \times p$ matrix and [B] is a $p \times n$ matrix, then the size of the resulting matrix [C] is an $m \times n$ matrix.

How does one calculate the elements of [C] matrix?

$$c_{ij} = \sum_{k=1}^{p} a_{ik} b_{kj}$$

$$= a_{i1}b_{1j} + a_{i2}b_{2j} + \dots\dots\dots + a_{ip}b_{pj}$$

for each $i = 1, 2, \dots m$ and $j = 1, 2, \dots n$.

To put it in simpler terms, the ith row and j^{th} column of the [C] matrix in [C] = [A][B] is calculated by multiplying the ith row of [A] by the jth column of [B] — that is,

$$c_{ij} = [a_{i1}\ a_{i2} \ldots\ldots a_{ip}] \begin{bmatrix} b_{1j} \\ b_{2j} \\ \vdots \\ \vdots \\ b_{pj} \end{bmatrix}$$

$$= a_{i1}\, b_{1j} + a_{i2}\, b_{2j} + \ldots\ldots + a_{ip}\, b_{pj}.$$

$$= \sum_{k=1}^{p} a_{ik} b_{kj} \ .$$

Example A.11

Given

$$[A] = \begin{bmatrix} 5 & 2 & 3 \\ 1 & 2 & 7 \end{bmatrix}$$

$$[B] = \begin{bmatrix} 3 & -2 \\ 5 & -8 \\ 9 & -10 \end{bmatrix},$$

find

$$[C] = [A][B] \ .$$

Solution

For example, the element c_{12} of the [C] matrix can be found by multiplying the first row of [A] by the second column of [B]:

$$c_{12} = [5 \quad 2 \quad 3] \begin{bmatrix} -2 \\ -8 \\ -10 \end{bmatrix}$$

$$= (5)(-2) + (2)(-8) + (3)(-10)$$

$$= -56.$$

Similarly, one can find the other elements of $[C]$ to give

$$[C] = \begin{bmatrix} 52 & -56 \\ 76 & -88 \end{bmatrix}.$$

What is a scalar product of a constant and a matrix?

If $[A]$ is an $n \times n$ matrix and k is a real number, then the scalar product of k and $[A]$ is another matrix $[B]$, where $b_{ij} = ka_{ij}$.

Example A.12

Let

$$[A] = \begin{bmatrix} 2.1 & 3 & 2 \\ 5 & 1 & 6 \end{bmatrix}.$$

Find 2 $[A]$.

Solution

$$[A] = \begin{bmatrix} 2.1 & 3 & 2 \\ 5 & 1 & 6 \end{bmatrix};$$

then,

$$2[A] = 2\begin{bmatrix} 2.1 & 3 & 2 \\ 5 & 1 & 6 \end{bmatrix}$$

$$= \begin{bmatrix} (2)(2.1) & (2)(3) & (2)(2) \\ (2)(5) & (2)(1) & (2)(6) \end{bmatrix}$$

$$= \begin{bmatrix} 4.2 & 6 & 4 \\ 10 & 2 & 12 \end{bmatrix}.$$

What is a linear combination of matrices?

If $[A_1]$, $[A_2]$,…,$[A_p]$ are matrices of the same size and k_1, k_2,…,k_p are scalars, then

$$k_1[A_1]+k_2[A_2]+........+k_p[A_p]$$

is called a linear combination of $[A_1]$, $[A_2]$,…,$[A_p]$.

Example A.13

If

$$[A_1]=\begin{bmatrix}5 & 6 & 2\\ 3 & 2 & 1\end{bmatrix},\ [A_2]=\begin{bmatrix}2.1 & 3 & 2\\ 5 & 1 & 6\end{bmatrix},\ [A_3]=\begin{bmatrix}0 & 2.2 & 2\\ 3 & 3.5 & 6\end{bmatrix},$$

then find

$$[A_1]+2[A_2]-0.5[A_3]\ .$$

Solution

$$[A_1]+2[A_2]-0.5[A_3]=\begin{bmatrix}5 & 6 & 2\\ 3 & 2 & 1\end{bmatrix}+2\begin{bmatrix}2.1 & 3 & 2\\ 5 & 1 & 6\end{bmatrix}-0.5\begin{bmatrix}0 & 2.2 & 2\\ 3 & 3.5 & 6\end{bmatrix}$$

$$=\begin{bmatrix}5 & 6 & 2\\ 3 & 2 & 1\end{bmatrix}+\begin{bmatrix}4.2 & 6 & 4\\ 10 & 2 & 12\end{bmatrix}-\begin{bmatrix}0 & 1.1 & 1\\ 1.5 & 1.75 & 3\end{bmatrix}$$

$$=\begin{bmatrix}9.2 & 10.9 & 5\\ 11.5 & 2.25 & 10\end{bmatrix}.$$

What are some of the rules of binary matrix operations?

Commutative law of addition: If $[A]$ and $[B]$ are $m \times n$ matrices, then

$$[A]+[B]=[B]+[A]\ .$$

Associate law of addition: If $[A]$, $[B]$, and $[C]$ all are $m \times n$ matrices, then

$$[A]+([B]+[C])=([A]+[B])+[C]\,.$$

Associate law of multiplication: If $[A]$, $[B]$, and $[C]$ are $m \times n$, $n \times p$, and $p \times r$ size matrices, respectively, then

$$[A]([B][C])=([A][B])[C]$$

and the resulting matrix size on both sides is $m \times r$.

Distributive law: If [A] and [B] are $m \times n$ size matrices and [C] and [D] are $n \times p$ size matrices, then

$$[A]([C]+[D])=[A][C]+[A][D]$$

$$([A]+[B])[C]=[A][C]+[B][C]$$

and the resulting matrix size on both sides is $m \times p$.

Example A.14

Illustrate the associative law of multiplication of matrices using

$$[A]=\begin{bmatrix}1 & 2\\ 3 & 5\\ 0 & 2\end{bmatrix},\quad [B]=\begin{bmatrix}2 & 5\\ 9 & 6\end{bmatrix},\quad [C]=\begin{bmatrix}2 & 1\\ 3 & 5\end{bmatrix}.$$

Solution

$$[B][C]=\begin{bmatrix}2 & 5\\ 9 & 6\end{bmatrix}\begin{bmatrix}2 & 1\\ 3 & 5\end{bmatrix}=\begin{bmatrix}19 & 27\\ 36 & 39\end{bmatrix}$$

$$[A][B][C]=\begin{bmatrix}1 & 2\\ 3 & 5\\ 0 & 2\end{bmatrix}\begin{bmatrix}19 & 27\\ 36 & 39\end{bmatrix}=\begin{bmatrix}91 & 105\\ 237 & 276\\ 72 & 78\end{bmatrix}$$

$$[A][B] = \begin{bmatrix} 1 & 2 \\ 3 & 5 \\ 0 & 2 \end{bmatrix} \begin{bmatrix} 2 & 5 \\ 9 & 6 \end{bmatrix} = \begin{bmatrix} 20 & 17 \\ 51 & 45 \\ 18 & 12 \end{bmatrix}$$

$$[A][B][C] = \begin{bmatrix} 20 & 17 \\ 51 & 45 \\ 18 & 12 \end{bmatrix} \begin{bmatrix} 2 & 1 \\ 3 & 5 \end{bmatrix}$$

$$= \begin{bmatrix} 91 & 105 \\ 237 & 276 \\ 72 & 78 \end{bmatrix}.$$

These illustrate the associate law of multiplication of matrices.

Is [*A*][*B*] = [*B*][*A*]?

First, both operations, $[A][B]$ and $[B][A]$, are only possible if $[A]$ and $[B]$ are square matrices of same size. Why? If $[A][B]$ exists, the number of columns of $[A]$ must be the same as the number of rows of $[B]$; if $[B][A]$ exists, the number of columns of $[B]$ must be the same as the number of rows of $[A]$.

Even then, in general, $[A][B] \neq [B][A]$.

Example A.15

Illustrate whether $[A][B] = [B][A]$ for the following matrices:

$$[A] = \begin{bmatrix} 6 & 3 \\ 2 & 5 \end{bmatrix}, \quad [B] = \begin{bmatrix} -3 & 2 \\ 1 & 5 \end{bmatrix}.$$

Solution

$$[A][B] = \begin{bmatrix} 6 & 3 \\ 2 & 5 \end{bmatrix} \begin{bmatrix} -3 & 2 \\ 1 & 5 \end{bmatrix}$$

$$= \begin{bmatrix} -15 & 27 \\ -1 & 29 \end{bmatrix}$$

$$[B][A] = \begin{bmatrix} -3 & 2 \\ 1 & 5 \end{bmatrix} \begin{bmatrix} 6 & 3 \\ 2 & 5 \end{bmatrix}$$

$$= \begin{bmatrix} -14 & 1 \\ 16 & 28 \end{bmatrix}$$

$$[A][B] \neq [B][A].$$

What is the transpose of a matrix?

Let $[A]$ be an $m \times n$ matrix. Then $[B]$ is the transpose of the $[A]$ if $b_{ji} = a_{ij}$ for all i and j. That is, the ith row and the jth column element of $[A]$ is the jth row and i^{th} column element of $[B]$. Note that $[B]$ would be an nxm matrix. The transpose of $[A]$ is denoted by $[A]^T$.

Example A.16

Find the transpose of

$$[A] = \begin{bmatrix} 25 & 20 & 3 & 2 \\ 5 & 10 & 15 & 25 \\ 6 & 16 & 7 & 27 \end{bmatrix}$$

Solution

The transpose of $[A]$ is

$$[A]^T = \begin{bmatrix} 25 & 5 & 6 \\ 20 & 10 & 16 \\ 3 & 15 & 7 \\ 2 & 25 & 27 \end{bmatrix}.$$

Note that the transpose of a row vector is a column vector and the transpose of a column vector is a row vector. Also, note that the transpose of a transpose of a matrix is the matrix — that is, $([A]^T)^T = [A]$. Also, $(A + B)^T = A^T + B^T$; $(cA)^T = cA^T$.

What is a symmetric matrix?

A square matrix $[A]$ with real elements, where $a_{ij} = a_{ji}$ for $i = 1,\ldots,n$ and $j = 1,\ldots,n$, is called a symmetric matrix. This is same as that if $[A] = [A]^T$, then $[A]$ is a symmetric matrix.

Example A.17

Give an example of a symmetric matrix.

Solution

A symmetric matrix is

$$[A]=\begin{bmatrix} 21.2 & 3.2 & 6 \\ 3.2 & 21.5 & 8 \\ 6 & 8 & 9.3 \end{bmatrix}$$

because $a_{12} = a_{21} = 3.2$; $a_{13} = a_{31} = 6$; and $a_{23} = a_{32} = 8$.

What is a skew-symmetric matrix?

A square matrix $[A]$ with real elements, where $a_{ij} = -a_{ji}$ for $i = 1,\ldots,n$ and $j = 1,\ldots,n$, is called a skew symmetric matrix. This is same as that if $[A] = -[A]^T$, then $[A]$ is a skew symmetric matrix.

Example A.18

Give an example of a skew-symmetric matrix.

Solution

A skew-symmetric matrix is

$$\begin{bmatrix} 0 & 1 & 2 \\ -1 & 0 & -5 \\ -2 & 5 & 0 \end{bmatrix}$$

because $a_{12} = -a_{21} = 1$; $a_{13} = -a_{31} = 2$; $a_{23} = -a_{32} = -5$. Because $a_{ii} = -a_{ii}$ only if $a_{ii} = 0$, all the diagonal elements of a skew-symmetric matrix must be zero.

Matrix algebra is used for solving systems of equations. Can you illustrate this concept?

Matrix algebra is used to solve a system of simultaneous linear equations. Let us illustrate with an example of three simultaneous linear equations:

$$25a + 5b + c = 106.8$$

$$64a + 8b + c = 177.2$$

$$144a + 12b + c = 279.2\ .$$

This set of equations can be rewritten in the matrix form as

$$\begin{bmatrix} 25a+ & 5b+ & c \\ 64a+ & 8b+ & c \\ 144a+ & 12b+ & c \end{bmatrix} = \begin{bmatrix} 106.8 \\ 177.2 \\ 279.2 \end{bmatrix}.$$

The preceding equation can be written as a linear combination as follows

$$a\begin{bmatrix} 25 \\ 64 \\ 144 \end{bmatrix} + b\begin{bmatrix} 5 \\ 8 \\ 12 \end{bmatrix} + c\begin{bmatrix} 1 \\ 1 \\ 1 \end{bmatrix} = \begin{bmatrix} 106.8 \\ 177.2 \\ 279.2 \end{bmatrix}$$

and, further using matrix multiplications, gives

$$\begin{bmatrix} 25 & 5 & 1 \\ 64 & 8 & 1 \\ 144 & 12 & 1 \end{bmatrix} \begin{bmatrix} a \\ b \\ c \end{bmatrix} = \begin{bmatrix} 106.8 \\ 177.2 \\ 279.2 \end{bmatrix}.$$

For a general set of m linear equations and n unknowns,

$$a_{11}x_1 + a_{22}x_2 + \cdots\cdots + a_{1n}x_n = c_1$$

$$a_{21}x_1 + a_{22}x_2 + \cdots\cdots + a_{2n}x_n = c_2$$

$$\cdots\cdots\cdots\cdots\cdots\cdots\cdots\cdots\cdots\cdots\cdots\cdots\cdots$$

$$\cdots\cdots\cdots\cdots\cdots\cdots\cdots\cdots\cdots\cdots\cdots\cdots\cdots$$

$$a_{m1}x_1 + a_{m2}x_2 + \ldots\ldots.. + a_{mn}x_n = c_m$$

can be rewritten in the matrix form as

$$\begin{bmatrix} a_{11} & a_{12} & \cdot & \cdot & a_{1n} \\ a_{21} & a_{22} & \cdot & \cdot & a_{2n} \\ \vdots & & & & \vdots \\ \vdots & & & & \vdots \\ a_{m1} & a_{m2} & \cdot & \cdot & a_{mn} \end{bmatrix} \begin{bmatrix} x_1 \\ x_2 \\ \cdot \\ \cdot \\ x_n \end{bmatrix} = \begin{bmatrix} c_1 \\ c_2 \\ \cdot \\ \cdot \\ c_m \end{bmatrix}.$$

Denoting the matrices by [*A*], [*X*], and [*C*], the system of equation is [*A*] [*X*] = [*C*], where [*A*] is called the *coefficient matrix*, [*C*] is called the *right-hand side vector*, and [*X*] is called the *solution vector*.

Sometimes [*A*] [*X*] = [*C*] systems of equations are written in the *augmented form* — that is,

$$[A \vdots C] = \begin{bmatrix} a_{11} & a_{12} & \cdots\cdots & a_{1n} \vdots c_1 \\ a_{21} & a_{22} & \cdots\cdots & a_{2n} \vdots c_2 \\ \vdots & & & \vdots \\ \vdots & & & \vdots \\ a_{m1} & a_{m2} & \cdots\cdots & a_{mn} \vdots c_n \end{bmatrix}.$$

Can you divide two matrices because that will help me find the solution vector for a general set of equations given by [*A*] [*X*] = [*C*]?

If [*A*][*B*]=[*C*] is defined, it might seem intuitive that $[A] = \frac{[C]}{[B]}$, but matrix division is not defined. However, an *inverse of a matrix* can be defined for certain types of square matrices. The inverse of a square matrix [*A*], if existing, is denoted by $[A]^{-1}$ such that $[A][A]^{-1} = [I] = [A]^{-1}[A]$.

In other words, let [*A*] be a square matrix. If [*B*] is another square matrix of the same size so that [*B*][*A*] = [*I*], then [*B*] is the inverse of [*A*]. [*A*] is then called *invertible* or *nonsingular*. If $[A]^{-1}$ does not exist, [*A*] is called *noninvertible* or *singular*.

Example A.19

Show whether

$$[B] = \begin{bmatrix} 3 & 2 \\ 5 & 3 \end{bmatrix}$$

is the inverse of

$$[A] = \begin{bmatrix} -3 & 2 \\ 5 & -3 \end{bmatrix}.$$

Solution

$$[B][A] = \begin{bmatrix} 3 & 2 \\ 5 & 3 \end{bmatrix} \begin{bmatrix} -3 & 2 \\ 5 & -3 \end{bmatrix}$$

$$= \begin{bmatrix} 1 & 0 \\ 0 & 1 \end{bmatrix}$$

$$= [I].$$

$[B][A] = [I]$, so $[B]$ is the inverse of $[A]$ and $[A]$ is the inverse of $[B]$. However, we can also show that

$$[A][B] = \begin{bmatrix} -3 & 2 \\ 5 & -3 \end{bmatrix} \begin{bmatrix} 3 & 2 \\ 5 & 3 \end{bmatrix}$$

$$= \begin{bmatrix} 1 & 0 \\ 0 & 1 \end{bmatrix}$$

$$= [I]$$

to show that $[A]$ is the inverse of $[B]$.

Can I use the concept of the inverse of a matrix to find the solution of a set of equations $[A][X] = [C]$?

Yes, if the number of equations is the same as the number of unknowns, the coefficient matrix $[A]$ is a square matrix.

Given $[A][X] = [C]$. Then, if $[A]^{-1}$ exists, multiplying both sides by $[A]^{-1}$:

$$[A]^{-1}\,[A][X] = [A]^{-1}\,[C]$$

$$[I][X] = [A]^{-1}[C]$$

$$[X] = [A]^{-1}\,[C].$$

This implies that if we are able to find $[A]^{-1}$, the solution vector of $[A][X] = [C]$ is simply a multiplication of $[A]^{-1}$ and the right-hand side vector, $[C]$.

How do I find the inverse of a matrix?

If $[A]$ is an $n \times n$ matrix, then $[A]^{-1}$ is an $n \times n$ matrix and, according to the definition of inverse of a matrix, $[A][A]^{-1} = [I]$.

Denoting,

$$[A]=\begin{bmatrix} a_{11} & a_{12} & \cdot & \cdot & a_{1n} \\ a_{21} & a_{22} & \cdot & \cdot & a_{2n} \\ \cdot & \cdot & \cdot & \cdot & \cdot \\ \cdot & \cdot & \cdot & \cdot & \cdot \\ a_{n1} & a_{n2} & \cdot & \cdot & a_{nn} \end{bmatrix}$$

$$[A]^{-1}=\begin{bmatrix} a'_{11} & a'_{12} & \cdot & \cdot & a'_{1n} \\ a'_{21} & a'_{22} & \cdot & \cdot & a'_{2n} \\ \cdot & \cdot & \cdot & \cdot & \cdot \\ \cdot & \cdot & \cdot & \cdot & \cdot \\ a'_{n1} & a'_{n2} & \cdot & \cdot & a'_{nm} \end{bmatrix}$$

$$[I]=\begin{bmatrix} 1 & 0 & \cdot & \cdot & \cdot & 0 \\ 0 & 1 & & & & 0 \\ 0 & & \cdot & & & \cdot \\ \cdot & & & 1 & & \cdot \\ \cdot & & & & \cdot & \cdot \\ 0 & \cdot & \cdot & \cdot & \cdot & 1 \end{bmatrix}.$$

Using the definition of matrix multiplication, the first column of the $[A]^{-1}$ matrix can then be found by solving:

$$\begin{bmatrix} a_{11} & a_{12} & \cdot & \cdot & a_{1n} \\ a_{21} & a_{22} & \cdot & \cdot & a_{2n} \\ \cdot & \cdot & \cdot & \cdot & \cdot \\ \cdot & \cdot & \cdot & \cdot & \cdot \\ a_{n1} & a_{n2} & \cdot & \cdot & a_{nn} \end{bmatrix}\begin{bmatrix} a'_{11} \\ a'_{21} \\ \cdot \\ \cdot \\ a'_{n1} \end{bmatrix}=\begin{bmatrix} 1 \\ 0 \\ \cdot \\ \cdot \\ 0 \end{bmatrix}.$$

Similarly, one can find the other columns of the $[A]^{-1}$ matrix by changing the right-hand side accordingly.

Example A.20

Solve the set of equations:

$$25a+5b+c=106.8$$

$$64a + 8b + c = 177.2$$

$$144a + 12b + c = 279.2\ .$$

Solution

In matrix form, the preceding three simultaneous linear equations are written as

$$\begin{bmatrix} 25 & 5 & 1 \\ 64 & 8 & 1 \\ 144 & 12 & 1 \end{bmatrix} \begin{bmatrix} a \\ b \\ c \end{bmatrix} = \begin{bmatrix} 106.8 \\ 177.2 \\ 279.2 \end{bmatrix}.$$

First, we will find the inverse of

$$[A] = \begin{bmatrix} 25 & 5 & 1 \\ 64 & 8 & 1 \\ 144 & 12 & 1 \end{bmatrix}$$

and then use the definition of inverse to find the coefficients a, b, c.

If

$$[A]^{-1} = \begin{bmatrix} a'_{11} & a'_{12} & a'_{13} \\ a'_{21} & a'_{22} & a'_{23} \\ a'_{31} & a'_{32} & a'_{33} \end{bmatrix}$$

is the inverse of $[A]$, then

$$\begin{bmatrix} 25 & 5 & 1 \\ 64 & 8 & 1 \\ 144 & 12 & 1 \end{bmatrix} \begin{bmatrix} a'_{11} & a'_{12} & a'_{13} \\ a'_{21} & a'_{22} & a'_{23} \\ a'_{31} & a'_{32} & a'_{33} \end{bmatrix} = \begin{bmatrix} 1 & 0 & 0 \\ 0 & 1 & 0 \\ 0 & 0 & 1 \end{bmatrix}$$

gives three sets of equations:

$$\begin{bmatrix} 25 & 5 & 1 \\ 64 & 8 & 1 \\ 144 & 12 & 1 \end{bmatrix} \begin{bmatrix} a'_{11} \\ a'_{21} \\ a'_{31} \end{bmatrix} = \begin{bmatrix} 1 \\ 0 \\ 0 \end{bmatrix}$$

$$\begin{bmatrix} 25 & 5 & 1 \\ 64 & 8 & 1 \\ 144 & 12 & 1 \end{bmatrix} \begin{bmatrix} a'_{12} \\ a'_{22} \\ a'_{32} \end{bmatrix} = \begin{bmatrix} 0 \\ 1 \\ 0 \end{bmatrix}$$

$$\begin{bmatrix} 25 & 5 & 1 \\ 64 & 8 & 1 \\ 144 & 12 & 1 \end{bmatrix} \begin{bmatrix} a'_{13} \\ a'_{23} \\ a'_{33} \end{bmatrix} = \begin{bmatrix} 0 \\ 0 \\ 1 \end{bmatrix}.$$

Solving the preceding three sets of equations separately gives

$$\begin{bmatrix} a'_{11} \\ a'_{21} \\ a'_{31} \end{bmatrix} = \begin{bmatrix} 0.04762 \\ -0.9524 \\ 4.571 \end{bmatrix}$$

$$\begin{bmatrix} a'_{12} \\ a'_{22} \\ a'_{32} \end{bmatrix} = \begin{bmatrix} -0.08333 \\ 1.417 \\ -5.000 \end{bmatrix}$$

$$\begin{bmatrix} a'_{13} \\ a'_{23} \\ a'_{33} \end{bmatrix} = \begin{bmatrix} 0.03571 \\ -0.4643 \\ 1.429 \end{bmatrix}.$$

Therefore,

$$[A]^{-1} = \begin{bmatrix} 0.04762 & -0.08333 & 0.03571 \\ -0.9524 & 1.417 & -0.4643 \\ 4.571 & -5.000 & 1.429 \end{bmatrix}.$$

Now, $[A][X] = [C]$, where

$$[X] = \begin{bmatrix} a \\ b \\ c \end{bmatrix}$$

$$[C] = \begin{bmatrix} 106.8 \\ 177.2 \\ 279.2 \end{bmatrix}.$$

Using the definition of $[A]^{-1}$,

$$[A]^{-1}[A][X] = [A]^{-1}[C]$$

$$[X] = [A]^{-1}[C]$$

$$= \begin{bmatrix} -0.04762 & -0.08333 & 0.03571 \\ -0.9524 & 1.417 & -0.4643 \\ 4.571 & -5.000 & 1.429 \end{bmatrix} \begin{bmatrix} 106.8 \\ 177.2 \\ 279.2 \end{bmatrix}$$

$$\begin{bmatrix} a \\ b \\ c \end{bmatrix} = \begin{bmatrix} 0.2900 \\ 19.70 \\ 1.050 \end{bmatrix}.$$

Computationally and algorithmically more efficient, a set of simultaneous linear equations, such as those given previously, can also be solved by using various numerical techniques. These techniques are explained completely in the source (http://numericalmethods.eng.usf.edu) of this appendix. Some of the common techniques of solving a set of simultaneous linear equations are

Matrix inverse method

Gaussian elimination method

Gauss–Siedel method

LU decomposition method

Key Terms

Matrix
Vector
Row vector
Column vector
Submatrix

Square matrix
Diagonal matrix
Identity matrix
Zero matrix
Equal matrices
Addition of matrices
Subtraction of matrices
Multiplication of matrices
Scalar product of matrices
Linear combination of matrices
Rules of binary matrix operation
Transpose of a matrix
Symmetric matrix
Skew symmetric matrix
Inverse of a matrix

Appendix B: Transformation of Stresses and Strains

Equation (2.100) and Equation (2.94) give the relationship between stresses/strains in the global (x,y) coordinate system and the local (1,2) coordinate system, respectively. Note that the transformation is independent of material properties and depends only on the angle between the x-axis and 1-axis, or the angle through which the coordinate system (1,2) is rotated anticlockwise.

B.1 Transformation of Stress

Consider that σ_x, σ_y, and τ_{xy} are the stresses on the rectangular element at a point O in a two-dimensional body (Figure 2.38). One now wants to find the values of the stresses σ_1, σ_2, and τ_{12} on another rectangular element but at the same point O on the body. To do so, make a cut at an angle θ normal to direction 1. Now the stresses in the local 1–2 coordinate system can be related to those in the global x–y coordinate system.

Summing the forces in the direction 1 gives,

$$\sigma_1 \overline{BC} - \tau_{xy}\overline{AB}\,\mathrm{Cos}\,\theta - \sigma_y \overline{AB}\,\mathrm{Sin}\,\theta - \tau_{xy}\overline{AC}\,\mathrm{Sin}\,\theta - \sigma_x \overline{AC}\,\mathrm{Cos}\,\theta = 0$$

$$\sigma_1 = \tau_{xy}\frac{\overline{AB}}{\overline{BC}}\mathrm{Cos}\,\theta + \sigma_y\frac{\overline{AB}}{\overline{BC}}\mathrm{Sin}\,\theta + \tau_{xy}\frac{\overline{AC}}{\overline{BC}}\mathrm{Sin}\,\theta + \sigma_x\frac{\overline{AC}}{\overline{BC}}\mathrm{Cos}\,\theta \;.$$

Now,

$$\mathrm{Sin}\,\theta = \frac{\overline{AB}}{\overline{BC}},$$

and

$$\mathrm{Cos}\,\theta = \frac{\overline{AC}}{\overline{BC}} \;;$$

we have

$$\sigma_1 \tau_{xy}\,\mathrm{Sin}\,\theta\,\mathrm{Cos}\,\theta + \sigma_y\,\mathrm{Sin}^2\theta + \tau_{xy}\,\mathrm{Cos}\,\theta\,\mathrm{Sin}\,\theta + \sigma_x\,\mathrm{Cos}^2\theta$$

$$\sigma_1 = \sigma_x\,\mathrm{Cos}^2\theta + \sigma_y\,\mathrm{Sin}^2\theta + 2\tau_{xy}\,\mathrm{Sin}\,\theta\,\mathrm{Cos}\,\theta \;. \tag{B.1}$$

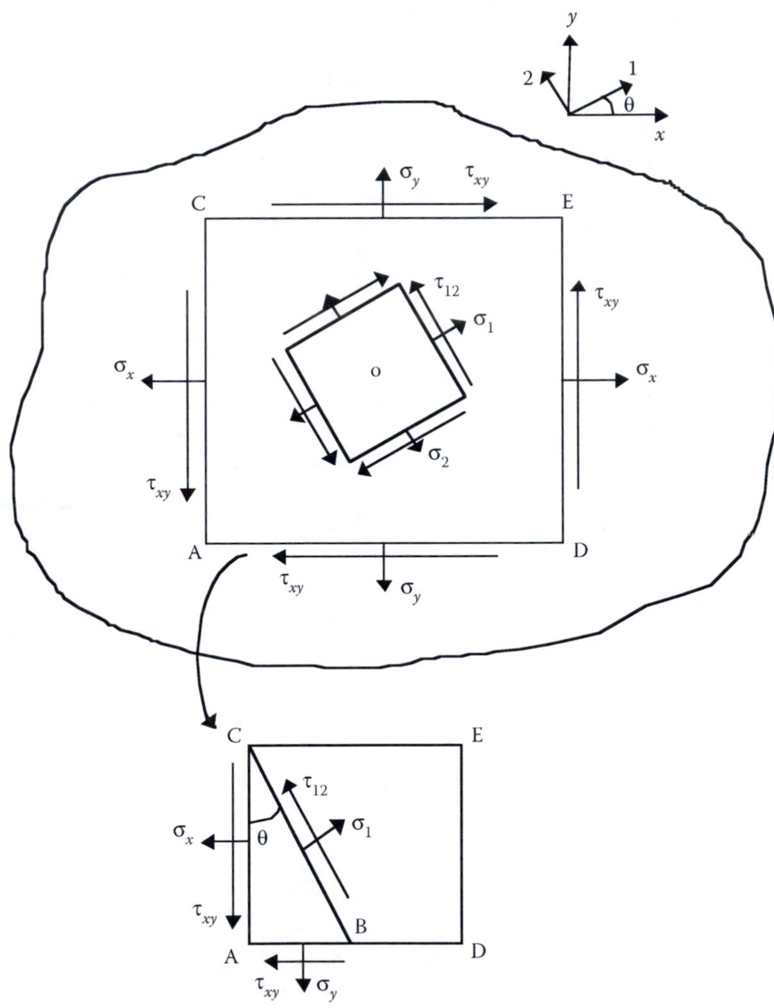

FIGURE 2.38
Free body diagrams for transformation of stresses between local and global axes.

Similarly, summing the forces in direction 2 gives

$$\tau_{12} = -\sigma_x \operatorname{Sin}\theta \operatorname{Cos}\theta + \sigma_y \operatorname{Sin}\theta \operatorname{Cos}\theta + \tau_{xy}(\operatorname{Cos}^2\theta - \operatorname{Sin}^2\theta)\ . \qquad \text{(B.2)}$$

By making a cut at an angle, θ, normal to direction 2,

$$\sigma_2 = \sigma_x \operatorname{Sin}^2\theta + \sigma_y \operatorname{Cos}^2\theta - 2\tau_{xy} \operatorname{Sin}\theta \operatorname{Cos}\theta\ . \qquad \text{(B.3)}$$

In matrix form, Equation (B.1), Equation (B.2), and Equation (B.3) relate the local stresses to global stresses as

$$\begin{bmatrix} \sigma_1 \\ \sigma_2 \\ \tau_{12} \end{bmatrix} = \begin{bmatrix} c^2 & s^2 & 2sc \\ s^2 & c^2 & -2sc \\ -sc & sc & c^2 - s^2 \end{bmatrix} \begin{bmatrix} \sigma_x \\ \sigma_y \\ \tau_{xy} \end{bmatrix} \tag{B.4}$$

where $c = Cos\ \theta$ and $s = Sin\ \theta$.

The 3×3 matrix in Equation (B.4) is called the transformation matrix $[T]$:

$$[T] = \begin{bmatrix} c^2 & s^2 & 2sc \\ s^2 & c^2 & -2sc \\ -sc & sc & c^2 - s^2 \end{bmatrix}. \tag{B.5}$$

By inverting (B.5),

$$[T]^{-1} = \begin{bmatrix} c^2 & s^2 & -2sc \\ s^2 & c^2 & 2sc \\ sc & -sc & c^2 - s^2 \end{bmatrix}. \tag{B.6}$$

This relates the global stresses to local stresses as

$$\begin{bmatrix} \sigma_x \\ \sigma_y \\ \tau_{xy} \end{bmatrix} = \begin{bmatrix} c^2 & s^2 & -2sc \\ s^2 & c^2 & 2sc \\ sc & -sc & c^2 - s^2 \end{bmatrix} \begin{bmatrix} \sigma_1 \\ \sigma_2 \\ \tau_{12} \end{bmatrix}. \tag{B.7}$$

B.2 Transformation of Strains

Consider an arbitrary line, AB, in direction 1 at an angle, θ, to the x-direction. Under loads, the line AB deforms to $A'B'$. By definition of normal strain along AB,

$$\varepsilon_1 = \frac{A'B' - AB}{AB} \tag{B.8}$$

$$= \frac{A'B'}{AB} - 1\,.$$

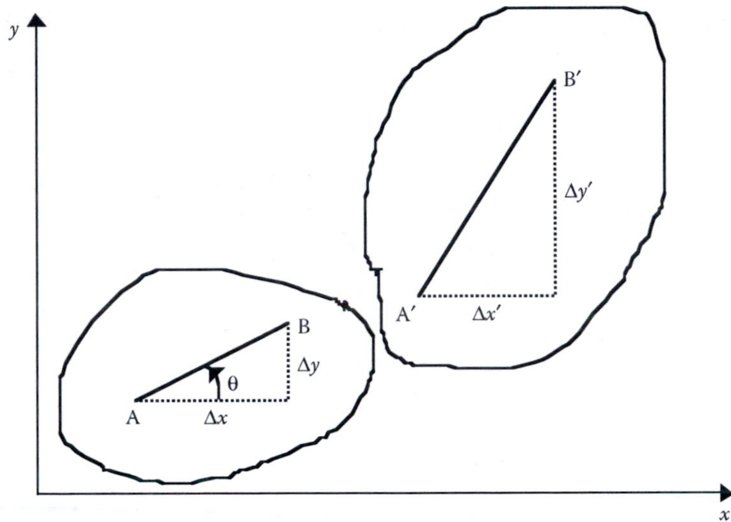

FIGURE 2.39
Line element for transformation of strains between local and global axes.

From Figure 2.39,

$$1 + \varepsilon_1 = \frac{A'B'}{AB}. \tag{B.9}$$

$$(AB)^2 = (\Delta x)^2 + (\Delta y)^2 \tag{B.10}$$

$$(A'B')^2 = (\Delta x')^2 + (\Delta y')^2 . \tag{B.11}$$

However, from definition of strain,

$$\Delta x' = \left(1 + \frac{\partial u}{\partial x}\right)\Delta x + \frac{\partial u}{\partial y}\Delta y \tag{B.12}$$

$$\Delta y' = \frac{\partial v}{\partial x}\,\Delta x + \left(1 + \frac{\partial v}{\partial y}\right)\Delta y . \tag{B.13}$$

Then, from Equation (B.11) through Equation (B.13),

$$(A'B')^2 = \left[\left(1+\frac{\partial u}{\partial x}\right)\Delta x + \frac{\partial u}{\partial y}(\Delta y)\right]^2 + \left[\frac{\partial v}{\partial x}\Delta x + \left(1+\frac{\partial v}{\partial y}\right)\Delta y\right]^2.$$

Neglecting products and squares of derivatives of strain,

$$(A'B')^2 = \left(1+2\frac{\partial u}{\partial x}\right)(\Delta x)^2 + \left(1+2\frac{\partial v}{\partial y}\right)(\Delta y)^2 + 2\left(\frac{\partial u}{\partial y}+\frac{\partial v}{\partial x}\right)\Delta x \Delta y. \quad \text{(B.14)}$$

From Equation (B.9),

$$(1+\varepsilon_1)^2 = \frac{(A'B')^2}{(AB)^2}$$

$$= \frac{\left(1+2\frac{\partial u}{\partial x}\right)(\Delta x)^2 + \left(1+2\frac{\partial v}{\partial y}\right)(\Delta y)^2 + 2\left(\frac{\partial u}{\partial y}+\frac{\partial v}{\partial x}\right)\Delta x \Delta y}{(\Delta x)^2 + (\Delta y)^2}$$

$$= \left(1+2\frac{\partial u}{\partial x}\right)\frac{(\Delta x)^2}{(\Delta x)^2+(\Delta y)^2} + \left(1+2\frac{\partial v}{\partial y}\right)\frac{(\Delta y)^2}{(\Delta x)^2+(\Delta y)^2}$$

$$+\ 2\left(\frac{\partial u}{\partial y}+\frac{\partial v}{\partial x}\right)\frac{\Delta x \Delta y}{(\Delta x)^2+(\Delta y)^2}$$

$$= \left(1+2\frac{\partial u}{\partial x}\right)\mathrm{Cos}^2\theta + \left(1+2\frac{\partial v}{\partial y}\right)\mathrm{Sin}^2\theta + 2\left(\frac{\partial u}{\partial y}+\frac{\partial v}{\partial x}\right)\mathrm{Sin}\,\theta\,\mathrm{Cos}\,\theta$$

$$(1+\varepsilon_1)^2 = (1+2\varepsilon_x)\mathrm{Cos}^2\theta + (1+2\varepsilon_y)\mathrm{Sin}^2\theta + 2\gamma_{xy}\,\mathrm{Sin}\,\theta\,\mathrm{Cos}\,\theta$$

$$1+\varepsilon_1^2+2\varepsilon_1 = 1+2\varepsilon_x\,\mathrm{Cos}^2\theta + 2\varepsilon_y\,\mathrm{Sin}^2\theta + 2\gamma_{xy}\,\mathrm{Sin}\,\theta\,\mathrm{Cos}\,\theta\,.$$

Neglecting again the squares of the strains,

$$\varepsilon_1 = \varepsilon_x\,\mathrm{Cos}^2\theta + \varepsilon_y\,\mathrm{Sin}^2\theta + \gamma_{xy}\,\mathrm{Sin}\,\theta\,\mathrm{Cos}\,\theta\,. \quad \text{(B.15)}$$

Similarly, one can take an arbitrary line in direction 2 and prove

$$\varepsilon_2 = \varepsilon_x \,\text{Sin}^2\theta + \varepsilon_y \,\text{Cos}^2\theta - \gamma_{xy}\,\text{Sin}\,\theta\,\text{Cos}\,\theta, \tag{B.16}$$

and, by taking two straight lines in direction 1 and 2 (perpendicular to each other), one can prove

$$\gamma_{12} = -2\varepsilon_x \,Sin\theta Cos\theta + 2\varepsilon_y \,\text{Sin}\,\theta\,\text{Cos}\,\theta + \gamma_{xy}(\text{Cos}^2\theta - \text{Sin}^2\theta)\,. \tag{B.17}$$

In matrix form, Equation (B.15), Equation (B.16), and Equation (B.17) relate the local strains to global strains

$$\begin{bmatrix} \varepsilon_1 \\ \varepsilon_2 \\ \gamma_{12}/2 \end{bmatrix} = \begin{bmatrix} c^2 & s^2 & 2sc \\ s^2 & c^2 & -2sc \\ -sc & sc & c^2 - s^2 \end{bmatrix} \begin{bmatrix} \varepsilon_x \\ \varepsilon_y \\ \gamma_{xy}/2 \end{bmatrix}, \tag{B.18}$$

where the 3×3 matrix in Equation (B.18) is the transformation matrix $[T]$ given in Equation (B.5).

Inverting Equation (B.18) gives

$$\begin{bmatrix} \varepsilon_x \\ \varepsilon_y \\ \gamma_{xy}/2 \end{bmatrix} = \begin{bmatrix} c^2 & s^2 & -2sc \\ s^2 & c^2 & 2sc \\ sc & -sc & c^2 - s^2 \end{bmatrix} \begin{bmatrix} \varepsilon_1 \\ \varepsilon_2 \\ \gamma_{12}/2 \end{bmatrix}, \tag{B.19}$$

where the 3×3 matrix in Equation (B.19) is the inverse of the transformation matrix given in Equation (B.6).

Key Terms

Transformation of stress
Transformation of strain
Free body diagram
Transformation matrix

3

Micromechanical Analysis of a Lamina

Chapter Objectives

- Develop concepts of volume and weight fraction (mass fraction) of fiber and matrix, density, and void fraction in composites.
- Find the nine mechanical and four hygrothermal constants: four elastic moduli, five strength parameters, two coefficients of thermal expansion, and two coefficients of moisture expansion of a unidirectional lamina from the individual properties of the fiber and the matrix, fiber volume fraction, and fiber packing.
- Discuss the experimental characterization of the nine mechanical and four hygrothermal constants.

3.1 Introduction

In Chapter 2, the stress–strain relationships, engineering constants, and failure theories for an angle lamina were developed using four elastic moduli, five strength parameters, two coefficients of thermal expansion (CTE), and two coefficients of moisture expansion (CME) for a unidirectional lamina. These 13 parameters can be found experimentally by conducting several tension, compression, shear, and hygrothermal tests on unidirectional lamina (laminates). However, unlike in isotropic materials, experimental evaluation of these parameters is quite costly and time consuming because they are functions of several variables: the individual constituents of the composite material, fiber volume fraction, packing geometry, processing, etc. Thus, the need and motivation for developing analytical models to find these parameters are very important. In this chapter, we will develop simple relationships for the these parameters in terms of the stiffnesses, strengths, coefficients of thermal and moisture expansion of the individual constituents of a composite, fiber volume fraction, packing geometry, etc. An understanding of this

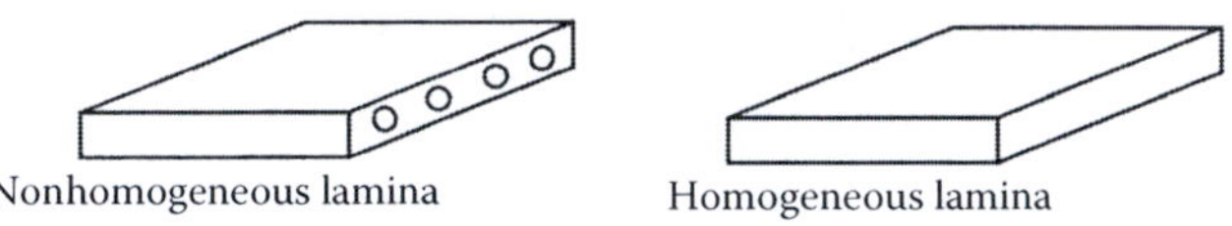

FIGURE 3.1
A nonhomogeneous lamina with fibers and matrix approximated as a homogeneous lamina.

relationship, called micromechanics of lamina, helps the designer to select the constituents of a composite material for use in a laminated structure.

Because this text is for a first course in composite materials, details will be explained only for the simple models based on the mechanics of materials approach and the semi-empirical approach. Results from other methods based on advanced topics such as elasticity are also explained for completeness.

As mentioned in Chapter 2, a unidirectional lamina is not homogeneous. However, one can assume the lamina to be homogeneous by focusing on the average response of the lamina to mechanical and hygrothermal loads (Figure 3.1). The lamina is simply looked at as a material whose properties are different in various directions, but not different from one location to another.

Also, the chapter focuses on a unidirectional continuous fiber-reinforced lamina. This is because it forms the basic building block of a composite structure, which is generally made of several unidirectional laminae placed at various angles. The modeling in the evaluation of the parameters is discussed first. This is followed by examples and experimental methods for finding these parameters.

3.2 Volume and Mass Fractions, Density, and Void Content

Before modeling the 13 parameters of a unidirectional composite, we introduce the concept of relative fraction of fibers by volume. This concept is critical because theoretical formulas for finding the stiffness, strength, and hygrothermal properties of a unidirectional lamina are a function of fiber volume fraction. Measurements of the constituents are generally based on their mass, so fiber mass fractions must also be defined. Moreover, defining the density of a composite also becomes necessary because its value is used in the experimental determination of fiber volume and void fractions of a composite. Also, the value of density is used in the definition of specific modulus and specific strength in Chapter 1.

3.2.1 Volume Fractions

Consider a composite consisting of fiber and matrix. Take the following symbol notations:

$v_{c,f,m}$ = volume of composite, fiber, and matrix, respectively
$\rho_{c,f,m}$ = density of composite, fiber, and matrix, respectively.

Now define the fiber volume fraction V_f and the matrix volume fraction V_m as

$$V_f = \frac{v_f}{v_c},$$

and

$$V_m = \frac{v_m}{v_c}. \tag{3.1a, b}$$

Note that the sum of volume fractions is

$$V_f + V_m = 1,$$

from Equation (3.1) as

$$v_f + v_m = v_c.$$

3.2.2 Mass Fractions

Consider a composite consisting of fiber and matrix and take the following symbol notation: $w_{c,f,m}$ = mass of composite, fiber, and matrix, respectively. The mass fraction (weight fraction) of the fibers (W_f) and the matrix (W_m) are defined as

$$W_f = \frac{w_f}{w_c}, \text{ and}$$

$$W_m = \frac{w_m}{w_c}. \tag{3.2a, b}$$

Note that the sum of mass fractions is

$$W_f + W_m = 1,$$

from Equation (3.2) as

$$w_f + w_m = w_c \cdot$$

From the definition of the density of a single material,

$$w_c = \rho_c v_c,$$

$$w_f = \rho_f v_f, \text{ and} \tag{3.3a–c}$$

$$w_m = \rho_m v_m.$$

Substituting Equation (3.3) in Equation (3.2), the mass fractions and volume fractions are related as

$$W_f = \frac{\rho_f}{\rho_c} V_f, \text{ and}$$

$$W_m = \frac{\rho_m}{\rho_c} V_m, \tag{3.4a, b}$$

in terms of the fiber and matrix volume fractions. In terms of individual constituent properties, the mass fractions and volume fractions are related by

$$W_f = \frac{\dfrac{\rho_f}{\rho_m}}{\dfrac{\rho_f}{\rho_m} V_f + V_m} V_f,$$

$$W_m = \frac{1}{\dfrac{\rho_f}{\rho_m}(1 - V_m) + V_m} V_m \cdot \tag{3.5a, b}$$

One should always state the basis of calculating the fiber content of a composite. It is given in terms of mass or volume. Based on Equation (3.4), it is evident that volume and mass fractions are not equal and that the mismatch between the mass and volume fractions increases as the ratio between the density of fiber and matrix differs from one.

3.2.3 Density

The derivation of the density of the composite in terms of volume fractions is found as follows. The mass of composite w_c is the sum of the mass of the fibers w_f and the mass of the matrix w_m as

$$w_c = w_f + w_m. \tag{3.6}$$

Substituting Equation (3.3) in Equation (3.6) yields

$$\rho_c v_c = \rho_f v_f + \rho_m v_m,$$

and

$$\rho_c = \rho_f \frac{v_f}{v_c} + \rho_m \frac{v_m}{v_c}. \tag{3.7}$$

Using the definitions of fiber and matrix volume fractions from Equation (3.1),

$$\rho_c = \rho_f V_f + \rho_m V_m. \tag{3.8}$$

Now, consider that the volume of a composite v_c is the sum of the volumes of the fiber v_f and matrix (v_m):

$$v_c = v_f + v_m. \tag{3.9}$$

The density of the composite in terms of mass fractions can be found as

$$\frac{1}{\rho_c} = \frac{W_f}{\rho_f} + \frac{W_m}{\rho_m}. \tag{3.10}$$

Example 3.1

A glass/epoxy lamina consists of a 70% fiber volume fraction. Use properties of glass and epoxy from Table 3.1* and Table 3.2, respectively, to determine the

* Table 3.1 and Table 3.2 give the typical properties of common fibers and matrices in the SI system of units, respectively. Note that fibers such as graphite and aramids are transversely isotropic, but matrices are generally isotropic. The typical properties of common fibers and matrices are again given in Table 3.3 and Table 3.4, respectively, in the USCS system of units.

TABLE 3.1

Typical Properties of Fibers (SI System of Units)

Property	Units	Graphite	Glass	Aramid
Axial modulus	GPa	230	85	124
Transverse modulus	GPa	22	85	8
Axial Poisson's ratio	—	0.30	0.20	0.36
Transverse Poisson's ratio	—	0.35	0.20	0.37
Axial shear modulus	GPa	22	35.42	3
Axial coefficient of thermal expansion	μm/m/°C	–1.3	5	–5.0
Transverse coefficient of thermal expansion	μm/m/°C	7.0	5	4.1
Axial tensile strength	MPa	2067	1550	1379
Axial compressive strength	MPa	1999	1550	276
Transverse tensile strength	MPa	77	1550	7
Transverse compressive strength	MPa	42	1550	7
Shear strength	MPa	36	35	21
Specific gravity	—	1.8	2.5	1.4

TABLE 3.2

Typical Properties of Matrices (SI System of Units)

Property	Units	Epoxy	Aluminum	Polyamide
Axial modulus	GPa	3.4	71	3.5
Transverse modulus	GPa	3.4	71	3.5
Axial Poisson's ratio	—	0.30	0.30	0.35
Transverse Poisson's ratio	—	0.30	0.30	0.35
Axial shear modulus	GPa	1.308	27	1.3
Coefficient of thermal expansion	μm/m/°C	63	23	90
Coefficient of moisture expansion	m/m/kg/kg	0.33	0.00	0.33
Axial tensile strength	MPa	72	276	54
Axial compressive strength	MPa	102	276	108
Transverse tensile strength	MPa	72	276	54
Transverse compressive strength	MPa	102	276	108
Shear strength	MPa	34	138	54
Specific gravity	—	1.2	2.7	1.2

1. Density of lamina
2. Mass fractions of the glass and epoxy
3. Volume of composite lamina if the mass of the lamina is 4 kg
4. Volume and mass of glass and epoxy in part (3)

Solution

1. From Table 3.1, the density of the fiber is

$$\rho_f = 2500\ kg/m^3.$$

TABLE 3.3

Typical Properties of Fibers (USCS System of Units)

Property	Units	Graphite	Glass	Aramid
Axial modulus	Msi	33.35	12.33	17.98
Transverse modulus	Msi	3.19	12.33	1.16
Axial Poisson's ratio	—	0.30	0.20	0.36
Transverse Poisson's ratio	—	0.35	0.20	0.37
Axial shear modulus	Msi	3.19	5.136	0.435
Axial coefficient of thermal expansion	μin./in./°F	–0.7222	2.778	–2.778
Transverse coefficient of thermal expansion	μin./in./°F	3.889	2.778	2.278
Axial tensile strength	ksi	299.7	224.8	200.0
Axial compressive strength	ksi	289.8	224.8	40.02
Transverse tensile strength	ksi	11.16	224.8	1.015
Transverse compressive strength	ksi	6.09	224.8	1.015
Shear strength	ksi	5.22	5.08	3.045
Specific gravity	—	1.8	2.5	1.4

TABLE 3.4

Typical Properties of Matrices (USCS System of Units)

Property	Units	Epoxy	Aluminum	Polyamide
Axial modulus	Msi	0.493	10.30	0.5075
Transverse modulus	Msi	0.493	10.30	0.5075
Axial Poisson's ratio	—	0.30	0.30	0.35
Transverse Poisson's ratio	—	0.30	0.30	0.35
Axial shear modulus	Msi	0.1897	3.915	0.1885
Coefficient of thermal expansion	μin./in./°F	35	12.78	50
Coefficient of moisture expansion	in./in./lb/lb	0.33	0.00	0.33
Axial tensile strength	ksi	10.44	40.02	7.83
Axial compressive strength	ksi	14.79	40.02	15.66
Transverse tensile strength	ksi	10.44	40.02	7.83
Transverse compressive strength	ksi	14.79	40.02	15.66
Shear strength	ksi	4.93	20.01	7.83
Specific gravity	—	1.2	2.7	1.2

From Table 3.2, the density of the matrix is

$$\rho_m = 1200\ kg/m^3.$$

Using Equation (3.8), the density of the composite is

$$\rho_c = (2500)(0.7) + (1200)(0.3)$$

$$= 2110\ kg/m^3.$$

2. Using Equation (3.4), the fiber and matrix mass fractions are

$$W_f = \frac{2500}{2110} \times 0.3$$

$$= 0.8294$$

$$W_m = \frac{1200}{2110} \times 0.3 \; .$$

$$= 0.1706$$

Note that the sum of the mass fractions,

$$W_f + W_m = 0.8294 + 0.1706$$

$$= 1.000.$$

3. The volume of composite is

$$v_c = \frac{w_c}{\rho_c}$$

$$= \frac{4}{2110}$$

$$= 1.896 \times 10^{-3} m^3 \; .$$

4. The volume of the fiber is

$$v_f = V_f v_c$$

$$= (0.7)(1.896 \times 10^{-3})$$

$$= 1.327 \times 10^{-3} m^3 \; .$$

The volume of the matrix is

$$v_m = V_m v_c$$

$$=(0.3)(0.1896 \times 10^{-3})$$

$$= 0.5688 \times 10^{-3}\ m^3 .$$

The mass of the fiber is

$$w_f = \rho_f v_f$$

$$= (2500)(1.327 \times 10^{-3})$$

$$= 3.318\ kg .$$

The mass of the matrix is

$$w_m = \rho_m v_m$$

$$= (1200)(0.5688 \times 10^{-3})$$

$$= 0.6826\ kg .$$

3.2.4 Void Content

During the manufacture of a composite, voids are introduced in the composite as shown in Figure 3.2. This causes the theoretical density of the composite to be higher than the actual density. Also, the void content of a

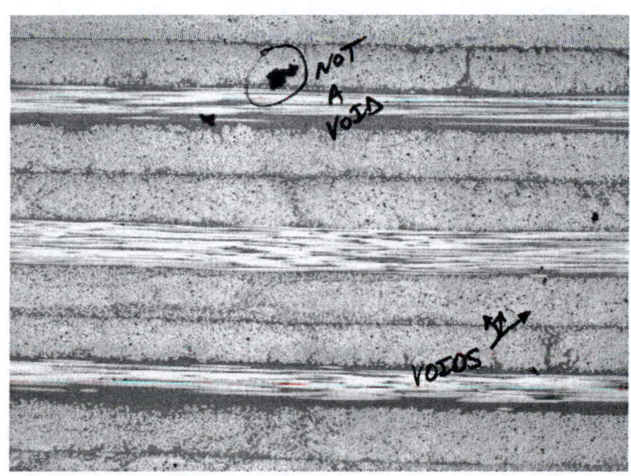

FIGURE 3.2
Photomicrographs of cross-section of a lamina with voids.

composite is detrimental to its mechanical properties. These detriments include lower

- Shear stiffness and strength
- Compressive strengths
- Transverse tensile strengths
- Fatigue resistance
- Moisture resistance

A decrease of 2 to 10% in the preceding matrix-dominated properties generally takes place with every 1% increase in the void content.[1]

For composites with a certain volume of voids V_v the volume fraction of voids V_v is defined as

$$V_v = \frac{v_v}{v_c}. \tag{3.11}$$

Then, the total volume of a composite (v_c) with voids is given by

$$v_c = v_f + v_m + v_v. \tag{3.12}$$

By definition of the experimental density ρ_{ce} of a composite, the actual volume of the composite is

$$v_c = \frac{w_c}{\rho_{ce}}, \tag{3.13}$$

and, by the definition of the theoretical density ρ_{ct} of the composite, the theoretical volume of the composite is

$$v_f + v_m = \frac{w_c}{\rho_{ct}}. \tag{3.14}$$

Then, substituting the preceding expressions (3.13) and (3.14) in Equation (3.12),

$$\frac{w_c}{\rho_{ce}} = \frac{w_c}{\rho_{ct}} + v_v .$$

The volume of void is given by

$$v_v = \frac{w_c}{\rho_{ce}}\left(\frac{\rho_{ct} - \rho_{ce}}{\rho_{ct}}\right). \tag{3.15}$$

Substituting Equation (3.13) and Equation (3.15) in Equation (3.11), the volume fraction of the voids is

$$\begin{aligned} V_v &= \frac{v_v}{v_c} \\ &= \frac{\rho_{ct} - \rho_{ce}}{\rho_{ct}}. \end{aligned} \tag{3.16}$$

Example 3.2

A graphite/epoxy cuboid specimen with voids has dimensions of $a \times b \times c$ and its mass is M_c. After it is put it into a mixture of sulfuric acid and hydrogen peroxide, the remaining graphite fibers have a mass M_f. From independent tests, the densities of graphite and epoxy are ρ_f and ρ_m, respectively. Find the volume fraction of the voids in terms of a, b, c, M_f, M_c, ρ_f, and ρ_m.

Solution

The total volume of the composite v_c is the sum total of the volume of fiber v_f, matrix v_m, and voids v_v:

$$v_c = v_f + v_m + v_v. \tag{3.17}$$

From the definition of density,

$$v_f = \frac{M_f}{\rho_f}, \tag{3.18a}$$

$$v_m = \frac{M_c - M_f}{\rho_m}. \tag{3.18b}$$

The specimen is a cuboid, so the volume of the composite is

$$v_c = abc. \tag{3.19}$$

Substituting Equation (3.18) and Equation (3.19) in Equation (3.17) gives

$$abc = \frac{M_f}{\rho_f} + \frac{M_c - M_f}{\rho_m} + v_v ,$$

and the volume fraction of voids then is

$$V_v = \frac{v_v}{abc} = 1 - \frac{1}{abc}\left[\frac{M_f}{\rho_f} + \frac{M_c - M_f}{\rho_m}\right] \tag{3.20}$$

Alternative Solution

The preceding problem can also be solved by using Equation (3.16). The theoretical density of the composite is

$$\rho_{ct} = \rho_f V_f' + \rho_m (1 - V_f') , \tag{3.21}$$

where V_f' is the theoretical fiber volume fraction given as

$$V_f' = \frac{volume\ of\ fibers}{volume\ of\ fibers + volume\ of\ matrix}$$

$$V_f' = \frac{\dfrac{M_f}{\rho_f}}{\dfrac{M_f}{\rho_f} + \dfrac{M_c - M_f}{\rho_m}} . \tag{3.22}$$

The experimental density of the composite is

$$\rho_{ce} = \frac{M_c}{abc} . \tag{3.23}$$

Substituting Equation (3.21) through Equation (3.23) in the definition of void volume fractions given by Equation (3.16),

$$V_v = 1 - \frac{1}{abc}\left[\frac{M_f}{\rho_f} + \frac{M_c - M_f}{\rho_m}\right]. \tag{3.24}$$

Experimental determination: the fiber volume fractions of the constituents of a composite are found generally by the burn or the acid digestion tests. These tests involve taking a sample of composite and weighing it. Then the density of the specimen is found by the liquid displacement method in which the sample is weighed in air and then in water. The density of the composite is given by

$$\rho_c = \frac{w_c}{w_c - w_i}\rho_w , \tag{3.25}$$

where

w_c = weight of composite
w_i = weight of composite when immersed in water
ρ_w = density of water (1000 kg/m^3 or 62.4 lb/ft^3)

For specimens that float in water, a sinker is attached. The density of the composite is then found by

$$\rho_c = \frac{w_c}{w_c + w_s - w_w}\rho_w , \tag{3.26}$$

where

w_c = weight of composite
w_s = weight of sinker when immersed in water
w_w = weight of sinker and specimen when immersed in water

The sample is then dissolved in an acid solution or burned.[2] Glass-based composites are burned, and carbon and aramid-based composites are digested in solutions. Carbon and aramid-based composites cannot be burned because carbon oxidizes in air above 300°C (572°F) and the aramid fiber can decompose at high temperatures. Epoxy-based composites can be digested by nitric acid or a hot mixture of ethylene glycol and potassium hydroxide; polyamide- and phenolic resin-based composites use mixtures of sulfuric acid and hydrogen peroxide. When digestion or burning is complete, the remaining fibers are washed and dried several times and then weighed. The fiber and matrix weight fractions can be found using Equation (3.2). The densities of the fiber and the matrix are known; thus, one can use Equation (3.4) to determine the volume fraction of the constituents of the composite and Equation (3.8) to calculate the theoretical density of the composite.

3.3 Evaluation of the Four Elastic Moduli

As shown in Section 2.4.3, there are four elastic moduli of a unidirectional lamina:

- Longitudinal Young's modulus, E_1
- Transverse Young's modulus, E_2
- Major Poisson's ratio, ν_{12}
- In-plane shear modulus, G_{12}

Three approaches for determining the four elastic moduli are discussed next.

3.3.1 Strength of Materials Approach

From a unidirectional lamina, take a representative volume element* that consists of the fiber surrounded by the matrix (Figure 3.3). This representative volume element (RVE) can be further represented as rectangular blocks. The fiber, matrix, and the composite are assumed to be of the same width, h, but of thicknesses t_f, t_m, and t_c, respectively. The area of the fiber is given by

$$A_f = t_f h \, . \tag{3.27a}$$

The area of the matrix is given by

$$A_m = t_m h, \tag{3.27b}$$

and the area of the composite is given by

$$A_c = t_c h. \tag{3.27c}$$

The two areas are chosen in the proportion of their volume fractions so that the fiber volume fraction is defined as

$$\begin{aligned} V_f &= \frac{A_f}{A_c} \\ &= \frac{t_f}{t_c}, \end{aligned} \tag{3.28a}$$

and the matrix fiber volume fraction V_m is

* A representative volume element (RVE) of a material is the smallest part of the material that represents the material as a whole. It could be otherwise intractable to account for the distribution of the constituents of the material.

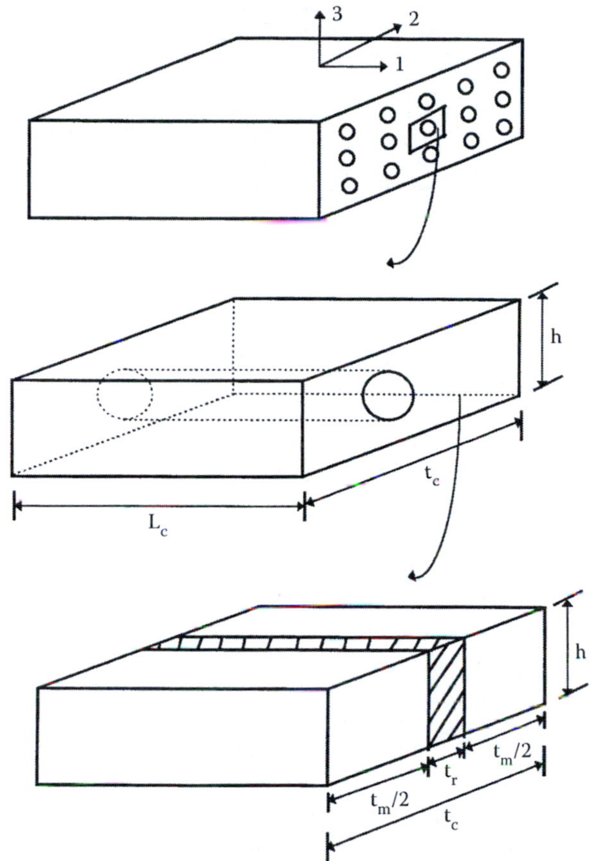

FIGURE 3.3
Representative volume element of a unidirectional lamina.

$$V_m = \frac{A_m}{A_c} = \frac{t_m}{t_c} = 1 - V_f. \tag{3.28b}$$

The following assumptions are made in the strength of materials approach model:

- The bond between fibers and matrix is perfect.
- The elastic moduli, diameters, and space between fibers are uniform.
- The fibers are continuous and parallel.

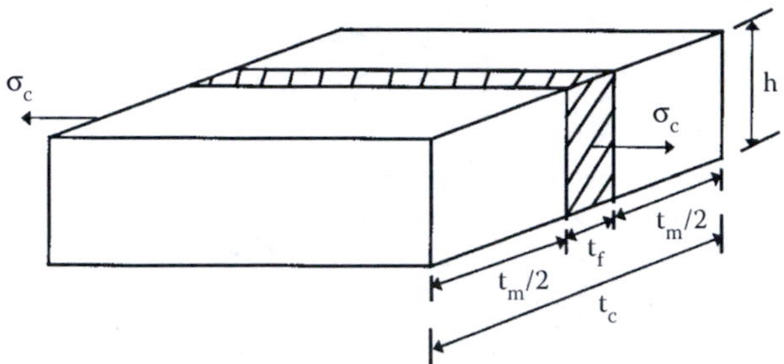

FIGURE 3.4
A longitudinal stress applied to the representative volume element to calculate the longitudinal Young's modulus for a unidirectional lamina.

- The fibers and matrix follow Hooke's law (linearly elastic).
- The fibers possess uniform strength.
- The composite is free of voids.

3.3.1.1 *Longitudinal Young's Modulus*

From Figure 3.4, under a uniaxial load F_c on the composite RVE, the load is shared by the fiber F_f and the matrix F_m so that

$$F_c = F_f + F_m. \tag{3.29}$$

The loads taken by the fiber, the matrix, and the composite can be written in terms of the stresses in these components and cross-sectional areas of these components as

$$F_c = \sigma_c A_c, \tag{3.30a}$$

$$F_f = \sigma_f A_f, \tag{3.30b}$$

$$F_m = \sigma_m A_m, \tag{3.30c}$$

where

$\sigma_{c,f,m}$ = stress in composite, fiber, and matrix, respectively
$A_{c,f,m}$ = area of composite, fiber, and matrix, respectively

Assuming that the fibers, matrix, and composite follow Hooke's law and that the fibers and the matrix are isotropic, the stress–strain relationship for each component and the composite is

$$\sigma_c = E_1 \varepsilon_c, \tag{3.31a}$$

$$\sigma_f = E_f \varepsilon_f, \tag{3.31b}$$

and

$$\sigma_m = E_m \varepsilon_m, \tag{3.31c}$$

where

$\varepsilon_{c,f,m}$ = strains in composite, fiber, and matrix, respectively
$E_{1,f,m}$ = elastic moduli of composite, fiber, and matrix, respectively

Substituting Equation (3.30) and Equation (3.31) in Equation (3.29) yields

$$E_1 \varepsilon_c A_c = E_f \varepsilon_f A_f + E_m \varepsilon_m A_m. \tag{3.32}$$

The strains in the composite, fiber, and matrix are equal ($\varepsilon_c = \varepsilon_f = \varepsilon_m$); then, from Equation (3.32),

$$E_1 = E_f \frac{A_f}{A_c} + E_m \frac{A_m}{A_c}. \tag{3.33}$$

Using Equation (3.28), for definitions of volume fractions,

$$E_1 = E_f V_f + E_m V_m. \tag{3.34}$$

Equation 3.34 gives the longitudinal Young's modulus as a weighted mean of the fiber and matrix modulus. It is also called the rule of mixtures.

The ratio of the load taken by the fibers F_f to the load taken by the composite F_c is a measure of the load shared by the fibers. From Equation (3.30) and Equation (3.31),

$$\frac{F_f}{F_c} = \frac{E_f}{E_1} V_f. \tag{3.35}$$

In Figure 3.5, the ratio of the load carried by the fibers to the load taken by the composite is plotted as a function of fiber-to-matrix Young's moduli ratio E_f/E_m for the constant fiber volume fraction V_f. It shows that as the fiber to matrix moduli ratio increases, the load taken by the fiber increases tremendously.

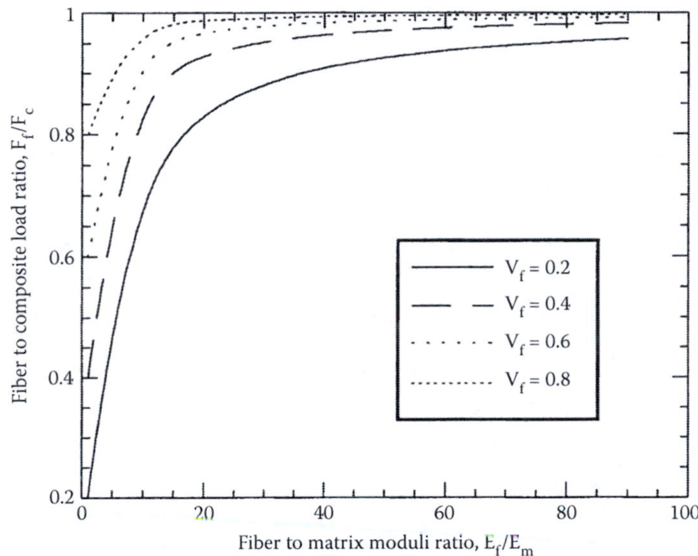

FIGURE 3.5
Fraction of load of composite carried by fibers as a function of fiber volume fraction for constant fiber to matrix moduli ratio.

Example 3.3

Find the longitudinal elastic modulus of a unidirectional glass/epoxy lamina with a 70% fiber volume fraction. Use the properties of glass and epoxy from Table 3.1 and Table 3.2, respectively. Also, find the ratio of the load taken by the fibers to that of the composite.

Solution

From Table 3.1, the Young's modulus of the fiber is

$$E_f = 85 \text{ GPa}.$$

From Table 3.2, the Young's modulus of the matrix is

$$E_m = 3.4 \text{ GPa}.$$

Using Equation (3.34), the longitudinal elastic modulus of the unidirectional lamina is

$$E_1 = (85)(0.7) + (3.4)(0.3)$$

$$= 60.52 \ GPa.$$

Using Equation (3.35), the ratio of the load taken by the fibers to that of the composite is

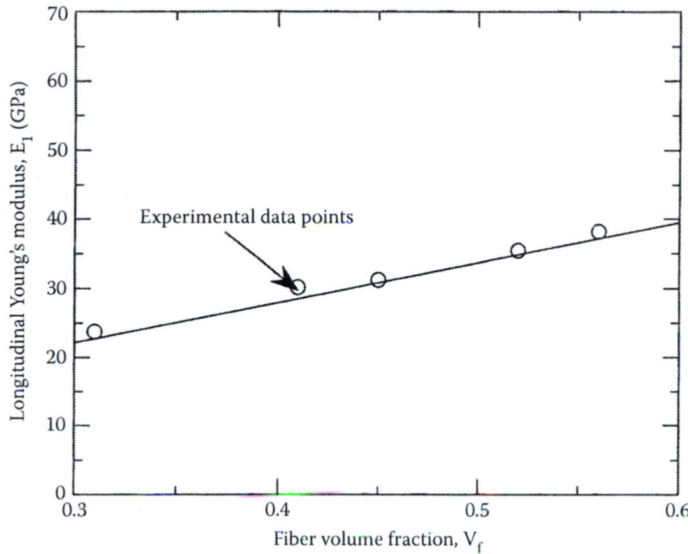

FIGURE 3.6
Longitudinal Young's modulus as function of fiber volume fraction and comparison with experimental data points for a typical glass/polyester lamina. (Experimental data points reproduced with permission of ASM International.)

$$\frac{F_f}{F_c} = \frac{85}{60.52}(0.7)$$

$$= 0.9831.$$

Figure 3.6 shows the linear relationship between the longitudinal Young's modulus of a unidirectional lamina and fiber volume fraction for a typical graphite/epoxy composite per Equation (3.34). It also shows that Equation (3.34) predicts results that are close to the experimental data points.[3]

3.3.1.2 Transverse Young's Modulus

Assume now that, as shown in Figure 3.7, the composite is stressed in the transverse direction. The fibers and matrix are again represented by rectangular blocks as shown. The fiber, the matrix, and composite stresses are equal. Thus,

$$\sigma_c = \sigma_f = \sigma_m, \tag{3.36}$$

where $\sigma_{c,f,m}$ = stress in composite, fiber, and matrix, respectively.

Now, the transverse extension in the composite Δ_c is the sum of the transverse extension in the fiber Δ_f, and that is the matrix, Δ_m.

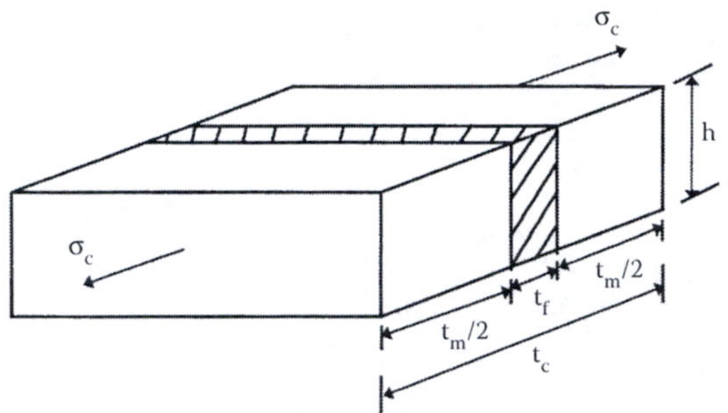

FIGURE 3.7
A transverse stress applied to a representative volume element used to calculate transverse Young's modulus of a unidirectional lamina.

$$\Delta_c = \Delta_f + \Delta_m. \tag{3.37}$$

Now, by the definition of normal strain,

$$\Delta_c = t_c \varepsilon_c, \tag{3.38a}$$

$$\Delta_f = t_f \varepsilon_f, \tag{3.38b}$$

and

$$\Delta_m = t_m \varepsilon_m, \tag{3.38c}$$

where

$t_{c,f,m}$ = thickness of the composite, fiber and matrix, respectively
$\varepsilon_{c,f,m}$ = normal transverse strain in the composite, fiber, and matrix, respectively

Also, by using Hooke's law for the fiber, matrix, and composite, the normal strains in the composite, fiber, and matrix are

$$\varepsilon_c = \frac{\sigma_c}{E_2}, \tag{3.39a}$$

$$\varepsilon_f = \frac{\sigma_f}{E_f}, \tag{3.39b}$$

and

$$\varepsilon_m = \frac{\sigma_m}{E_m}. \tag{3.39c}$$

Substituting Equation (3.38) and Equation (3.39) in Equation (3.37) and using Equation (3.36) gives

$$\frac{1}{E_2} = \frac{1}{E_f}\frac{t_f}{t_c} + \frac{1}{E_m}\frac{t_m}{t_c}. \tag{3.40}$$

Because the thickness fractions are the same as the volume fractions as the other two dimensions are equal for the fiber and the matrix (see Equation 3.28):

$$\frac{1}{E_2} = \frac{V_f}{E_f} + \frac{V_m}{E_m}. \tag{3.41}$$

Equation (3.41) is based on the weighted mean of the compliance of the fiber and the matrix.

Example 3.4

Find the transverse Young's modulus of a glass/epoxy lamina with a fiber volume fraction of 70%. Use the properties of glass and epoxy from Table 3.1 and Table 3.2, respectively.

Solution

From Table 3.1, the Young's modulus of the fiber is

$$E_f = 85 \text{ GPa}.$$

From Table 3.2, the Young's modulus of the matrix is

$$E_m = 3.4 \text{ GPa}.$$

Using Equation (3.41), the transverse Young's modulus, E_2, is

$$\frac{1}{E_2} = \frac{0.7}{85} + \frac{0.3}{3.4},$$

$$E_2 = 10.37\ GPa.$$

Figure 3.8 plots the transverse Young's modulus as a function of fiber volume fraction for constant fiber-to-matrix elastic moduli ratio, E_f/E_m. For metal and ceramic matrix composites, the fiber and matrix elastic moduli are of the same order. (For example, for a SiC/aluminum metal matrix composite, $E_f/E_m = 4$ and for a SiC/CAS ceramic matrix composite, $E_f/E_m = 2$). The transverse Young's modulus of the composite in such cases changes more smoothly as a function of the fiber volume fraction.

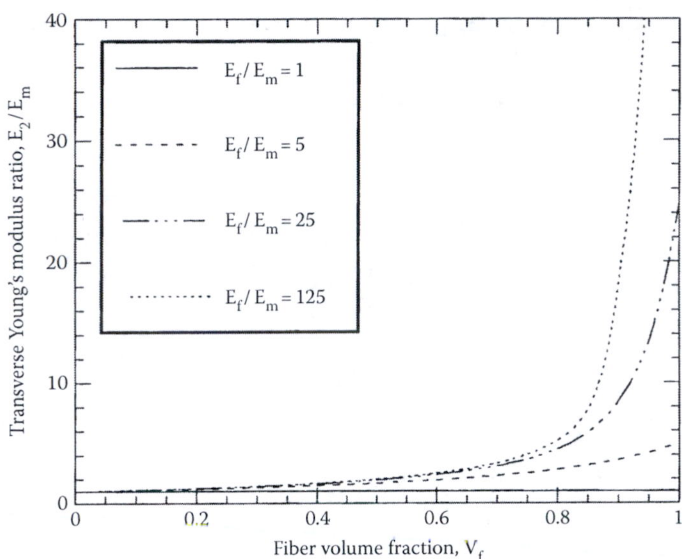

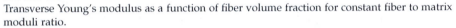

FIGURE 3.8
Transverse Young's modulus as a function of fiber volume fraction for constant fiber to matrix moduli ratio.

For polymeric composites, the fiber-to-matrix moduli ratio is very high. (For example, for a glass/epoxy polymer matrix composite, $E_f/E_m = 25$). The transverse Young's modulus of the composite in such cases changes appreciably only for large fiber volume fractions. Figure 3.8 shows that, for high E_f/E_m ratios, the contribution of the fiber modulus only increases substantially for a fiber volume fraction greater than 80%. These fiber volume fractions are not practical and in many cases are physically impossible due to the geometry of fiber packing. Figure 3.9 shows various possibilities of fiber packing. Note that the ratio of the diameter, d, to fiber spacing, s, d/s varies with geometrical packing. For circular fibers with square array packing (Figure 3.9a),

$$\frac{d}{s} = \left(\frac{4V_f}{\pi} \right)^{1/2}. \tag{3.42a}$$

This gives a maximum fiber volume fraction of 78.54% as $s \geq d$. For circular fibers with hexagonal array packing (Figure 3.9b),

$$\frac{d}{s} = \left(\frac{2\sqrt{3}V_f}{\pi} \right)^{1/2}. \tag{3.42b}$$

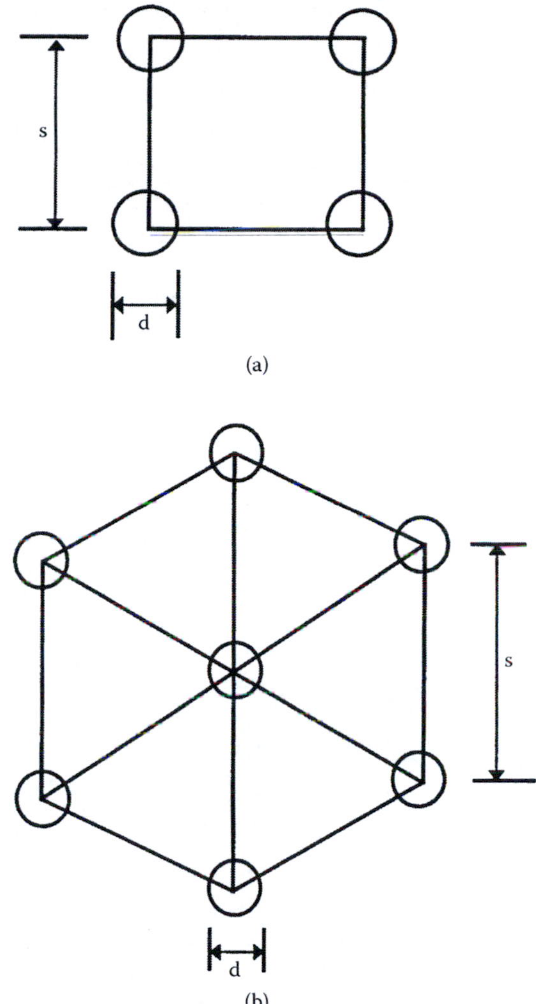

FIGURE 3.9
Fiber to fiber spacing in (a) square packing geometry and (b) hexagonal packing geometry.

This gives a maximum fiber volume fraction of 90.69% because $s \geq d$. These maximum fiber volume fractions are not practical to use because the fibers touch each other and thus have surfaces where the matrix cannot wet out the fibers.

In Figure 3.10, the transverse Young's modulus is plotted as a function of fiber volume fraction using Equation (3.41) for a typical boron/epoxy lamina. Also given are the experimental data points.[4] In Figure 3.10, the experimental and analytical results are not as close to each other as they are for the longitudinal Young's modulus in Figure 3.6.

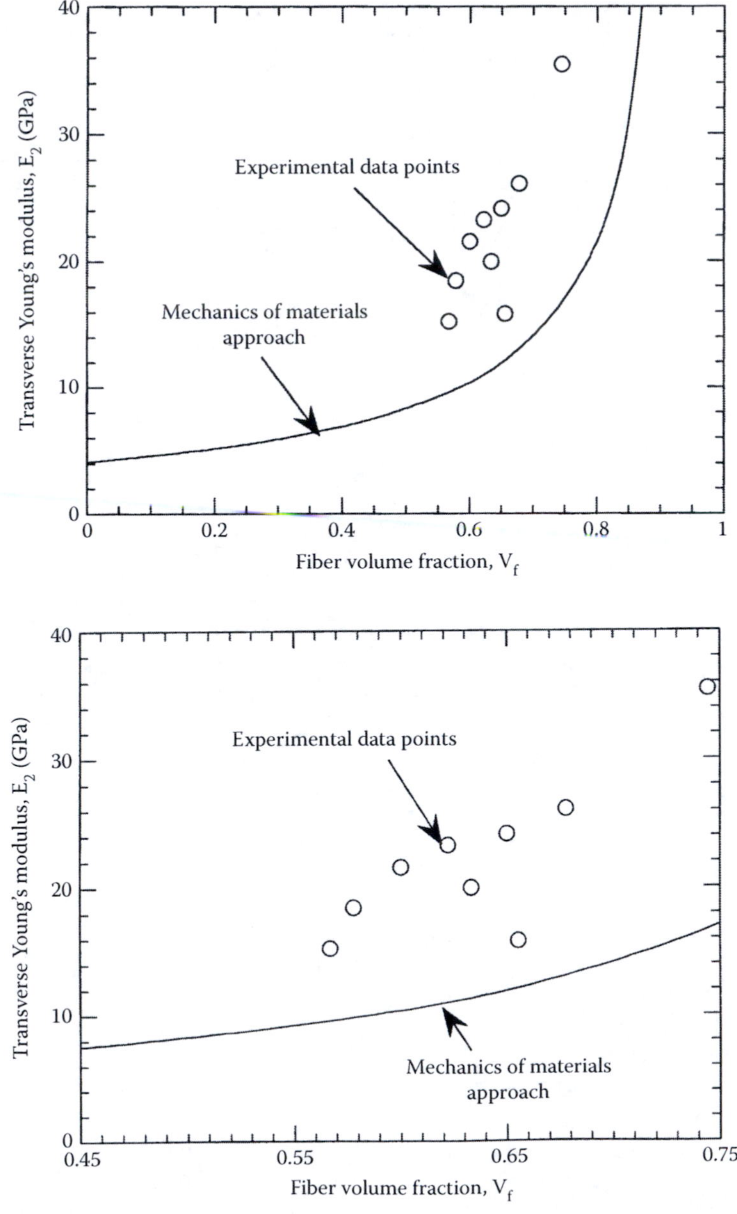

FIGURE 3.10
Theoretical values of transverse Young's modulus as a function of fiber volume fraction for a Boron/Epoxy unidirectional lamina (E_f = 414 GPa, ν_f = 0.2, E_m = 4.14 GPa, ν_m = 0.35) and comparison with experimental values. Figure (b) zooms figure (a) for fiber volume fraction between 0.45 and 0.75. (Experimental data from Hashin, Z., NASA tech. rep. contract no. NAS1-8818, November 1970.)

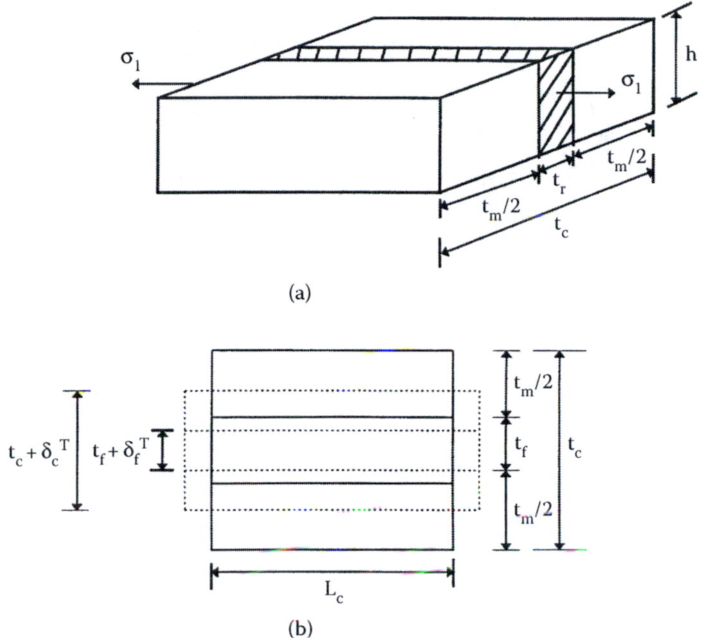

FIGURE 3.11
A longitudinal stress applied to a representative volume element to calculate Poisson's ratio of unidirectional lamina.

3.3.1.3 Major Poisson's Ratio

The major Poisson's ratio is defined as the negative of the ratio of the normal strain in the transverse direction to the normal strain in the longitudinal direction, when a normal load is applied in the longitudinal direction. Assume a composite is loaded in the direction parallel to the fibers, as shown in Figure 3.11. The fibers and matrix are again represented by rectangular blocks. The deformations in the transverse direction of the composite (δ_c^T) is the sum of the transverse deformations of the fiber (δ_f^T) and the matrix (δ_m^T) as

$$\delta_c^T = \delta_f^T + \delta_m^T. \tag{3.43}$$

Using the definition of normal strains,

$$\varepsilon_f^T = \frac{\delta_f^T}{t_f}, \tag{3.44a}$$

$$\varepsilon_m^T = \frac{\delta_m^T}{t_m}, \tag{3.44b}$$

and

$$\varepsilon_c^T = \frac{\delta_c^T}{t_c}, \tag{3.44c}$$

where $\varepsilon_{c,f,m}$ = transverse strains in composite, fiber, and matrix, respectively.
Substituting Equation (3.44) in Equation (3.43),

$$t_c \varepsilon_c^T = t_f \varepsilon_f^T + t_m \varepsilon_m^T. \tag{3.45}$$

The Poisson's ratios for the fiber, matrix, and composite, respectively, are

$$\nu_f = -\frac{\varepsilon_f^T}{\varepsilon_f^L}, \tag{3.46a}$$

$$\nu_m = -\frac{\varepsilon_m^T}{\varepsilon_m^L}, \tag{3.46b}$$

and

$$\nu_{12} = -\frac{\varepsilon_c^T}{\varepsilon_c^L}. \tag{3.46c}$$

Substituting in Equation (3.45),

$$-t_c \nu_{12} \varepsilon_c^L = -t_f \nu_f \varepsilon_f^L - t_m \nu_m \varepsilon_m^L, \tag{3.47}$$

where
$\nu_{12,f,m} \upsilon_{12,f,m}$ = Poisson's ratio of composite, fiber, and matrix, respectively
$\varepsilon_{c,f,m}^L$ = longitudinal strains of composite, fiber and matrix, respectively

However, the strains in the composite, fiber, and matrix are assumed to be the equal in the longitudinal direction ($\varepsilon_c^L = \varepsilon_f^L = \varepsilon_m^L$), which, from Equation (3.47), gives

$$t_c \nu_{12} = t_f \nu_f + t_m \nu_m,$$

$$\nu_{12} = \nu_f \frac{t_f}{t_c} + \nu_m \frac{t_m}{t_c}. \tag{3.48}$$

Because the thickness fractions are the same as the volume fractions, per Equation (3.28),

$$\nu_{12} = \nu_f V_f + \nu_m V_m. \tag{3.49}$$

Example 3.5

Find the major and minor Poisson's ratio of a glass/epoxy lamina with a 70% fiber volume fraction. Use the properties of glass and epoxy from Table 3.1 and Table 3.2, respectively.

Solution

From Table 3.1, the Poisson's ratio of the fiber is

$$\nu_f = 0.2.$$

From Table 3.2, the Poisson's ratio of the matrix is

$$\nu_m = 0.3.$$

Using Equation (3.49), the major Poisson's ratio is

$$\nu_{12} = (0.2)(0.7) + (0.3)(0.3)$$
$$= 0.230.$$

From Example 3.3, the longitudinal Young's modulus is

$$E_1 = 60.52 \text{ GPa}$$

and, from Example 3.4, the transverse Young's modulus is

$$E_2 = 10.37 \text{ GPa}.$$

Then, the minor Poisson's ratio from Equation (2.83) is

$$\nu_{21} = \nu_{12} \frac{E_2}{E_1}$$
$$= 0.230\left(\frac{10.37}{60.52}\right)$$
$$= 0.03941.$$

3.3.1.4 In-Plane Shear Modulus

Apply a pure shear stress τ_c to a lamina as shown in Figure 3.12. The fibers and matrix are represented by rectangular blocks as shown. The resulting

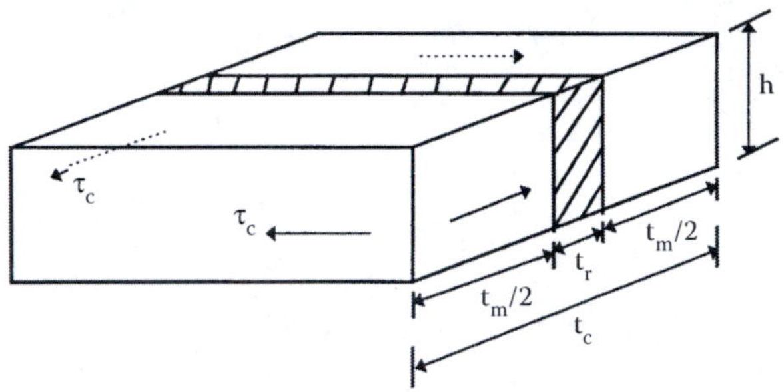

FIGURE 3.12
An in-plane shear stress applied to a representative volume element for finding in-plane shear modulus of a unidirectional lamina.

shear deformations of the composite δ_c the fiber δ_f, and the matrix δ_m are related by

$$\delta_c = \delta_f + \delta_m \,. \tag{3.50}$$

From the definition of shear strains,

$$\delta_c = \gamma_c t_c \,, \tag{3.51a}$$

$$\delta_f = \gamma_f t_f \,, \tag{3.51b}$$

and

$$\delta_m = \gamma_m t_m \,, \tag{3.51c}$$

where

$\gamma_{c,f,m}$ = shearing strains in the composite, fiber, and matrix, respectively
$t_{c,f,m}$ = thickness of the composite, fiber, and matrix, respectively.

From Hooke's law for the fiber, the matrix, and the composite,

$$\gamma_c = \frac{\tau_c}{G_{12}}, \tag{3.52a}$$

$$\gamma_f = \frac{\tau_f}{G_f}, \tag{3.52b}$$

and

$$\gamma_m = \frac{\tau_m}{G_m}, \tag{3.52c}$$

where $G_{12,f,m}$ = shear moduli of composite, fiber, and matrix, respectively.

From Equation (3.50) through Equation (3.52),

$$\frac{\tau_c}{G_{12}} t_c = \frac{\tau_f}{G_f} t_f + \frac{\tau_m}{G_m} t_m. \tag{3.53}$$

The shear stresses in the fiber, matrix, and composite are assumed to be equal ($\tau_c = \tau_f = \tau_m$), giving

$$\frac{1}{G_{12}} = \frac{1}{G_f}\frac{t_f}{t_c} + \frac{1}{G_m}\frac{t_m}{t_c}. \tag{3.54}$$

Because the thickness fractions are equal to the volume fractions, per Equation (3.28),

$$\frac{1}{G_{12}} = \frac{V_f}{G_f} + \frac{V_m}{G_m}. \tag{3.55}$$

Example 3.6

Find the in-plane shear modulus of a glass/epoxy lamina with a 70% fiber volume fraction. Use properties of glass and epoxy from Table 3.1 and Table 3.2, respectively.

Solution

The glass fibers and the epoxy matrix have isotropic properties. From Table 3.1, the Young's modulus of the fiber is

$$E_f = 85 \text{ GPa}$$

and the Poisson's ratio of the fiber is

$$\nu_f = 0.2.$$

The shear modulus of the fiber

$$G_f = \frac{E_f}{2(1+\nu_f)}$$

$$= \frac{85}{2(1+0.2)}$$

$$= 35.42 \; GPa.$$

From Table 3.2, the Young's modulus of the matrix is

$$E_m = 3.4 \text{ GPa}$$

and the Poisson's ratio of the fiber is

$$\nu_m = 0.3.$$

The shear modulus of the matrix is

$$G_m = \frac{E_m}{2(1+\nu_m)}$$

$$= \frac{3.40}{2(1+0.3)}$$

$$= 1.308 \; GPa.$$

From Equation (3.55), the in-plane shear modulus of the unidirectional lamina is

$$\frac{1}{G_{12}} = \frac{0.70}{35.42} + \frac{0.30}{1.308}$$

$$G_{12} = 4.014 \; GPa.$$

Figure 3.13a and Figure 3.13b show the analytical values from Equation (3.55) of the in-plane shear modulus as a function of fiber volume fraction for a typical glass/epoxy lamina. Experimental values[4] are also plotted in the same figure.

3.3.2 Semi-Empirical Models

The values obtained for transverse Young's modulus and in-plane shear modulus through Equation (3.41) and Equation (3.55), respectively, do not agree well with the experimental results shown in Figure 3.10 and Figure 3.13. This establishes a need for better modeling techniques. These techniques include numerical methods, such as finite element and finite difference, and boundary element methods, elasticity solution, and variational principal models.[5] Unfortunately, these models are available only as complicated equations or in graphical form. Due to these difficulties, semi-empirical models have been developed for design purposes. The most useful of these models include those of Halphin and Tsai[6] because they can be used over a wide range of elastic properties and fiber volume fractions.

Halphin and Tsai[6] developed their models as simple equations by curve fitting to results that are based on elasticity. The equations are semi-empirical in nature because involved parameters in the curve fitting carry physical meaning.

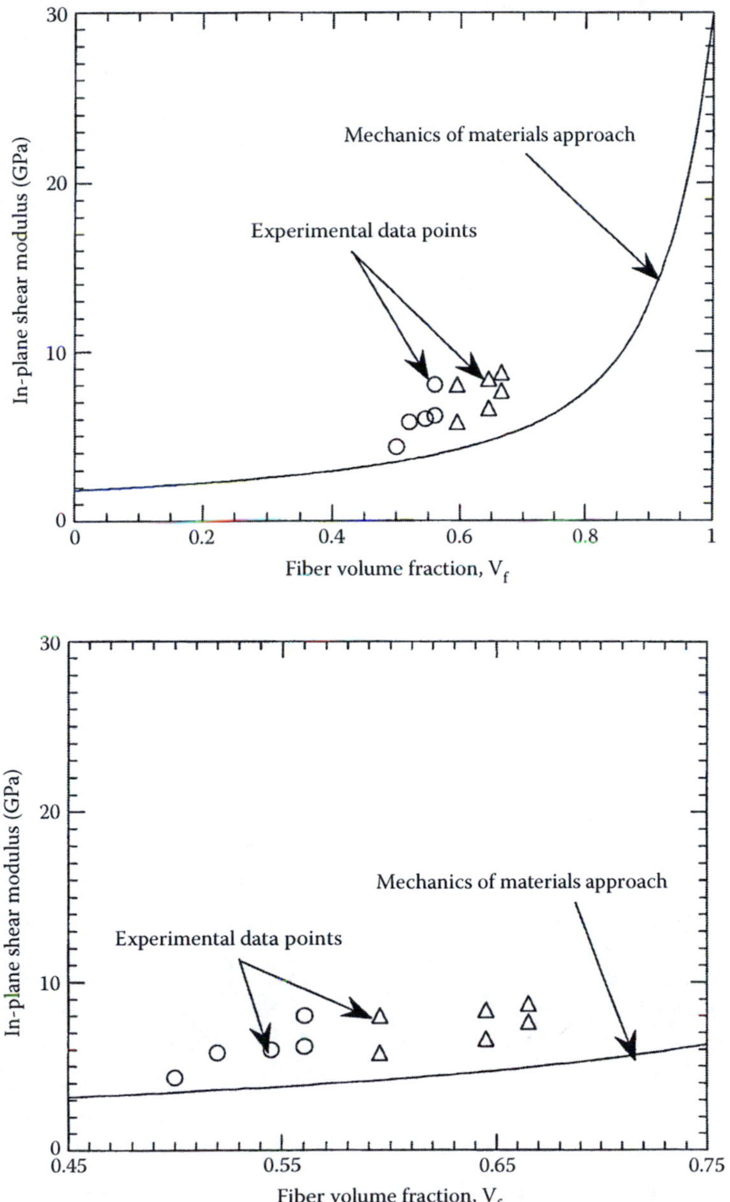

FIGURE 3.13
Theoretical values of in-plane shear modulus as a function of fiber volume fraction and comparison with experimental values for a unidirectional glass/epoxy lamina (G_f = 30.19 GPa, G_m = 1.83 GPa). Figure (b) zooms figure (a) for fiber volume fraction between 0.45 and 0.75. (Experimental data from Hashin, Z., NASA tech. rep. contract No. NAS1-8818, November 1970.)

3.3.2.1 Longitudinal Young's Modulus

The Halphin–Tsai equation for the longitudinal Young's modulus, E_1, is the same as that obtained through the strength of materials approach — that is,

$$E_1 = E_f V_f + E_m V_m. \tag{3.56}$$

3.3.2.2 Transverse Young's Modulus

The transverse Young's modulus, E_2, is given by[6]

$$\frac{E_2}{E_m} = \frac{1+\xi\eta V_f}{1-\eta V_f}, \tag{3.57}$$

where

$$\eta = \frac{(E_f / E_m)-1}{(E_f / E_m)+\xi}. \tag{3.58}$$

The term ξ is called the reinforcing factor and depends on the following:

- Fiber geometry
- Packing geometry
- Loading conditions

Halphin and Tsai[6] obtained the value of the reinforcing factor ξ by comparing Equation (3.57) and Equation (3.58) to the solutions obtained from the elasticity solutions. For example, for a fiber geometry of circular fibers in a packing geometry of a square array, $\xi = 2$. For a rectangular fiber cross-section of length a and width b in a hexagonal array, $\xi = 2(a/b)$, where b is in the direction of loading.[6] The concept of direction of loading is illustrated in Figure 3.14.

Example 3.7

Find the transverse Young's modulus for a glass/epoxy lamina with a 70% fiber volume fraction. Use the properties for glass and epoxy from Table 3.1 and Table 3.2, respectively. Use Halphin–Tsai equations for a circular fiber in a square array packing geometry.

Solution

Because the fibers are circular and packed in a square array, the reinforcing factor $\xi = 2$. From Table 3.1, the Young's modulus of the fiber is $E_f = 85$ GPa.

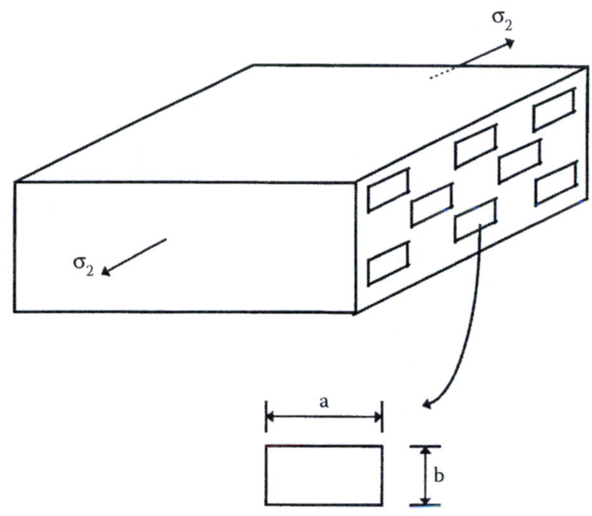

FIGURE 3.14
Concept of direction of loading for calculation of transverse Young's modulus by Halphin–Tsai equations.

From Table 3.2, the Young's modulus of the matrix is E_m = 3.4 GPa.
From Equation (3.58),

$$\eta = \frac{(85/3.4) - 1}{(85/3.4) + 2}$$

$$= 0.8889.$$

From Equation (3.57), the transverse Young's modulus of the unidirectional lamina is

$$\frac{E_2}{3.4} = \frac{1 + 2(0.8889)(0.7)}{1 - (0.8889)(0.7)}$$

$$E_2 = 20.20\ GPa.$$

For the same problem, from Example 3.4, this value of E_2 was found to be 10.37 GPa by the mechanics of materials approach.

Figure 3.15a and Figure 3.15b show the transverse Young's modulus as a function of fiber volume fraction for a typical boron/epoxy composite. The Halphin–Tsai equations (3.57) and the mechanics of materials approach Equation (3.41) curves are shown and compared to experimental data points.

As mentioned previously, the parameters ξ and η have a physical meaning. For example,

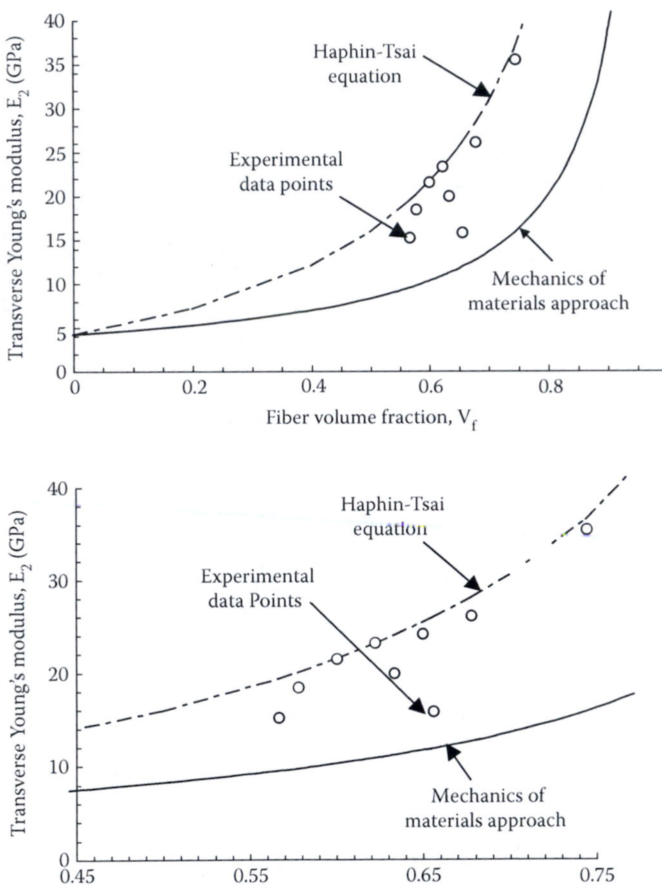

FIGURE 3.15
Theoretical values of transverse Young's modulus as a function of fiber volume fraction and comparison with experimental values for boron/epoxy unidirectional lamina (E_f = 414 GPa, ν_f = 0.2, E_m = 4.14 GPa, ν_m = 0.35). Figure (b) zooms figure (a) for fiber volume fraction between 0.45 and 0.75. (Experimental data from Hashin, Z., NASA tech. rep. contract no. NAS1-8818, November 1970.)

$E_f/E_m = 1$ implies $\eta = 0$, (homogeneous medium)

$E_f/E_m \to \infty$ implies $\eta = 1$ (rigid inclusions)

$E_f/E_m \to 0$ implies $\eta = -\frac{1}{\xi}$ (voids)

3.3.2.3 *Major Poisson's Ratio*

The Halphin–Tsai equation for the major Poisson's ratio, ν_{12}, is the same as that obtained using the strength of materials approach — that is,

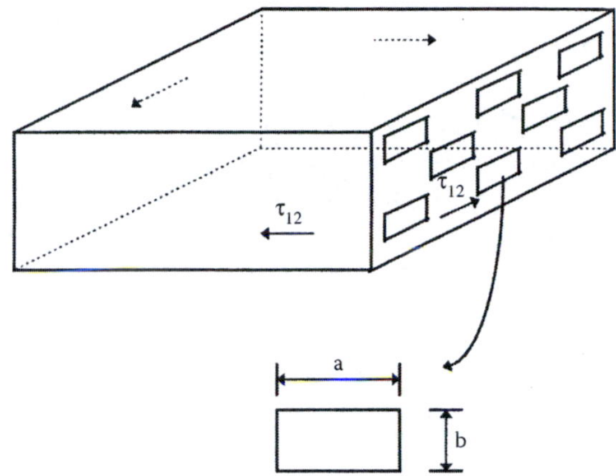

FIGURE 3.16
Concept of direction of loading to calculate in-plane shear modulus by Halphin–Tsai equations.

$$\nu_{12} = \nu_f V_f + \nu_m V_m. \tag{3.59}$$

3.3.2.4 *In-Plane Shear Modulus*

The Halphin–Tsai[6] equation for the in-plane shear modulus, G_{12}, is

$$\frac{G_{12}}{G_m} = \frac{1+\xi\eta V_f}{1-\eta V_f}, \tag{3.60}$$

where

$$\eta = \frac{(G_f / G_m) - 1}{(G_f / G_m) + \xi}. \tag{3.61}$$

The value of the reinforcing factor, ξ, depends on fiber geometry, packing geometry, and loading conditions. For example, for circular fibers in a square array, $\xi = 1$. For a rectangular fiber cross-sectional area of length a and width b in a hexagonal array, $\xi = \sqrt{3}\log_e(a/b)$, where a is the direction of loading. The concept of the direction of loading[7] is given in Figure 3.16.

The value of $\xi = 1$ for circular fibers in a square array gives reasonable results only for fiber volume fractions of up to 0.5. For example, for a typical glass/epoxy lamina with a fiber volume fraction of 0.75, the value of in-plane shear modulus using the Halphin–Tsai equation with $\xi = 1$ is 30% lower than that given by elasticity solutions. Hewitt and Malherbe[8] suggested choosing a function,

$$\xi = 1 + 40 V_f^{10}. \tag{3.62}$$

Example 3.8

Using Halphin–Tsai equations, find the shear modulus of a glass/epoxy composite with a 70% fiber volume fraction. Use the properties of glass and epoxy from Table 3.1 and Table 3.2, respectively. Assume that the fibers are circular and are packed in a square array. Also, get the value of the shear modulus by using Hewitt and Malherbe's[8] formula for the reinforcing factor.

Solution

For Halphin–Tsai's equations with circular fibers in a square array, the reinforcing factor $\xi = 1$. From Example 3.6, the shear modulus of the fiber is

$$G_f = 35.42 \text{ GPa}$$

and the shear modulus of the matrix is

$$G_m = 1.308 \text{ GPa}.$$

From Equation (3.61),

$$\eta = \frac{(35.42/1.308) - 1}{(35.42/1.308) + 1}$$

$$= 0.9288.$$

From Equation (3.60), the in-plane shear modulus is

$$\frac{G_{12}}{1.308} = \frac{1 + (1)(0.9288)(0.7)}{1 - (0.9288)(0.7)}$$

$$G_{12} = 6.169 \; GPa.$$

For the same problem, the value of G_{12} = 4.013 GPa was found by the mechanics of materials approach in Example 3.5.

Because the volume fraction is greater than 50%, Hewitt and Mahelbre[8] suggested a reinforcing factor (Equation 3.62):

$$\xi = 1 + 40V_f^{10}$$

$$= 1 + 40(0.7)^{10}.$$

$$= 2.130$$

Then, from Equation (3.61),

$$\eta = \frac{(35.42/1.308)-1}{(35.42/1.308)+2.130}.$$

$$= 0.8928$$

From Equation (3.60), the in-plane shear modulus is

$$\frac{G_{12}}{1.308} = \frac{1+(2.130)(0.8928)(0.7)}{1-(0.8928)(0.7)}.$$

$$G = 8.130\ GPa$$

Figure 3.17a and Figure 3.17b show the in-plane shear modulus as a function of fiber volume fraction for a typical glass/epoxy composite. The Halphin–Tsai equation (3.60) and the mechanics of materials approach, Equation (3.55) are shown and compared to the experimental[4] data points.

3.3.3 Elasticity Approach

In addition to the strength of materials and semi-empirical equation approaches, expressions for the elastic moduli based on elasticity are also available. Elasticity accounts for equilibrium of forces, compatibility, and Hooke's law relationships in three dimensions; the strength of materials approach may not satisfy compatibility and/or account for Hooke's law in three dimensions, and semi-empirical approaches are just as the name implies — partly empirical.

The elasticity models described here are called composite cylinder assemblage (CCA) models.[4,9–12] In a CCA model, one assumes the fibers are circular in cross-section, spread in a periodic arrangement, and continuous, as shown in Figure 3.18. Then the composite can be considered to be made of repeating elements called the representative volume elements (RVEs). The RVE is considered to represent the composite and respond the same as the whole composite does.

The RVE consists of a composite cylinder made of a single inner solid cylinder (fiber) bonded to an outer hollow cylinder (matrix) as shown in Figure 3.19. The radius of the fiber, a, and the outer radius of the matrix, b, are related to the fiber volume fraction, V_f, as

$$V_f = \frac{a^2}{b^2}. \tag{3.63}$$

Appropriate boundary conditions are applied to this composite cylinder based on the elastic moduli being evaluated.

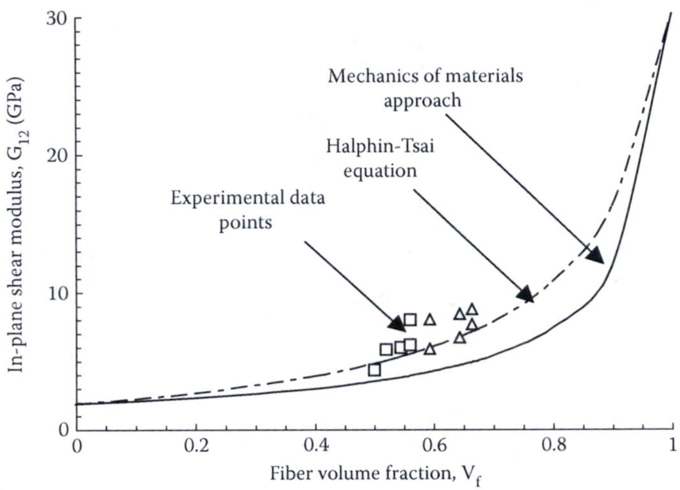

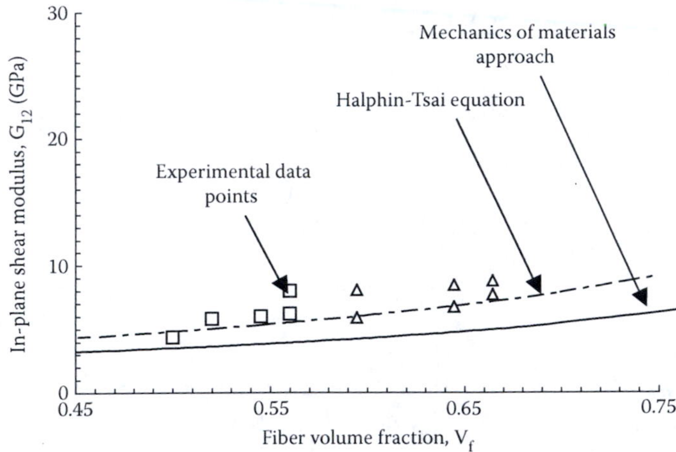

FIGURE 3.17
Theoretical values of in-plane shear modulus as a function of fiber volume fraction compared with experimental values for unidirectional glass/epoxy lamina ($G_f = 30.19$ GPa, $G_m = 1.83$ GPa). Figure (b) zooms figure (a) for fiber volume fraction between 0.45 and 0.75. (Experimental data from Hashin, Z., NASA tech. rep. contract No. NAS1-8818, November 1970.)

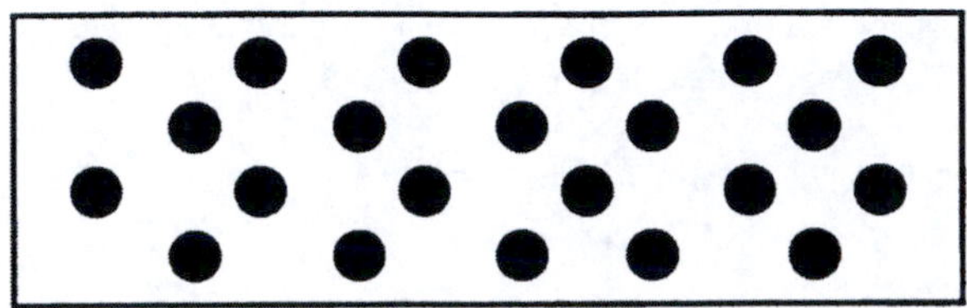

FIGURE 3.18
Periodic arrangement of fibers in a cross-section of a lamina.

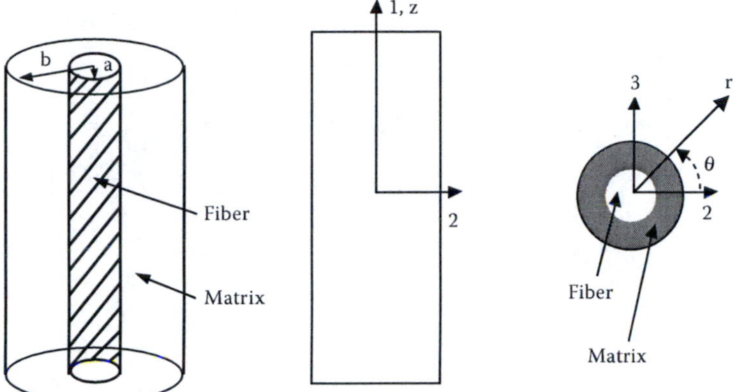

FIGURE 3.19
Composite cylinder assemblage (CCA) model used for predicting elastic moduli of unidirectional composites.

3.3.3.1 Longitudinal Young's Modulus

To find the elastic moduli along the fibers, we will apply an axial load, *P*, in direction 1 (Figure 3.19). The axial stress, σ_1, in direction 1 then is

$$\sigma_1 = \frac{P}{\pi b^2}. \tag{3.64}$$

Now, in terms of Hooke's law,

$$\sigma_1 = E_1 \in_1 \tag{3.65}$$

where

E_1 = longitudinal Young's modulus
$\in_1$ = axial strain in direction 1

Thus, from Equation (3.64) and Equation (3.65), we have

$$E_1 \in_1 = \frac{P}{\pi b^2}$$

$$E_1 = \frac{P}{\pi b^2 \in_1}. \tag{3.66}$$

To find E_1 in terms of elastic moduli of the fiber and the matrix, and the geometrical parameters such as fiber volume fraction, we need to relate the axial load, P, and the axial strain, $\in_1$, in these terms.

Assuming the response of a cylinder is axisymmetric, the equilibrium equation in the radial direction is given by[13]

$$\frac{d\sigma_r}{dr} + \frac{\sigma_r - \sigma_\theta}{r} = 0 \,, \tag{3.67}$$

where

σ_r = radial stress,
σ_θ = hoop stress.

The normal stress–normal strain relationships in polar coordinates, r–θ–z, for an isotropic material with Young's modulus, E, and Poisson's ratio, ν, are given by

$$\begin{bmatrix} \sigma_r \\ \sigma_\theta \\ \sigma_z \end{bmatrix} = \begin{bmatrix} \frac{E(1-\nu)}{(1-2\nu)(1+\nu)} & \frac{\nu E}{(1-2\nu)(1+\nu)} & \frac{\nu E}{(1-2\nu)(1+\nu)} \\ \frac{\nu E}{(1-2\nu)(1+\nu)} & \frac{E(1-\nu)}{(1-2\nu)(1+\nu)} & \frac{\nu E}{(1-2\nu)(1+\nu)} \\ \frac{\nu E}{(1-2\nu)(1+\nu)} & \frac{\nu E}{(1-2\nu)(1+\nu)} & \frac{E(1-\nu)}{(1-2\nu)(1+\nu)} \end{bmatrix} \begin{bmatrix} \in_r \\ \in_\theta \\ \in_z \end{bmatrix} . \tag{3.68}$$

The shear stresses and shear strains are zero in the r–θ–z coordinate system for axisymmetric response.

The strain displacement equations for axisymmetric response are

$$\in_r = \frac{du}{dr} \tag{3.69a}$$

$$\in_\theta = \frac{u}{r} \tag{3.69b}$$

$$\in_z = \frac{dw}{dz} \,, \tag{3.69c}$$

where

u = displacement in radial direction,
w = displacement in axial direction.

Substituting the strain-displacement equations (3.69a-c) in the stress–strain equations (3.68) and noting that $\epsilon_z = \epsilon_1$ everywhere gives

$$\begin{bmatrix} \sigma_r \\ \sigma_\theta \\ \sigma_z \end{bmatrix} = \begin{bmatrix} \frac{E(1-\nu)}{(1-2\nu)(1+\nu)} & \frac{\nu E}{(1-2\nu)(1+\nu)} & \frac{\nu E}{(1-2\nu)(1+\nu)} \\ \frac{\nu E}{(1-2\nu)(1+\nu)} & \frac{E(1-\nu)}{(1-2\nu)(1+\nu)} & \frac{\nu E}{(1-2\nu)(1+\nu)} \\ \frac{\nu E}{(1-2\nu)(1+\nu)} & \frac{\nu E}{(1-2\nu)(1+\nu)} & \frac{E(1-\nu)}{(1-2\nu)(1+\nu)} \end{bmatrix} \begin{bmatrix} \frac{du}{dr} \\ \frac{u}{r} \\ \epsilon_1 \end{bmatrix}, \tag{3.70}$$

which is rewritten for simplicity as

$$\begin{bmatrix} \sigma_r \\ \sigma_\theta \\ \sigma_z \end{bmatrix} = \begin{bmatrix} C_{11} & C_{12} & C_{12} \\ C_{12} & C_{11} & C_{12} \\ C_{12} & C_{12} & C_{11} \end{bmatrix} \begin{bmatrix} \frac{du}{dr} \\ \frac{u}{r} \\ \epsilon_1 \end{bmatrix}, \tag{3.71}$$

where the constants of the stiffness matrix are

$$C_{11} = \frac{E(1-\nu)}{(1-2\nu)(1+\nu)} \tag{3.72a}$$

$$C_{12} = \frac{\nu E}{(1-2\nu)(1+\nu)} . \tag{3.72b}$$

Substituting Equation (3.71) in the equilibrium equation (3.67) gives

$$\frac{d^2u}{dr^2} + \frac{1}{r}\frac{du}{dr} - \frac{u}{r^2} = 0 . \tag{3.73}$$

The solution to the linear ordinary differential equation is found by assuming that

$$u = \sum_{n=-\infty}^{\infty} A_n r^n \; . \tag{3.74}$$

Substituting Equation (3.74) in Equation (3.73) gives

$$\sum_{n=-\infty}^{\infty} n(n-1)A_n r^{n-2} + \frac{1}{r}\sum_{n=-\infty}^{\infty} nA_n r^{n-1} - \frac{1}{r^2}\sum_{n=-\infty}^{\infty} A_n r^n = 0$$

$$\sum_{n=-\infty}^{\infty} \left[n(n-1) + n - 1 \right] A_n r^{n-2} = 0$$

$$\sum_{n=-\infty}^{\infty} (n^2 - 1) A_n r^{n-2} = 0$$

$$\sum_{n=-\infty}^{\infty} (n-1)(n+1) A_n r^{n-2} = 0 \; . \tag{3.75}$$

The preceding expression (3.75) requires that

$$A_n = 0,\ n = -\infty, \ldots, \infty, \text{ except for } n = 1 \text{ and } n = -1. \tag{3.76}$$

Therefore, the form of the radial displacement is

$$u = A_1 r + \frac{A_{-1}}{r} \; . \tag{3.77}$$

To keep the terminology simple, assume that the form of the radial displacement with different names for the constants,

$$u = Ar + \frac{B}{r} \; . \tag{3.78}$$

The preceding equations are valid for a cylinder with an axisymmetric response. Thus, the radial displacement, u_f and u_m, in the fiber and matrix cylinders, respectively, can be assumed of the form

$$u_f = A_f r + \frac{B_f}{r}, 0 \le r \le a, \tag{3.79}$$

$$u_m = A_m r + \frac{B_m}{r}, a \le r \le b. \tag{3.80}$$

However, because the fiber is a solid cylinder and the radial displacement u_f is finite, $B_f = 0$; otherwise, the radial displacement of the fiber u_f would be infinite. Thus,

$$u_f = A_f r, 0 \le r \le a, \tag{3.81}$$

$$u_m = A_m r + \frac{B_m}{r}, a \le r \le b . \tag{3.82}$$

Differentiating Equation (3.81) and Equation (3.82) gives

$$\frac{du_f}{dr} = A_f \tag{3.83a}$$

$$\frac{du_m}{dr} = A_m - \frac{B_m}{r^2} . \tag{3.83b}$$

Using Equation (3.83a) and Equation (3.83b) in Equation (3.70), the stress–strain relationships for the fiber are

$$\begin{bmatrix} \sigma_r^f \\ \sigma_\theta^f \\ \sigma_z^f \end{bmatrix} = \begin{bmatrix} C_{11}^f & C_{12}^f & C_{12}^f \\ C_{12}^f & C_{11}^f & C_{12}^f \\ C_{12}^f & C_{12}^f & C_{11}^f \end{bmatrix} \begin{bmatrix} A_f \\ A_f \\ \epsilon_1 \end{bmatrix}, \tag{3.84}$$

where the stiffness constants of the fiber are

$$C_{11}^f = \frac{E_f(1-\nu_f)}{(1-2\nu_f)(1+\nu_f)}$$

$$C_{12}^f = \frac{\nu_f E_f}{(1-2\nu_f)(1+\nu_f)} \tag{3.85}$$

and the stress–strain relationships for the matrix are

$$\begin{bmatrix} \sigma_r^m \\ \sigma_\theta^m \\ \sigma_z^m \end{bmatrix} = \begin{bmatrix} C_{11}^m & C_{12}^m & C_{12}^m \\ C_{12}^m & C_{11}^m & C_{12}^m \\ C_{12}^m & C_{12}^m & C_{11}^m \end{bmatrix} \begin{bmatrix} A_m - \dfrac{B_m}{r^2} \\ A_m + \dfrac{B_m}{r^2} \\ \epsilon_1 \end{bmatrix}, \tag{3.86}$$

where the stiffness constants of the matrix are

$$C_{11}^m = \frac{E_m\left(1-\nu_m\right)}{\left(1-2\nu_m\right)\left(1+\nu_m\right)} \tag{3.87a}$$

$$C_{12}^m = \frac{\nu_m E_m}{\left(1-2\nu_m\right)\left(1+\nu_m\right)} . \tag{3.87b}$$

How do we now solve for the unknown constants A_f, A_m, B_m, and ε_1? The following four boundary and interface conditions will allow us to do that:

1. The radial displacement is continuous at the interface, $r = a$,

$$u_f\left(r=a\right) = u_m\left(r=a\right) . \tag{3.88}$$

Then, from Equation (3.81) and Equation (3.82),

$$A_f a = A_m a + \frac{B_m}{a} . \tag{3.89}$$

2. The radial stress is continuous at $r = a$:

$$\left(\sigma_r^f\right)\left(r=a\right) = \left(\sigma_r^m\right)\left(r=a\right) . \tag{3.90}$$

Then, from Equation (3.84) and Equation (3.86),

$$C_{11}^f A_f + C_{12}^f A_f + C_{12}^f \epsilon_1 = C_{11}^m\left(A_m - \frac{B_m}{a^2}\right) + C_{12}^m\left(A_m + \frac{B_m}{a^2}\right) + C_{12}^m \epsilon_1 . \tag{3.91}$$

3. Because the surface at $r = b$ is traction free, the radial stress on the outside of matrix, $r = b$, is zero:

$$\left(\sigma_r^m\right)\left(r=b\right)=0\ . \tag{3.92}$$

Then, Equation (3.84) gives

$$C_{11}^m\left(A_m-\frac{B_m}{b^2}\right)+C_{12}^m\left(A_m+\frac{B_m}{b^2}\right)+C_{12}^m\in_1=0\ . \tag{3.93}$$

4. The overall axial load over the fiber-matrix cross-sectional area in direction 1 is the applied load, P, then

$$\int_A \sigma_z dA = P \tag{3.94}$$

$$\int_0^b\int_0^{2\pi}\sigma_z r dr d\theta = P\ .$$

Because the axial normal stress, σ_z, is independent of θ,

$$\int_0^b \sigma_z 2\pi r dr = P\ . \tag{3.95}$$

Now,

$$\sigma_z=\sigma_z^f, 0\le r\le a$$

$$=\sigma_z^m, a\le r\le b\ . \tag{3.96}$$

Then, from Equation (3.84) and Equation (3.86),

$$\begin{aligned}&\int_0^a\left(C_{12}^f A_f + C_{12}^f A_f + C_{11}^f \in_1\right)2\pi r dr + \\ &\int_a^b\left(C_{12}^m\left(A_m-\frac{B_m}{r^2}\right)+C_{12}^m\left(A_m+\frac{B_m}{r^2}\right)+C_{11}^m\in_1\right)2\pi r dr = P\end{aligned}\ . \tag{3.97}$$

Solving Equation (3.89), Equation (3.91), Equation (3.93), and Equation (3.97), we get the solution to A_f, A_m, B_m, and ε_1.

Using the resulting solution for $\in_1$, and using Equation (3.66),

$$E_1 = \frac{P}{\pi b^2 \in_1}$$

$$= E_f V_f + E_m \left(1 - V_f\right)$$

$$- \frac{2 E_m E_f V_f \left(\nu_f - \nu_m\right)^2 \left(1 - V_f\right)}{E_f \left(2\nu_m^2 V_f - \nu_m + V_f \nu_m - V_f - 1\right) + E_m \left(-1 - 2V_f \nu_f^2 + \nu_f - V_f \nu_f + 2\nu_f^2 + V_f\right)} \tag{3.98}$$

Although the preceding expression can be written in a compact form by using definitions of shear and bulk modulus* of the material, we avoid doing so because results given in Equation (3.98) can now be found symbolically by computational systems such as Maple.[14] Note that the first two terms of Equation (3.98) represent the mechanics of materials approach result given by Equation (3.34).

Example 3.9

Find the longitudinal Young's modulus for a glass/epoxy lamina with a 70% fiber volume fraction. Use the properties for glass and epoxy from Table 3.1 and Table 3.2, respectively. Use equations obtained using the elasticity model.

Solution

From Table 3.1, the Young's modulus of fiber is

$$E_f = 85 \text{ GPa};$$

the Poisson's ratio of the fiber is

$$\nu_f = 0.2.$$

From Table 3.2, the Young's modulus of matrix is

$$E_m = 3.4 \text{ GPa}$$

* Bulk modulus of an elastic body is defined as the slope of the applied hydrostatic pressure vs. volume dilation curve. Hydrostatic stress is defined as $\sigma_{xx} = \sigma_{yy} = \sigma_{zz} = -p$, $\tau_{xy} = 0$, $\tau_{yz} = 0$, $\tau_{zx} = 0$ and volume dilation, D_v, is defined as the sum of resulting normal strains. $D_v = \varepsilon_x + \varepsilon_y + \varepsilon_z$. The bulk modulus, K, is used for finding volume changes in a given body subjected to hydrostatic pressure.

and the Poisson's ratio of the matrix is

$$\nu_m = 0.3.$$

Using Equation (3.98), the longitudinal Young's modulus

$$E_1 = (85\times10^9)(0.7) + (3.4\times10^9)(1-0.7)$$

$$-\frac{2(3.4\times10^9)(85\times10^9)(0.7)\left(0.2-0.3\right)^2\left(1-0.7\right)}{\begin{pmatrix}(85\times10^9)\left(2(0.3)^2(0.7)-0.3+(0.7)(0.3)-0.7-1\right)+\\(3.4\times10^9)\left(-1-2(0.7)(0.2)^2+0.2-(0.7)(0.2)+2(0.2)^2+0.7\right)\end{pmatrix}}$$

$$= 60.53\times10^9 \text{ Pa}$$

$$= 60.53 \text{ GPa}.$$

For the same problem, the longitudinal Young's modulus was found to be 60.52 GPa from the mechanics of materials approach as well as the Halphin–Tsai equations.

3.3.3.2 Major Poisson's Ratio

In Section 3.3.3.1, we solved the problems of an axially loaded cylinder. This same problem can be used to determine the axial Poisson's ratio, ν_{12}, because of the definition of major Poisson's ratio as

$$\nu_{12} = -\frac{\epsilon_r}{\epsilon_1}, \tag{3.99}$$

when a body is only under an axial load in direction 1.

From the definition of radial strain from Equation (3.69a) that, at $r = b$,

$$\epsilon_r\,(r=b) = \frac{u_m\left(b\right)}{b}, \tag{3.100}$$

the major Poisson's ratio is

$$\nu_{12} = -\frac{\dfrac{u_m\left(r=b\right)}{b}}{\epsilon_1}. \tag{3.101}$$

Using Equation (3.101),

$$\nu_{12} = -\frac{\left(A_m + \dfrac{B_m}{b^2}\right)}{\epsilon_1}. \tag{3.102}$$

Using the solution obtained in Section 3.3.3.1 for A_m, B_m, and ϵ_1 by solving Equation (3.89), Equation (3.91), Equation (3.93), and Equation (3.97), we get

$$\nu_{12} = \nu_f V_f + \nu_m V_m$$
$$+\frac{V_f V_m\left(\nu_f - \nu_m\right)\left(2E_f\nu_m^2 + \nu_m E_f - E_f + E_m - E_m\nu_f - 2E_m\nu_f^2\right)}{\left(2\nu_m^2 V_f - \nu_m + \nu_m V_f - 1 - V_f\right)E_f + \left(2\nu_f^2 - V_f\nu_f - 2V_f\nu_f^2 + V_f + \nu_f - 1\right)E_m} \tag{3.103}$$

Although the preceding expression can be written in a compact form by using definitions of shear and bulk modulus of the material, we avoid doing so because results given in Equation (3.103) can be found symbolically by computational systems such as Maple.[14] Note that the first two terms of Equation (3.103) are the same as the mechanics of materials approach result given by Equation (3.34).

Example 3.10

Find the major Poisson's ratio for a glass/epoxy lamina with a 70% fiber volume fraction. Use the properties for glass and epoxy from Table 3.1 and Table 3.2, respectively. Use equations obtained using the elasticity model.

Solution

Using Equation (3.103), the major Poisson's ratio is

$$\nu_{12} = (0.2)(0.7) + (0.3)(0.3)$$
$$+\frac{(0.7)(0.3)\left(0.2-0.3\right)\begin{pmatrix}(2)(85\times10^9)(0.3)^2 + (0.3)(85\times10^9) - 85\times10^9 + 3.4\times10^9\\ -(3.4\times10^9)(0.2) - (2)(3.4\times10^9)(0.2)^2\end{pmatrix}}{\begin{pmatrix}\left((2)(0.3)^2(0.7) - 0.3 + (0.3)(0.7) - 1 - (0.7)\right)(85\times10^9) +\\ \left(2(0.3)^2 - (0.7)(0.2) - (2)(0.7)(0.2)^2 + 0.7 + 0.2 - 1\right)(3.4\times10^9)\end{pmatrix}}$$
$$= 0.2238.$$

For the same problem, the major Poisson's ratio was found to be 0.2300 from the mechanics of materials approach as well as the Halphin–Tsai equations.

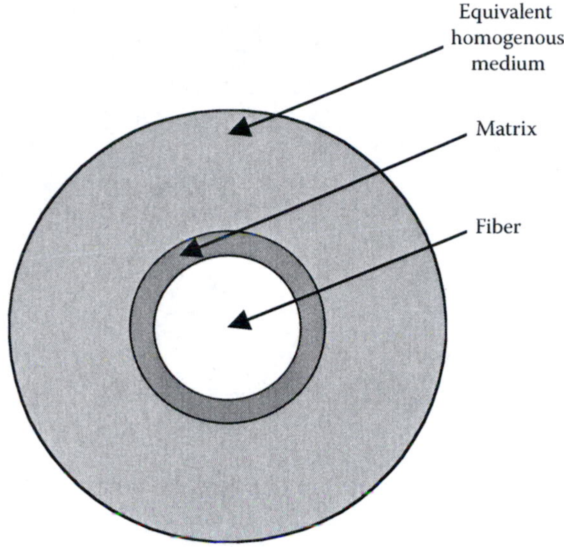

FIGURE 3.20
Three-phase model of a composite.

3.3.3.3 Transverse Young's Modulus

The CCA model only gives lower and upper bounds of the transverse Young's modulus of the composite. However, for the sake of completeness, we will summarize the result from a three-phase model. This model (Figure 3.20), however, yields an exact solution[12] for the transverse shear modulus, G_{23}. However, the transverse Young's modulus can be found as follows.

Assuming that the resulting composite properties are transversely isotropic (a valid assumption for hexagonally arranged fibers; 2–3 plane is isotropic),

$$E_2 = 2(1 + \nu_{23})G_{23}\,, \tag{3.104}$$

where ν_{23} = transverse Poisson's ratio.

The transverse Poisson's ratio, ν_{23}, is given by[15]

$$\nu_{23} = \frac{K^* - mG_{23}}{K^* + mG_{23}}\,, \tag{3.105}$$

where

$$m = 1 + 4K^* \frac{\nu_{12}^2}{E_1}\,. \tag{3.106}$$

The bulk modulus, K^*, of the composite under longitudinal plane strain is

$$K^* = \frac{K_m\left(K_f + G_m\right)V_m + K_f\left(K_m + G_m\right)V_f}{\left(K_f + G_m\right)V_m + \left(K_m + G_m\right)V_f}. \quad (3.107)$$

The bulk modulus K_f of the fiber under longitudinal plane strain is

$$K_f = \frac{E_f}{2\left(1+\nu_f\right)\left(1-2\nu_f\right)}. \quad (3.108)$$

The bulk modulus K_m of the matrix under longitudinal plane strain is

$$K_m = \frac{E_m}{2\left(1+\nu_m\right)\left(1-2\nu_m\right)}. \quad (3.109)$$

To derive the solution for G_{23} for use in Equation (3.104) is out of scope of this book; however, for the sake of completeness, the final solution is given next. Based on the three-phase model (Figure 3.20) where the fiber is surrounded by matrix, which is then surrounded by a homogeneous material equivalent to the composite, the transverse shear modulus, G_{23}, is given by the acceptable solution of the quadratic equation:[12]

$$A\left(\frac{G_{23}}{G_m}\right)^2 + 2B\left(\frac{G_{23}}{G_m}\right) + C = 0, \quad (3.110)$$

where

$$A = 3V_f\left(1-V_f\right)^2\left(\frac{G_f}{G_m}-1\right)\left(\frac{G_f}{G_m}+\eta_f\right)$$

$$+\left[\frac{G_f}{G_m}\eta_m + \eta_f\eta_m - \left(\frac{G_f}{G_m}\eta_m - \eta_f\right)V_f^3\right]\left[V_f\eta_m\left(\frac{G_f}{G_m}-1\right) - \left(\frac{G_f}{G_m}\eta_m + 1\right)\right]$$

$$B = -3V_f\left(1-V_f\right)^2\left(\frac{G_f}{G_m}-1\right)\left(\frac{G_f}{G_m}+\eta_f\right)$$

$$+\frac{1}{2}\left[\eta_m\frac{G_f}{G_m} + \left(\frac{G_f}{G_m}-1\right)V_f + 1\right]\left[\left(\eta_m - 1\right)\left(\frac{G_f}{G_m}+\eta_f\right)\right.$$

$$\left. -2\left(\frac{G_f}{G_m}\eta_m - \eta_f\right)V_f^3\right] + \frac{V_f}{2}\left(\eta_m + 1\right)\left(\frac{G_f}{G_m}-1\right)\left[\frac{G_f}{G_m} + \eta_f + \left(\frac{G_f}{G_m}\eta_m - \eta_f\right)V_f^3\right]$$

$$C = 3V_f\left(1 - V_f\right)^2\left(\frac{G_f}{G_m} - 1\right)\left(\frac{G_f}{G_m} + \eta_f\right)$$

$$+\left[\eta_m \frac{G_f}{G_m} + \left(\frac{G_f}{G_m} - 1\right)V_f + 1\right]\left[\frac{G_f}{G_m} + \eta_f + \left(\frac{G_f}{G_m}\eta_m - \eta_f\right)V_f^3\right] \quad (3.111)$$

$$\eta_m = 3 - 4\nu_m \ ,$$

$$\eta_f = 3 - 4\nu_f \ . \quad (3.112)$$

Then, using Equation (3.104) through Equation (3.109), we get the transverse Young's modulus, E_2.

Example 3.11

Find the transverse Young's modulus for a glass/epoxy lamina with a 70% fiber volume fraction. Use the properties for glass and epoxy from Table 3.1 and Table 3.2, respectively. Use equations obtained using the elasticity model.

Solution

From Equation (3.112),

$$\eta_f = 3 - 4\left(0.2\right)$$

$$= 2.2$$

$$\eta_m = 3 - 4\left(0.3\right)$$

$$= 1.8 \ .$$

From Equation (3.108) and Equation (3.109),

$$K_f = \frac{85 \times 10^9}{2\left(1 + 0.2\right)\left(1 - 2 \times 0.2\right)}$$

$$= 59.03 \times 10^9\, Pa$$

$$K_m = \frac{3.4 \times 10^9}{2(1+0.3)(1-2\times 0.3)}$$

$$= 3.269 \times 10^9 \, Pa \; .$$

From Equation (3.107),

$$K^* = \frac{\begin{pmatrix} 3.269\times 10^9\left(59.03\times 10^9 + 1.308\times 10^9\right)(0.3) + \\ 59.03\times 10^9\left(3.269\times 10^9 + 1.308\times 10^9\right)(0.7) \end{pmatrix}}{\left(59.03\times 10^9 + 1.308\times 10^9\right)(0.3) + \left(3.269\times 10^9 + 1.308\times 10^9\right)(0.7)}$$

$$= 11.66 \times 10^9 \, Pa \; .$$

The three constants of the quadratic Equation (3.110) are given by Equation (3.111) as

$$A = 3(0.7)(1-0.7)^2\left(\frac{35.42\times 10^9}{1.308\times 10^9} - 1\right)\left(\frac{35.42\times 10^9}{1.308\times 10^9} + 2.2\right)$$

$$+\left[\frac{35.42\times 10^9}{1.308\times 10^9}(1.8) + 2.2\times 1.8 - \left(\frac{35.42\times 10^9}{1.308\times 10^9}(1.8) - 2.2\right)0.7^3\right]$$

$$\left[(0.7)(1.8)\left(\frac{35.42\times 10^9}{1.308\times 10^9} - 1\right) - \left(\frac{35.42\times 10^9}{1.308\times 10^9}(1.8) + 1\right)\right]$$

$$= -476.0 \; .$$

$$B = -3(0.7)(1-0.7)^2\left(\frac{35.42\times 10^9}{1.308\times 10^9} - 1\right)\left(\frac{35.42\times 10^9}{1.308\times 10^9} + 2.2\right)$$

$$+\frac{1}{2}\left[1.8\left(\frac{35.42\times 10^9}{1.308\times 10^9}\right) + \left(\frac{35.42\times 10^9}{1.308\times 10^9} - 1\right)(0.7) + 1\right]$$

$$\left[(1.8-1)\left(\frac{35.42\times10^9}{1.308\times10^9}+2.2\right)-2\left(\frac{35.42\times10^9}{1.308\times10^9}(1.8)-2.2\right)0.7^3\right]+\frac{0.7}{2}$$

$$\left[(1.8+1)\left(\frac{35.42\times10^9}{1.308\times10^9}-1\right)+\left(\frac{35.42\times10^9}{1.308\times10^9}+1.8+\frac{35.42\times10^9}{1.308\times10^9}(1.8)-2.2\right)0.7^3\right]$$

$$=723.0.$$

$$C=3(0.7)(1-0.7)^2\left(\frac{35.42\times10^9}{1.308\times10^9}-1\right)\left(\frac{35.42\times10^9}{1.308\times10^9}+2.2\right)$$

$$+\left[1.8\frac{35.42\times10^9}{1.308\times10^9}+\left(\frac{35.42\times10^9}{1.308\times10^9}-1\right)(0.7)+1\right]$$

$$\left[\frac{35.42\times10^9}{1.308\times10^9}+2.2+\left(\frac{35.42\times10^9}{1.308\times10^9}(1.8)-2.2\right)0.7^3\right]$$

$$=3222.$$

Substituting values of *A*, *B*, and *C* in Equation (3.110),

$$-476.0\left(\frac{G_{23}}{1.308\times10^9}\right)^2+2(723.0)\left(\frac{G_{23}}{1.308\times10^9}\right)+3222=0$$

$$-278.4\times10^{-18}G_{23}^2+1106\times10^{-9}G_{23}+3222=0$$

gives $G_{23} = 5.926 \times 10^9$ Pa, -1.953×10^9 Pa. Thus, the acceptable solution is

$$G_{23}=5.926\times10^9\,Pa\,.$$

From Equation (3.106),

$$m=1+4(11.66\times10^9)\frac{0.2238^2}{60.53\times10^9}$$

$$=1.039\,.$$

From Equation (3.105),

$$\nu_{23} = \frac{11.66 \times 10^9 - 1.039\left(5.926 \times 10^9\right)}{11.66 \times 10^9 + 1.039\left(5.926 \times 10^9\right)}$$

$$= 0.3089 \, .$$

From Equation (3.104),

$$E_2 = 2\left(1 + 0.3089\right)\left(5.926 \times 10^9\right)$$

$$= 15.51 \times 10^9 \, Pa$$

$$= 15.51 \; GPa.$$

For the same problem, the transverse Young's modulus was found to be 10.37 GPa from the mechanics of materials approach and 20.20 GPa from the Halphin–Tsai equations.

Figure 3.21a and Figure 3.21b show the transverse Young's modulus as a function of fiber volume fraction for a typical boron/epoxy unidirectional lamina. The elasticity equation (3.104), Halphin–Tsai equation (3.60), and the mechanics of materials approach (Equation 3.55) are shown and compared to the experimental data points.

3.3.3.4 Axial Shear Modulus

To find the axial shear modulus, G_{12}, of a unidirectional composite, we consider the same concentric cylinder model (Figure 3.19). Consider a long fiber of radius, a, and shear modulus, G_f, surrounded by a long concentric cylinder of matrix of outer radius, b, and shear modulus, G_m. The composite cylinder (Figure 3.19) is subjected to a shear strain, γ_{12}^0 , in the 1–2 plane.

Following the derivation,[4,12,16] the normal displacements in the 1, 2, 3 direction for the fiber or matrix are assumed of the following form:

$$u_1 = -\frac{\gamma_{12}^0}{2} x_2 + F\left(x_2, x_3\right)$$

$$u_2 = \frac{\gamma_{12}^0}{2} x_1$$

$$u_3 = 0 \, , \qquad \text{(3.113a, b, c)}$$

where γ_{12}^0 is the applied shear strain to the boundary.

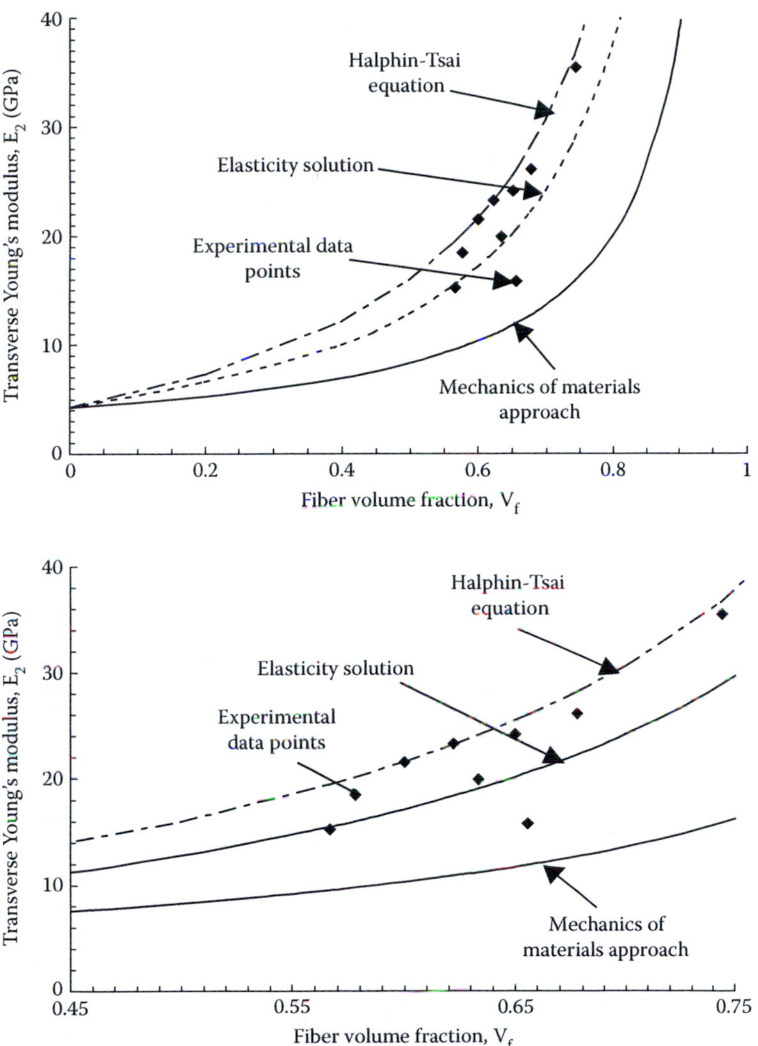

FIGURE 3.21
Theoretical values of transverse Young's modulus as a function of fiber volume fraction and comparison with experimental values for boron/epoxy unidirectional lamina (E_f = 414 GPa, ν_f = 0.2, E_m = 4.14 GPa, ν_m = 0.35). Figure (b) zooms figure (a) for fiber volume fraction between 0.45 and 0.75. (Experimental data from Hashin, Z., NASA tech. rep. contract No. NAS1-8818, November 1970.)

The preceding assumption of the form of the displacements is based on a semi-inverse method[17] that is beyond the scope of this book. Individual expressions for displacement of the fiber and matrix will be shown later in the derivation.

From the strain-displacement[13] equations and the expressions for the displacement field in Equation (3.113a, b, c),

$$\epsilon_{11} = \frac{\partial u_1}{\partial x_1}$$

$$= 0$$

$$\epsilon_{22} = \frac{\partial u_2}{\partial x_2}$$

$$= 0$$

$$\epsilon_{33} = \frac{\partial u_3}{\partial x_3}$$

$$= 0$$

$$\gamma_{23} = \frac{\partial u_2}{\partial x_3} + \frac{\partial u_3}{\partial x_2}$$

$$= 0$$

$$\gamma_{12} = \frac{\partial u_1}{\partial x_2} + \frac{\partial u_2}{\partial x_1}$$

$$= \frac{\partial F}{\partial x_2}$$

$$\gamma_{31} = \frac{\partial u_1}{\partial x_3} + \frac{\partial u_3}{\partial x_1}$$

$$= \frac{\partial F}{\partial x_3}. \tag{3.114 a–f}$$

Because all normal strains in the 1, 2, and 3 directions are zero, all the normal stresses in 1, 2, 3 directions are also zero. Also, $\tau_{23} = 0$ because $\gamma_{23} = 0$.

Using Equation (3.114e) and Equation (3.114f), the only possible nonzero stresses are

$$\tau_{12} = G\gamma_{12}$$

$$= G\frac{\partial F}{\partial x_2} \tag{3.115a}$$

$$\tau_{13} = G\gamma_{13}$$

$$= G\frac{\partial F}{\partial x_3}, \tag{3.115b}$$

where G is the shear modulus of the material.

The equilibrium condition derived from the fact that the sum of the forces in direction 1 is zero gives[13]

$$\frac{\partial \sigma_1}{\partial x_1} + \frac{\partial \tau_{12}}{\partial x_2} + \frac{\partial \tau_{13}}{\partial x_3} = 0 . \tag{3.116}$$

With Equation (3.115a) and Equation (3.115b) and $\sigma_1 = 0$, the equilibrium equation (3.116) reduces it to

$$\frac{\partial^2 F}{\partial x_2^2} + \frac{\partial^2 F}{\partial x_3^2} = 0 . \tag{3.117}$$

Converting Equation (3.117) to polar coordinates needs the following:

$$x_2 = rCos\theta , \tag{3.118}$$

$$x_3 = rSin\theta \tag{3.119}$$

give

$$r^2 = x_2^2 + x_3^2 \tag{3.120a}$$

$$\theta = \tan^{-1}\frac{x_3}{x_2} . \tag{3.120b}$$

From Equation (3.118), Equation (3.119), and Equation (3.120a, b),

$$2r\frac{\partial r}{\partial x_2} = 2x_2$$

$$\frac{\partial r}{\partial x_2} = \frac{x_2}{r}$$

$$= Cos\theta \tag{3.121a}$$

$$2r\frac{\partial r}{\partial x_3} = 2x_3$$

$$\frac{\partial r}{\partial x_3} = \frac{x_3}{r}$$

$$= Sin\theta \tag{3.121b}$$

$$\frac{\partial \theta}{\partial x_2} = \frac{1}{1+\left(\frac{x_3}{x_2}\right)^2}\left(-\frac{x_3}{x_2^2}\right)$$

$$= -\frac{x_3}{x_2^2 + x_3^2}$$

$$= -\frac{rSin\theta}{r^2}$$

$$= -\frac{Sin\theta}{r} \tag{3.121c}$$

$$\frac{\partial \theta}{\partial x_3} = \frac{1}{1+\left(\frac{x_3}{x_2}\right)^2}\left(\frac{1}{x_2}\right)$$

$$= \frac{x_2}{x_2^2 + x_3^2}$$

$$= \frac{rCos\theta}{r}$$

$$= \frac{Cos\theta}{r} \,. \tag{3.121d}$$

Now, using the chain rule for derivatives,

$$\frac{\partial F}{\partial x_2} = \frac{\partial F}{\partial r}\frac{\partial r}{\partial x_2} + \frac{\partial F}{\partial \theta}\frac{\partial \theta}{\partial x_2} \,, \tag{3.122}$$

and using Equation (3.121a) and Equation (3.121c),

$$\frac{\partial F}{\partial x_2} = Cos\theta \frac{\partial F}{\partial r} - \frac{Sin\theta}{r}\frac{\partial F}{\partial \theta} \,. \tag{3.123}$$

Repeating a similar chain of rule of derivatives on Equation (3.122),

$$\frac{\partial^2 F}{\partial x_2^2} = Cos^2\theta \frac{\partial^2 F}{\partial r^2} + Sin^2\theta\left(\frac{1}{r}\frac{\partial F}{\partial r} + \frac{1}{r^2}\frac{\partial^2 F}{\partial \theta^2}\right) - 2Sin\theta Cos\theta \frac{\partial}{\partial r}\left(\frac{1}{r}\frac{\partial F}{\partial \theta}\right) . \tag{3.124a}$$

Similarly,

$$\frac{\partial^2 F}{\partial x_3^2} = Sin^2\theta \frac{\partial^2 F}{\partial r^2} + Cos^2\theta\left(\frac{1}{r}\frac{\partial F}{\partial r} + \frac{1}{r^2}\frac{\partial^2 F}{\partial \theta^2}\right) + 2Sin\theta Cos\theta \frac{\partial}{\partial r}\left(\frac{1}{r}\frac{\partial F}{\partial \theta}\right) . \tag{3.124b}$$

Substituting Equation (3.124a) and Equation (3.124b) in Equation (3.117) yields

$$\frac{\partial^2 F}{\partial r^2} + \frac{1}{r}\frac{\partial F}{\partial r} + \frac{1}{r^2}\frac{\partial^2 F}{\partial \theta^2} = 0 \,. \tag{3.125}$$

The solution of Equation (3.125) is given by

$$F(r,\theta) = \left(Ar + \frac{B}{r}\right)Cos\theta \tag{3.126}$$

only after noting that the complete solution to Equation (3.125) is of the form

$$F(r,\theta) = A_0 + \sum_{n=1}^{\infty} \left(A_n r^n + B_n r^{-n}\right)\left[C_n Sin(n\theta) + D_n Cos(n\theta)\right], \quad (3.127)$$

but that the surface, $r = b$, of the composite cylinder is only subjected to displacements:

$$u_{1m}(r=b) = \frac{\gamma_{12}^0}{2} x_2 \Big|_{r=b}$$

$$= \frac{\gamma_{12}^0}{2} bCos\theta \quad (3.128)$$

$$u_{2m}(r=b) = \frac{\gamma_{12}^0}{2} x_1 \Big|_{r=b}$$

$$= \frac{\gamma_{12}^0}{2} bSin\theta \quad (3.129)$$

$$u_{3m}(r=b) = 0\,. \quad (3.130)$$

Thus, the function $F(r,\theta)$ of Equation (3.126) for the fiber F_f and matrix F_m is given by

$$F_f(r,\theta) = \left(A_1 r + \frac{B_1}{r}\right)Cos\theta \quad (3.131)$$

$$F_m(r,\theta) = \left(A_2 r + \frac{B_2}{r}\right)Cos\theta\,. \quad (3.132)$$

How do we find A_1, B_1, A_2, and B_2? The following boundary and interface conditions are applied to find these four unknowns:

1. The axial displacements of the fiber u_{1f} and the matrix u_{1m} at the interface, $r = a$, are continuous:

$$u_{1f}(r=a) = u_{1m}(r=a)\,. \quad (3.133)$$

Now, from Equation (3.113a),

$$u_{1f} = -\frac{\gamma_{12}^0}{2} x_2 + F_f\left(x_2, x_3\right)$$

$$= -\frac{\gamma_{12}^0}{2} rCos\theta + F_f\left(rCos\theta, rSin\theta\right). \tag{3.134}$$

At $r = a$,

$$u_{1f}\left(r = a\right) = -\frac{\gamma_{12}^0}{2} aCos\theta + \left(A_1 a + \frac{B_1}{a}\right)Cos\theta. \tag{3.135}$$

Similarly, from Equations (3.313a), (3.318), and (3.131),

$$u_{1m}\left(r = a\right) = -\frac{\gamma_{12}^0}{2} aCos\theta + \left(A_2 a + \frac{B_2}{a}\right)Cos\theta. \tag{3.136}$$

Equating Equation (3.135) and Equation (3.136) per Equation (3.133) gives

$$A_1 a + \frac{B_1}{a} = A_2 a + \frac{B_2}{a}. \tag{3.137}$$

2. The displacement of the fiber u_{1f} is given by Equations (3.313a), (3.318), and (3.131) as

$$u_{1f} = -\frac{\gamma_{12}^0}{2} rCos\theta + \left(A_1 r + \frac{B_1}{r}\right)Cos\theta. \tag{3.138}$$

Because $r = 0$ is a point on the fiber and displacement in the fiber is finite,

$$B_1 = 0. \tag{3.139}$$

3. The shear stress in the fiber τ_{1rf} and that in the matrix τ_{1rm} are continuous at the interface $r = a$:

$$\tau_{1r,f}\left(r = a\right) = \tau_{1r,m}\left(r = a\right). \tag{3.140}$$

First, we need to derive an expression for τ_{1r} from transforming stresses between 1–r and 1–3 coordinates:

$$\tau_{1r} = Cos\theta\, \tau_{12} + Sin\theta\, \tau_{13}\ . \tag{3.141}$$

Using Equation (3.115a, b) in Equation (3.141),

$$\tau_{1r} = Cos\theta\, G\, \frac{\partial F}{\partial x_2} + Sin\theta\, G\, \frac{\partial F}{\partial x_3} \tag{3.142}$$

$$= G\left(Cos\theta \frac{\partial F}{\partial x_2} + Sin\theta \frac{\partial F}{\partial x_3} \right). \tag{3.143}$$

Substituting Equation (3.121a) and Equation (3.121b) in Equation (3.143),

$$\tau_{1r} = G\left(\frac{\partial x_2}{\partial r}\frac{\partial F}{\partial x_2} + \frac{\partial x_3}{\partial r}\frac{\partial F}{\partial x_3} \right),$$

gives

$$\tau_{1r} = G\frac{\partial F}{\partial r}\ . \tag{3.144}$$

Thus, in the fiber from Equation (3.131)

$$\tau_{1r,f} = G_f\, \frac{\partial F_f}{\partial r}$$

$$= G_f\left(A_1 - \frac{B_1}{r^2} \right) Cos\theta \tag{3.145}$$

and in the matrix from Equation (3.132),

$$\tau_{1r,m} = G_m\, \frac{\partial F_m}{\partial r}$$

$$= G_m\left(A_2 - \frac{B_2}{r^2}\right)Cos\theta \,. \tag{3.146}$$

Equating Equation (3.145) and Equation (3.146) at $r = a$, per Equation (3.140), gives

$$G_f\left(A_1 - \frac{B_1}{a^2}\right) = G_m\left(A_2 - \frac{B_2}{a^2}\right). \tag{3.147}$$

4. The displacement due to the applied shear strain of γ_{12}^0 at the boundary $r = b$ of the composite cylinder is given by

$$u_{1m}\left(r = b\right) = \frac{\gamma_{12}^0}{2}x_2\Big|_{r=b} \tag{3.148a}$$

$$= \frac{\gamma_{12}^0}{2}bCos\theta \,. \tag{3.148b}$$

Based on Equation (3.113a) and Equation (3.132),

$$u_{1m}\left(r = b\right) = -\frac{\gamma_{12}^0}{2}x_2 + F_m\left(x_2, x_3\right)\Big|_{r=b}$$

$$= -\frac{\gamma_{12}^0}{2}bCos\theta + \left(A_2 b + \frac{B_2}{b}\right)Cos\theta \,. \tag{3.149}$$

From Equation (3.148b) and Equation (3.149), we get

$$A_2 b + \frac{B_2}{b} = \gamma_{12}^0 b \,. \tag{3.150}$$

Solving the three simultaneous equations (Equation 3.137, Equation 3.147, and Equation 3.150) to find A_1, A_2, and B_2, ($B_1 = 0$ from Equation 3.139), we get

$$A_1 = \frac{2G_m}{G_m\left(1+V_f\right)+G_f\left(1-V_f\right)}\gamma_{12}^0 \tag{3.151}$$

$$A_2 = \frac{\left(G_f+G_m\right)}{G_m\left(1+V_f\right)+G_f\left(1-V_f\right)}\gamma_{12}^0 \tag{3.152}$$

$$B_2 = -\frac{a^2\left(-G_m+G_f\right)}{G_m\left(1+V_f\right)+G_f\left(1-V_f\right)}\gamma_{12}^0, \tag{3.153}$$

where, from Equation (3.63), the fiber volume fraction V_f is substituted for $\frac{a^2}{b^2}$.

The shear modulus G_{12} can be now be found as

$$G_{12} \equiv \frac{\tau_{12,m}\,|_{r=b}}{\gamma_{12}^0}, \tag{3.154}$$

where

$$\tau_{12,m}\,|_{r=b} = \text{ shear stress at } r = b$$

because, based on Equation (3.115a),

$$\tau_{12,m} = G_m\frac{\partial F_m}{\partial x_2}$$

$$= G_m\left(\frac{\partial F_m}{\partial r}\frac{\partial r}{\partial x_2}+\frac{\partial F_m}{\partial \theta}\frac{\partial \theta}{\partial x_2}\right). \tag{3.155}$$

Using Equation (3.121a) and Equation (3.121b),

$$\tau_{12,m} = G_m\left[\frac{\partial F_m}{\partial r}Cos\theta+\frac{\partial F_m}{\partial \theta}\left(-\frac{Sin\theta}{r}\right)\right]. \tag{3.156}$$

Using Equation (3.131) and Equation (3.132) in Equation (3.156) gives

$$\tau_{12,m} = G_m\left[\left(A_2 - \frac{B_2}{r^2}\right)Cos\theta Cos\theta + \left(A_2 r + \frac{B_2}{r}\right)(-Sin\theta)\cdot\left(-\frac{Sin\theta}{r}\right)\right]$$

$$= G_m\left[\left(A_2 - \frac{B_2}{r^2}\right)Cos^2\theta + \left(A_2 + \frac{B_2}{r^2}\right)Sin^2\theta\right]. \quad (3.157)$$

At $r = b$, $\theta = 0$

$$\tau_{12,m}\Big|_{r=b,\theta=0} = G_m\left(A_2 - \frac{B_2}{b^2}\right). \quad (3.158)$$

Substituting values of A_2 and B_2 from Equation (3.152) and Equation (3.153), respectively, in Equation (3.158) yields

$$\tau_{12,m}\Big|_{r=b,\theta=0} = G_m\left[\frac{G_f\left(1+V_f\right)+G_m\left(1-V_f\right)}{G_f\left(1-V_f\right)+G_m\left(1+V_f\right)}\right]\gamma_{12}^0 \quad (3.159)$$

and the shear modulus, G_{12}, can be found as

$$G_{12} \equiv \frac{\tau_{12,m}\big|_{r=b,\theta=0}}{\gamma_{12}^0}.$$

This gives

$$G_{12} = G_m\left[\frac{G_f\left(1+V_f\right)+G_m\left(1-V_f\right)}{G_f\left(1-V_f\right)+G_m\left(1+V_f\right)}\right]. \quad (3.160)$$

Example 3.12

Find the shear modulus, G_{12}, for a glass/epoxy composite with 70% fiber volume fraction. Use the properties for glass and epoxy from Table 3.1 and Table 3.2, respectively. Use the equations obtained using the elasticity model.

Solution

From Example 3.6, G_f = 35.42 GPa and G_m = 1.308 GPa. Using Equation (3.160), the in-plane shear modulus is

$$G_{12} = 1.308 \times 10^9 \left[\frac{(35.42 \times 10^9)(1+0.7) + (1.308 \times 10^9)(1-0.7)}{(35.42 \times 10^9)(1-0.7) + (1.308 \times 10^9)(1+0.7)} \right]$$

$$= 6.169 \times 10^9 \ Pa$$

$$= 6.169 \ GPa.$$

For the same problem, the shear modulus, G_{12}, is found to be 4.014 GPa from the mechanics of materials approach and 6.169 GPa from the Halphin-Tsai equations.

Figure 3.22a and Figure 3.22b show the in-plane shear modulus as a function of fiber volume fraction for a typical glass/epoxy unidirectional lamina. The elasticity equation (3.160), Halphin-Tsai equation (3.60), and the mechanics of materials approach (Equation 3.55) are shown and compared to the experimental data points.

A comparison of the elastic moduli from the mechanics of materials approach, the Halphin-Tsai equations, and elasticity models (Example 3.3 through Example 3.11) is given in Table 3.5.

3.3.4 Elastic Moduli of Lamina with Transversely Isotropic Fibers

Glass, aramids, and graphite are the three most common types of fibers used in composites; among these, aramids and graphite are transversely isotropic. From the definition of transversely isotropic materials in Chapter 2, such fibers have five elastic moduli.

If L represents the longitudinal direction along the length of the fiber and T represents the plane of isotropy (Figure 3.23) perpendicular to the longitudinal direction, the five elastic moduli of the transversely isotropic fiber are

E_{fL} = longitudinal Young's modulus

E_{fT} = Young's modulus in plane of isotropy

ν_{fL} = Poisson's ratio characterizing the contraction in the plane of isotropy when longitudinal tension is applied

ν_{fT} = Poisson's ratio characterizing the contraction in the longitudinal direction when tension is applied in the plane of isotropy

G_{fT} = in-plane shear modulus in the plane perpendicular to the plane of isotropy

The elastic moduli using strength of materials approach for lamina with transversely isotropic fibers[18] are

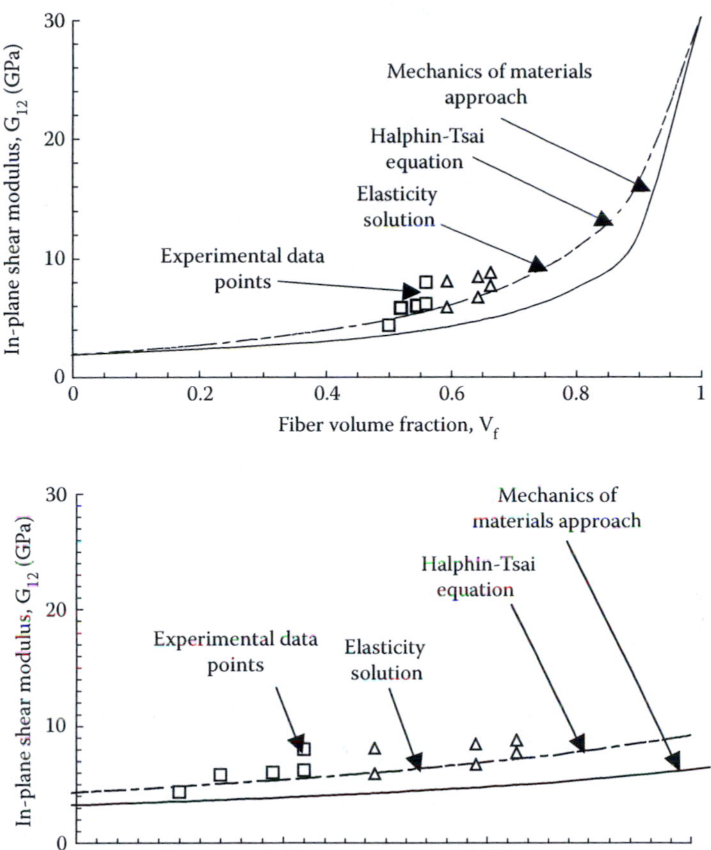

FIGURE 3.22
Theoretical values of in-plane shear modulus as a function of fiber volume fraction compared with experimental values for unidirectional glass/epoxy lamina ($G_f = 30.19$ GPa, $G_m = 1.83$ GPa). Figure (b) zooms figure (a) for fiber volume fraction between 0.45 and 0.75. (Experimental data from Hashin, Z., NASA tech. rep. contract No. NAS1-8818, November 1970.)

TABLE 3.5
Comparison of Predicted Elastic Moduli

Method	E_1 (GPa)	E_2 (GPa)	ν_{12}	G_{12} (GPa)
Mechanics of materials	60.52	10.37	0.2300	4.014
Halphin–Tsai	60.52	20.20	0.2300	6.169
Elasticity	60.53	15.51	0.2238	6.169[a]

[a] The Halphin–Tsai equations and the elasticity model equations give the same value for the shear modulus. Can you show that this is not a coincidence?

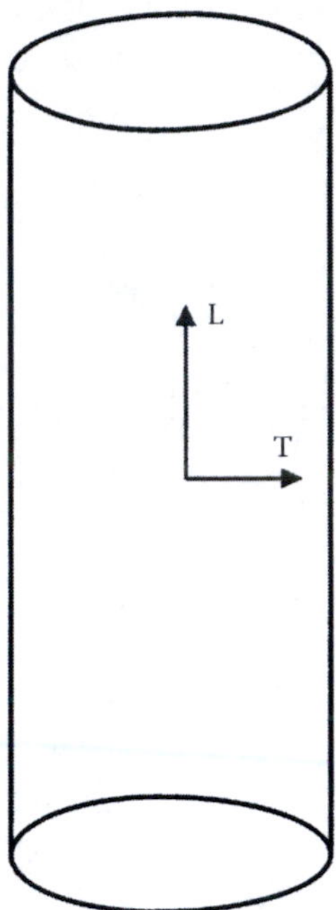

FIGURE 3.23
Longitudinal and transverse direction in a transversely isotropic fiber.

$$E_1 = E_{fL}V_f + E_mV_m,$$

$$\frac{1}{E_2} = \frac{V_f}{E_{fT}} + \frac{V_m}{E_m},$$

$$\nu_{12} = \nu_{fT}V_f + \nu_mV_m,$$

and

$$\frac{1}{G_{12}} = \frac{V_f}{G_{fT}} + \frac{V_m}{G_m}. \tag{3.161a–d}$$

The preceding expressions are similar to those of a lamina with isotropic fibers. The only difference is that appropriate transverse or longitudinal properties of the fiber are used. In composites such as carbon–carbon, the matrix is also transversely isotropic. In that case, the preceding equations cannot be used and are given elsewhere.[15,19]

3.4 Ultimate Strengths of a Unidirectional Lamina

As shown in Chapter 2, one needs to know five ultimate strength parameters for a unidirectional lamina:

- Longitudinal tensile strength $(\sigma_1^T)_{ult}$
- Longitudinal compressive strength $(\sigma_1^C)_{ult}$
- Transverse tensile strength $(\sigma_2^T)_{ult}$
- Transverse compressive strength $(\sigma_2^C)_{ult}$
- In-plane shear strength $(\tau_{12})_{ult}$

In this section, we will see whether and how these parameters can be found from the individual properties of the fiber and matrix by using the mechanics of materials approach. The strength parameters for a unidirectional lamina are much harder to predict than the stiffnesses because the strengths are more sensitive to the material and geometric nonhomogeneities, fiber–matrix interface, fabrication process, and environment. For example, a weak interface between the fiber and matrix may result in premature failure of the composite under a transverse tensile load, but may increase its longitudinal tensile strength. For these reasons of sensitivity, some theoretical and empirical models are available for some of the strength parameters. Eventually, the experimental evaluation of these strengths becomes important because it is direct and reliable. These experimental techniques are also discussed in this section.

3.4.1 Longitudinal Tensile Strength

A simple mechanics of materials approach model is presented (Figure 3.24). Assume that

- Fiber and matrix are isotropic, homogeneous, and linearly elastic until failure.
- The failure strain for the matrix is higher than for the fiber, which is the case for polymeric matrix composites. For example, glass fibers fail at strains of 3 to 5%, but an epoxy fails at strains of 9 to 10%.

Now, if

$(\sigma_f)_{ult}$ = ultimate tensile strength of fiber

E_f = Young's modulus of fiber

$(\sigma_m)_{ult}$ = ultimate tensile strength of matrix

E_m = Young's modulus of matrix

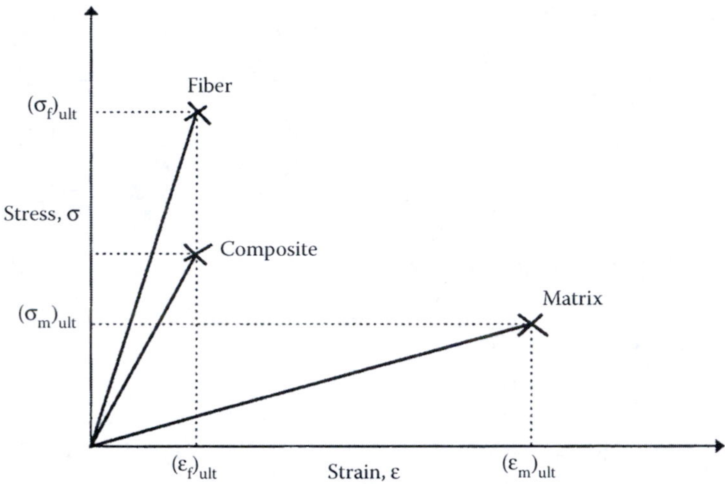

FIGURE 3.24
Stress–strain curve for a unidirectional composite under uniaxial tensile load along fibers.

then the ultimate failure strain of the fiber is

$$(\varepsilon_f)_{ult} = \frac{(\sigma_f)_{ult}}{E_f}, \tag{3.162}$$

and the ultimate failure strain of the matrix is

$$(\varepsilon_m)_{ult} = \frac{(\sigma_m)_{ult}}{E_m}. \tag{3.163}$$

Because the fibers carry most of the load in polymeric matrix composites, it is assumed that, when the fibers fail at the strain of $(\varepsilon_f)_{ult}$, the whole composite fails. Thus, the composite tensile strength is given by

$$(\sigma_1^T)_{ult} = (\sigma_f)_{ult} V_f + (\varepsilon_f)_{ult} E_m (1 - V_f). \tag{3.164}$$

Once the fibers have broken, can the composite take more load? The stress that the matrix can take alone is given by (σ_{mult}) $(1 - V_f)$. Only if this stress is greater than $(\sigma_1^T)_{ult}$ (Equation 3.164), is it possible for the composite to take more load. The volume fraction of fibers for which this is possible is called the minimum fiber volume fraction, $(V_f)_{minimum}$, and is

$$(\sigma_m)_{ult}\left[1-(V_f)_{minimum}\right] > (\sigma_f)_{ult}(V_f)_{minimum} + (\varepsilon_f)_{ult}E_m\left[1-(V_f)_{minimum}\right]$$

$$(V_f)_{minimum} < \frac{(\sigma_m)_{ult} - E_m(\varepsilon_f)_{ult}}{(\sigma_f)_{ult} - E_m(\varepsilon_f)_{ult} + (\sigma_m)_{ult}}. \tag{3.165}$$

It is also possible that, by adding fibers to the matrix, the composite will have lower ultimate tensile strength than the matrix. In that case, the fiber volume fraction for which this is possible is called the critical fiber volume fraction, $(V_f)_{critical}$, and is

$$(\sigma_m)_{ult} > (\sigma_f)_{ult}(V_f)_{critical} + (\varepsilon_f)_{ult}E_m\left[1-(V_f)_{critical}\right]$$

$$(V_f)_{critical} < \frac{(\sigma_m)_{ult} - E_m(\varepsilon_f)_{ult}}{(\sigma_f)_{ult} - E_m(\varepsilon_f)_{ult}}. \tag{3.166}$$

Example 3.13

Find the ultimate tensile strength for a glass/epoxy lamina with a 70% fiber volume fraction. Use the properties for glass and epoxy from Table 3.1 and Table 3.2, respectively. Also, find the minimum and critical fiber volume fractions.

Solution

From Table 3.1,

$$E_f = 85\ GPa$$

$$(\sigma_f)_{ult} = 1550\ MPa.$$

Thus,

$$(\varepsilon_f)_{ult} = \frac{1550 \times 10^6}{85 \times 10^9}$$

$$= 0.1823 \times 10^{-1}.$$

From Table 3.2,

$$E_m = 3.4\ GPa$$

$$(\sigma_m)_{ult} = 72\ MPa.$$

Thus,

$$(\varepsilon_m)_{ult} = \frac{72 \times 10^6}{3.4 \times 10^9}$$

$$= 0.2117 \times 10^{-1}.$$

Applying Equation (3.164), the ultimate longitudinal tensile strength is

$$(\sigma_1^T)_{ult} = (1550 \times 10^6)(0.7) + (0.1823 \times 10^{-1})(3.4 \times 10^9)(1 - 0.7)$$

$$= 1104\ MPa.$$

Applying Equation (3.165), the minimum fiber volume fraction is

$$(V_f)_{miminum} = \frac{72 \times 10^6 - (3.4 \times 10^9)(0.1823 \times 10^{-1})}{1550 \times 10^6 - (3.4 \times 10^9)(0.1823 \times 10^{-1}) + 72 \times 10^6}$$

$$= 0.6422 \times 10^{-2}$$

$$= 0.6422\%.$$

This implies that, if the fiber volume fraction is less than 0.6422%, the matrix can take more loading after all the fibers break. Applying Equation (3.166), the critical fiber volume fraction is

$$(V_f)_{critical} = \frac{72 \times 10^6 - (3.4 \times 10^9)(0.1823 \times 10^{-1})}{1550 \times 10^6 - (3.4 \times 10^9)(0.1823 \times 10^{-1})}$$

$$= 0.6732 \times 10^{-2}$$

$$= 0.6732\%.$$

This implies that, if the fiber volume fraction were less than 0.6732%, the composite longitudinal tensile strength would be less than that of the matrix.

Experimental evaluation: The general test method recommended for tensile strength is the ASTM test method for tensile properties of fiber–resin composites (D3039) (Figure 3.25). A tensile test geometry (Figure 3.26) to find

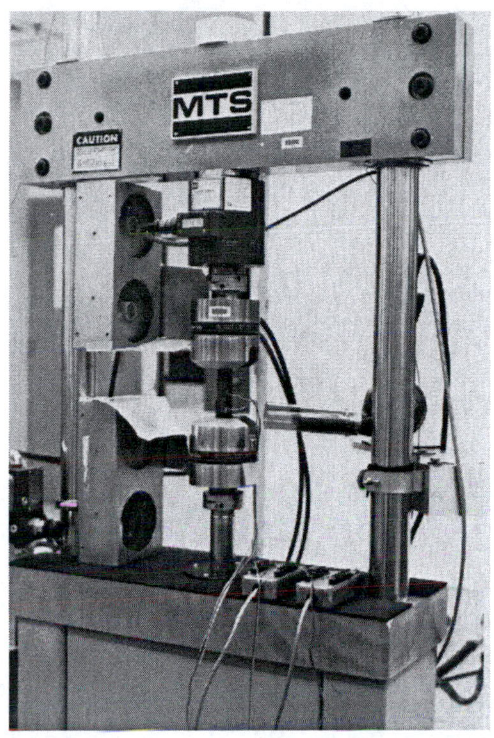

FIGURE 3.25
Tensile coupon mounted in the test frame for finding the tensile strengths of a unidirectional lamina. (Photo courtesy of Dr. R.Y. Kim, University of Dayton Research Institute, Dayton, OH.)

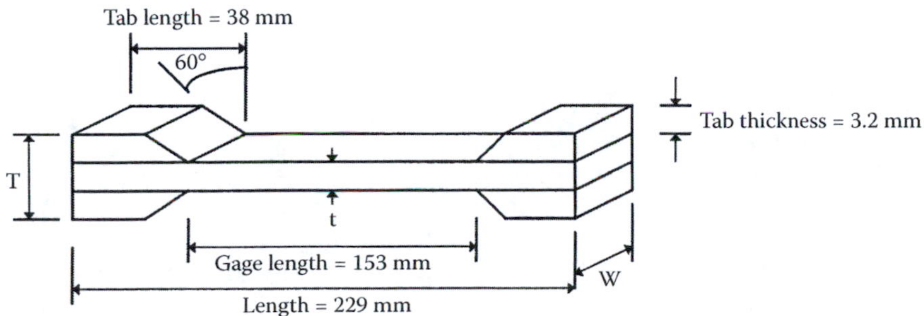

FIGURE 3.26
Geometry of a longitudinal tensile strength specimen.

the longitudinal tensile strength consists of six to eight 0° plies that are 12.5 mm (1/2 in.) wide and 229 mm (10 in.) long. The specimen is mounted with strain gages in the longitudinal and transverse directions. Tensile stresses are applied on the specimen at a rate of about 0.5 to 1 mm/min (0.02 to 0.04 in./min). A total of 40 to 50 data points for stress and strain is taken until a specimen fails. The stress in the longitudinal direction is plotted as a function

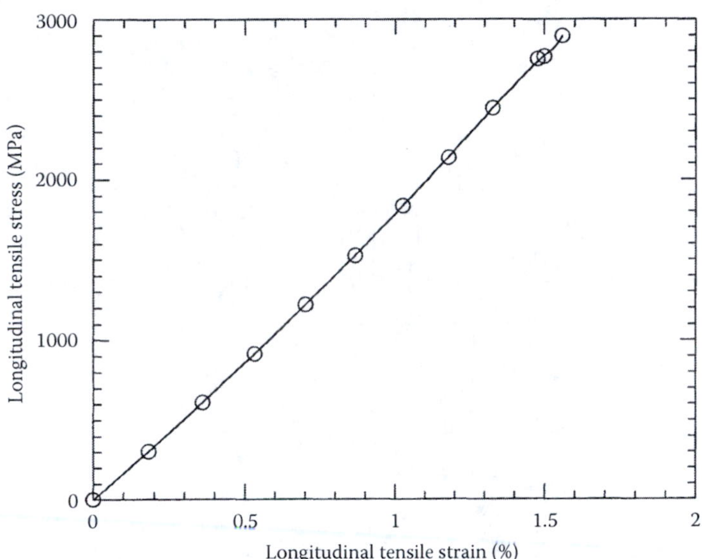

FIGURE 3.27
Stress–strain curve for a $[0]_8$ laminate under a longitudinal tensile load. (Data courtesy of Dr. R.Y. Kim, University of Dayton Research Institute, Dayton, OH).

of longitudinal strain, as shown in Figure 3.27. The data are reduced using linear regression. The longitudinal Young's modulus is the initial slope of the σ_1 vs. ε_1 curve.

From Figure 3.27, the following values are obtained:

$$E_1 = 187.5\ GPa,$$

$$(\sigma_1^T)_{ult} = 2896\ MPa,$$

$$(\varepsilon_1^T)_{ult} = 1.560\%.$$

Discussion: Failure of a unidirectional ply under a longitudinal tensile load takes place with

1. Brittle fracture of fibers
2. Brittle fracture of fibers with pullout
3. Fiber pullout with fiber–matrix debonding

The three failure modes are shown in Figure 3.28. The mode of failure depends on the fiber–matrix bond strength and fiber volume fraction.[20] For low fiber volume fractions, $0 < V_f < 0.40$, a typical glass/epoxy composite

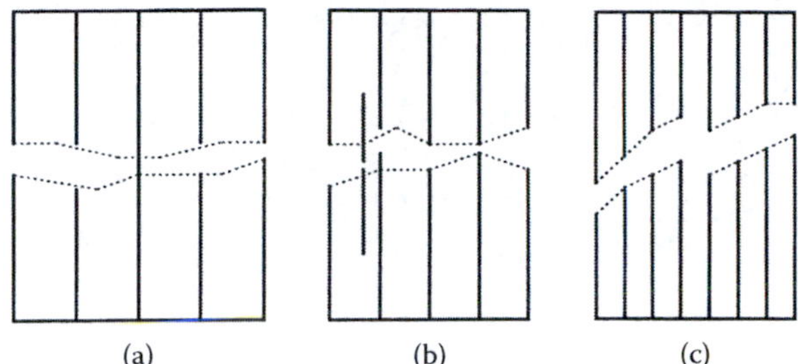

FIGURE 3.28
Modes of failure of unidirectional lamina under a longitudinal tensile load.

exhibits a mode (1) type failure. For intermediate fiber volume fractions, $0.4 < V_f < 0.65$, mode (2) type failure occurs. For high fiber volume fractions, $V_f > 0.65$, it exhibits mode (3) type of failure.

3.4.2 Longitudinal Compressive Strength

The model used for calculating the longitudinal tensile strength for a unidirectional lamina cannot also be used for its longitudinal compressive strength because the failure modes are different. Three typical failure modes are shown in Figure 3.29:

- Fracture of matrix and/or fiber–matrix bond due to tensile strains in the matrix and/or bond
- Microbuckling of fibers in shear or extensional mode
- Shear failure of fibers

Ultimate tensile strains in matrix failure mode: A mechanics of materials approach model based on the failure of the composite in the transverse direction due to transverse tensile strains is given next.[20] Assuming that one is applying a longitudinal compressive stress of magnitude σ_1, then the magnitude of longitudinal compressive strain is given by

$$|\varepsilon_1| = \frac{|\sigma_1|}{E_1}. \tag{3.167}$$

Because the major Poisson's ratio is ν_{12}, the transverse strain is tensile and is given by

$$|\varepsilon_2| = \nu_{12}\frac{|\sigma_1|}{E_1}. \tag{3.168}$$

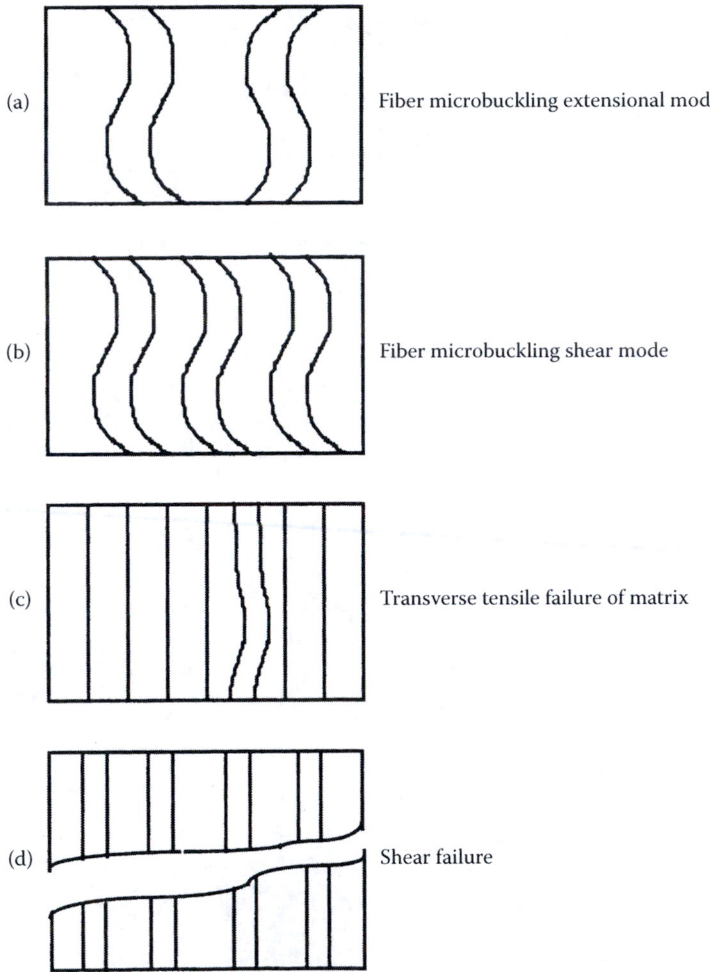

FIGURE 3.29
Modes of failure of a unidirectional lamina under a longitudinal compressive load.

Using maximum strain failure theory, if the transverse strain exceeds the ultimate transverse tensile strain, $(\varepsilon_2^T)_{ult}$, the lamina is considered to have failed in the transverse direction. Thus,

$$(\sigma_1^c)_{ult} = \frac{E_1(\varepsilon_2^T)_{ult}}{\nu_{12}}. \tag{3.169}$$

The value of the longitudinal modulus, E_1, and the major Poisson's ratio, υ_{12}, can be found from Equation (3.34) and Equation (3.49), respectively. However, for the value of $(\varepsilon_2^T)_{ult}$, one can use the empirical formula,

$$(\varepsilon_2^T)_{ult} = (\varepsilon_m^T)_{ult}(1 - V_f^{1/3}), \tag{3.170}$$

or the mechanics of materials formula,

$$(\varepsilon_2^T)_{ult} = (\varepsilon_m^T)_{ult}\left[\frac{d}{s}\left(\frac{E_m}{E_f} - 1\right) + 1\right], \tag{3.171}$$

where
$(\varepsilon_m^T)_{ult}$ = ultimate tensile strain of the matrix
d = diameter of the fibers
s = center-to-center spacing between the fibers

Equation (3.170) and Equation (3.171) will be discussed later in Section 3.4.3.

Shear/extensional fiber microbuckling failure mode: local buckling models for calculating longitudinal compressive strengths have been developed.[21,22] Because these results are based on advanced topics, only the final expressions are given:

$$(\sigma_1^c)_{ult} = \min[S_1^c, S_2^c], \tag{3.172}$$

where

$$S_1^c = 2\left[V_f + (1 - V_f)\frac{E_m}{E_f}\right]\sqrt{\frac{V_f E_m E_f}{3(1 - V_f)}}, \tag{3.173a}$$

and

$$S_2^c = \frac{G_m}{1 - V_f}. \tag{3.173b}$$

Note that the extensional mode buckling stress (S_1^c) is higher than the shear mode buckling stress (S_2^c) for most cases. Extensional mode buckling is prevalent only in low fiber volume fraction composites.

Shear stress failure of fibers mode: A unidirectional composite may fail due to direct shear failure of fibers. In this case, the rule of mixtures gives the shear strength of the unidirectional composite as

$$(\tau_{12})_{ult} = (\tau_f)_{ult} V_f + (\tau_m)_{ult} V_m, \tag{3.174}$$

where

$(\tau_f)_{ult}$ = ultimate shear strength of the fiber
$(\tau_m)_{ult}$ = ultimate shear strength of the matrix

The maximum shear stress in a lamina under a longitudinal compressive load σ_1^c is $(\sigma_1^c)/2$ at 45° to the loading axis. Thus,

$$(\sigma_1^c)_{ult} = 2\left[(\tau_f)_{ult}V_f + (\tau_m)_{ult}V_m\right]. \tag{3.175}$$

Three models based on each of the preceding failure modes were discussed to find the magnitude of the ultimate longitudinal compressive strength. One may caution that these models are not found to match the experimental results as is partially evident in the comparison of experimental and predicted values[23] of longitudinal compressive strength given in Table 3.6. Comparison with other equations (3.169) and (3.175) is not available because the properties of constituents are not given in the reference,[23] although fiber buckling is the most probable mode of failure in advanced polymer matrix composites.

Several factors may contribute to this discrepancy, including

- Irregular spacing of fibers causing premature failure in matrix-rich areas
- Less than perfect bonding between the fibers and the matrix
- Poor alignment of fibers
- Not accounting for Poisson's ratio mismatch between the fiber and the matrix
- Not accounting for the transversely isotropic nature of fibers such as aramids and graphite

In addition, there is controversy concerning the techniques used in measuring compressive strengths.

TABLE 3.6

Comparison of Experimental and Predicted Values of Longitudinal Compressive Strength of Unidirectional Laminae[a]

Material	Experimental strength	Equation (3.78a) (MPa)	Equation (3.78b) (MPa)
Glass/polyester	600–1000	8700	2200
Type I carbon/epoxy	700–900	22,800	2900
Kevlar 49/epoxy	240–290	13,200	2900

[a] $V_f = 0.50$.

Source: Hull, D., *Introduction to Composite Materials*, Cambridge University Press, 1981, Table 7.2. Reprinted with the permission of Cambridge University Press.

Example 3.14

Find the longitudinal compressive strength of a glass/epoxy lamina with a 70% fiber volume fraction. Use the properties of glass and epoxy from Table 3.1 and Table 3.2, respectively. Assume that fibers are circular and are in a square array.

Solution

From Table 3.1, the Young's modulus for the fiber is

$$E_f = 85\ GPa$$

and the Poisson's ratio of the fiber is

$$\nu_f = 0.20.$$

The ultimate tensile strength of the fiber is

$$(\sigma_f)_{ult} = 1550\ MPa$$

and the ultimate shear strength of the fiber is

$$(\tau_f)_{ult} = 35\ MPa.$$

From Table 3.2, the Young's modulus of the matrix is

$$E_m = 3.4\ GPa$$

and the Poisson's ratio of the matrix is

$$\nu_m = 0.30.$$

The ultimate normal strength of the matrix is

$$(\sigma_m)_{ult} = 72\ MPa$$

and the ultimate shear strength of the matrix is

$$(\tau_m)_{ult} = 34\ MPa.$$

From Example 3.3, the longitudinal Young's modulus of the unidirectional lamina is

$$E_1 = 60.52\ GPa.$$

From Example 3.5, the major Poisson's ratio of the unidirectional lamina is

$$\nu_{12} = 0.23.$$

Using Equation (3.42a), the fiber diameter to fiber spacing ratio is

$$\frac{d}{s} = \left[\frac{4(0.7)}{\pi}\right]^{1/2}$$

$$= 0.9441.$$

The ultimate tensile strain of the matrix is

$$(\varepsilon_m)_{ult} = \frac{72 \times 10^6}{3.40 \times 10^9}$$

$$= 0.2117 \times 10^{-1}.$$

Using the transverse ultimate tensile strain failure mode formula (3.171),

$$(\varepsilon_2^T)_{ult} = 0.2117 \times 10^{-1}\left[0.9441\left(\frac{3.4 \times 10^9}{85 \times 10^9} - 1\right) + 1\right]$$

$$= 0.1983 \times 10^{-2}.$$

From the empirical Equation (3.170),

$$(\varepsilon_2^T)_{ult} = (0.2117 \times 10^{-1})(1 - 0.7^{1/3})$$

$$= 0.2373 \times 10^{-2}.$$

Using the lesser of these two values of ultimate transverse tensile strain, $(\varepsilon_2^T)_{ult}$, and Equation (3.169),

$$(\sigma_1^C)_{ult} = \frac{(60.52 \times 10^9)(0.1983 \times 10^{-2})}{0.23}$$

$$= 521.8 \; MPa.$$

Using shear/extensional fiber microbuckling failure mode formulas (3.173a),

$$S_1^C = 2\left[0.7 + (1 - 0.7)\frac{3.4 \times 10^9}{85 \times 10^9}\right]\sqrt{\frac{(0.7)(3.4 \times 10^9)(85 \times 10^9)}{3(1 - 0.7)}}$$

$$= 21349 \; MPa.$$

From Example 3.6, the shear modulus of the matrix is

$$G_m = 1.308\ GPa.$$

Using Equation (3.173b),

$$S_2^C = \frac{1.308 \times 10^9}{1 - 0.7}$$

$$= 4360\ MPa.$$

Thus, from Equation (3.172), the ultimate longitudinal compressive strength is

$$(\sigma_1^C)_{ult} = \min(21349, 4360) = 4360\ MPa.$$

Using shear stress failure of fibers mode, the ultimate longitudinal compressive strength from Equation (3.175) is

$$(\sigma_1^C)_{ult} = 2\left[(35 \times 10^6)(0.7) + (34 \times 10^6)(0.3)\right]$$

$$= 69.4\ MPa.$$

Taking the minimum value of the preceding, the ultimate longitudinal compressive strength is predicted as

$$(\sigma_1^C)_{ult} = 69.4\ MPa$$

Experimental evaluation: The compressive strength of a lamina has been found by several different methods. A highly recommended method is the IITRI (Illinois Institute of Technology Research Institute), compression test.[24] Figure 3.30 shows the (ASTM D3410 Celanese) IITRI fixture mounted in a test frame. A specimen (Figure 3.31) consists generally of 16 to 20 plies of 0° lamina that are 6.4 mm (1/4 in.) wide and 127 mm (5 in.) long. Strain gages are mounted in the longitudinal direction on both faces of the specimen to check for parallelism of the edges and ends. The specimen is compressed at a rate of 0.5 to 1 mm/min (0.02 to 0.04 in./min). A total of 40 to 50 data points for stress and strain are taken until the specimen fails. The stress in the longitudinal direction is plotted as a function of longitudinal strain and is shown for a typical graphite/epoxy lamina in Figure 3.32. The data are reduced using linear regression and the modulus is the initial slope of the stress–strain curve. From Figure 3.32, the following values are obtained:

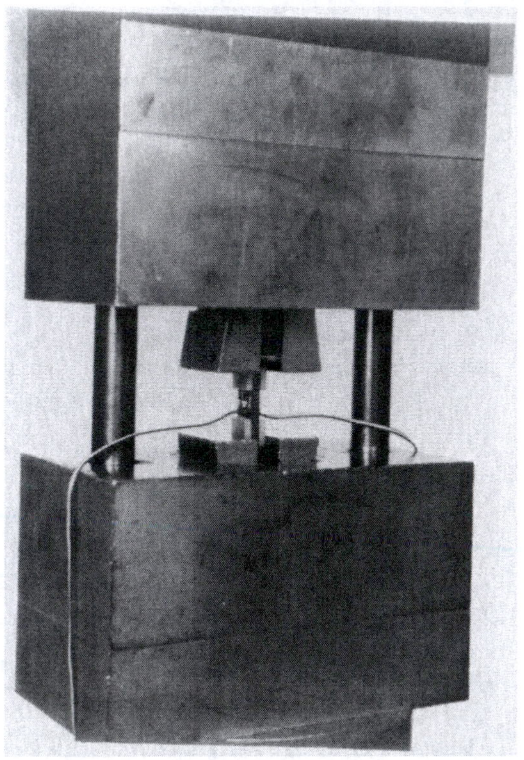

FIGURE 3.30
IITRI fixture mounted in a test frame for finding the compressive strengths of a lamina. (Data reprinted with permission from *Experimental Characterization of Advanced Composites*, Carlsson, L.A. and Pipes, R.B., Technomic Publishing Co., Inc., 1987, p. 76. Copyright CRC Press, Boca Raton, FL.)

$$E_1^c = 199 \; GPa,$$

$$(\sigma_1^c)_{ult} = 1908 \; MPa, \text{ and}$$

$$(\varepsilon_1^c)_{ult} = 0.9550\%.$$

3.4.3 Transverse Tensile Strength

A mechanics of materials approach model for finding the transverse tensile strength of a unidirectional lamina is given next.[25] Assumptions used in the model include

- A perfect fiber–matrix bond
- Uniform spacing of fibers

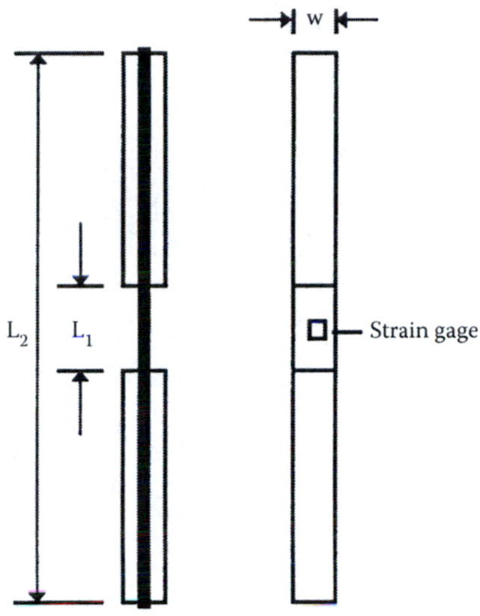

Specimen dimensions		
L_1, mm	L_2, mm	w^*, mm
12.7±1	12.7±1.5	12.7±0.1 or 6.4 ± 0.1

FIGURE 3.31
Geometry of a longitudinal compressive strength specimen. (Data reprinted with permission from *Experimental Characterization of Advanced Composites*, Carlsson, L.A. and Pipes, R.B., Technomic Publishing Co., Inc., 1987, p. 76. Copyright CRC Press, Boca Raton, FL.)

- The fiber and matrix follow Hooke's law
- There are no residual stresses

Assume a plane model of a composite as shown by the shaded portion in Figure 3.33. In this case,

s = distance between center of fibers
d = diameter of fibers

The transverse deformations of the fiber, δ_f, the matrix, δ_m, and the composite, δ_c, are related by

$$\delta_c = \delta_f + \delta_m. \tag{3.176}$$

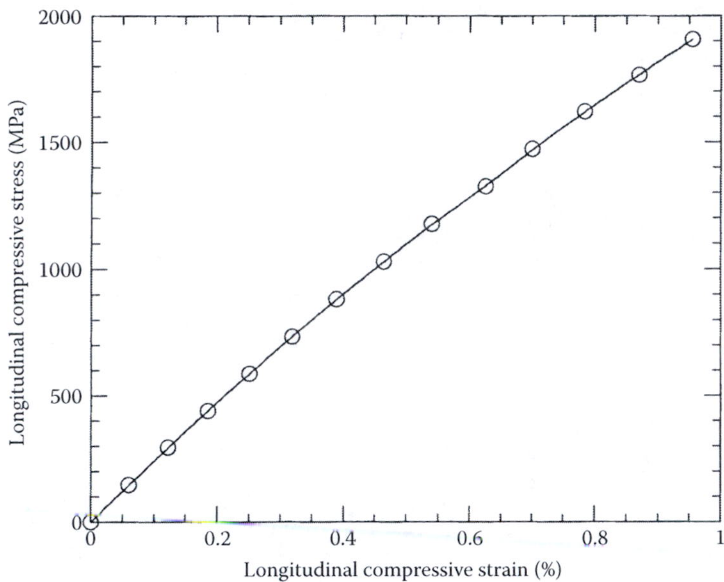

FIGURE 3.32
Stress–strain curve for a $[0]_{24}$ graphite/epoxy laminate under a longitudinal compressive load. (Data courtesy of Dr. R.Y. Kim, University of Dayton Research Institute, Dayton, OH.)

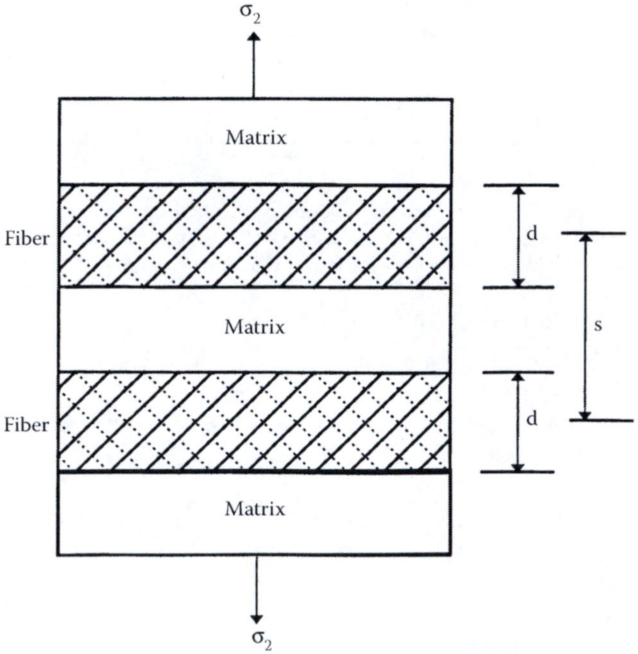

FIGURE 3.33
Representative volume element to calculate transverse tensile strength of a unidirectional lamina.

Now, by the definition of strain, the deformations are related to the transverse strains,

$$\delta_c = s\varepsilon_c \,, \tag{3.177a}$$

$$\delta_f = d\varepsilon_f \,, \tag{3.177b}$$

$$\delta_m = (s-d)\varepsilon_m \,, \tag{3.177c}$$

where $\varepsilon_{c,f,m}$ = the transverse strain in the composite, fiber, and matrix, respectively.

Substituting the expressions in Equation (3.82) in Equation (3.81), we get

$$\varepsilon_c = \frac{d}{s}\varepsilon_f + \left(1-\frac{d}{s}\right)\varepsilon_m. \tag{3.178}$$

Now, under transverse loading, one assumes that the stresses in the fiber and matrix are equal (see derivation of transverse Young's modulus in Section 3.2.1.2). Then, the strains in the fiber and matrix are related through Hooke's law as

$$E_f\varepsilon_f = E_m\varepsilon_m. \tag{3.179}$$

Substituting the expression for the transverse strain in the fiber, ε_f, in Equation (3.178), the transverse strain in the composite

$$\varepsilon_c = \left[\frac{d}{s}\frac{E_m}{E_f} + \left(1-\frac{d}{s}\right)\right]\varepsilon_m. \tag{3.180}$$

If one assumes that the transverse failure of the lamina is due to the failure of the matrix, then the ultimate transverse failure strain is

$$(\varepsilon_2^T)_{ult} = \left[\frac{d}{s}\frac{E_m}{E_f} + \left(1-\frac{d}{s}\right)\right](\varepsilon_m^T)_{ult}, \tag{3.181}$$

where $(\varepsilon_m^T)_{ult}$ = ultimate tensile failure strain of the matrix.

The ultimate transverse tensile strength is then given by

$$(\sigma_2^T)_{ult} = E_2(\varepsilon_2^T)_{ult}, \tag{3.182}$$

where $(\varepsilon_2^T)_{ult}$ is given by Equation (3.181). The preceding expression assumes that the fiber is perfectly bonded to the matrix. If the adhesion between the fiber and matrix is poor, the composite transverse strength will be further reduced.

Example 3.15

Find the ultimate transverse tensile strength for a unidirectional glass/epoxy lamina with a 70% fiber volume fraction. Use properties of glass and epoxy from Table 3.1 and Table 3.2, respectively. Assume that the fibers are circular and arranged in a square array.

Solution

From Example 3.14, the ultimate transverse tensile strain of the lamina

$$(\varepsilon_2^T)_{ult} = 0.1983 \times 10^{-2}$$

is the lower estimate from using Equation (3.170) and Equation (3.171).

From Example 3.4, the transverse Young's modulus of the lamina is E_2 = 10.37 GPa.

Using Equation (3.182), the ultimate transverse tensile strength of the lamina is

$$(\sigma_2^T)_{ult} = (10.37 \times 10^9)(0.1983 \times 10^{-2})$$
$$= 20.56\ MPa.$$

Experimental evaluation: The procedure for finding the transverse tensile strength is the same as for finding the longitudinal tensile strength. Only the specimen dimensions differ. The standard width of the specimen is 25.4 mm (1 in.) and 8 to 16 plies are used. This is mainly done to increase the amount of load required to break the specimen. Figure 3.34 shows the typical stress–strain curve for a 90° graphite/peek laminate. From Figure 3.34, the following data are obtained:

$$E_2 = 9.963 \text{ GPa}$$

$$(\sigma_2^T)_{ult} = 53.28\ MPa$$

$$(\varepsilon_2^T)_{ult} = 0.5355\%$$

Discussion: Predicting transverse tensile strength is quite complicated. Under a transverse tensile load, factors other than the individual properties of the fiber and matrix are important. These include the bond strength between the fiber and the matrix, the presence of voids, and the presence

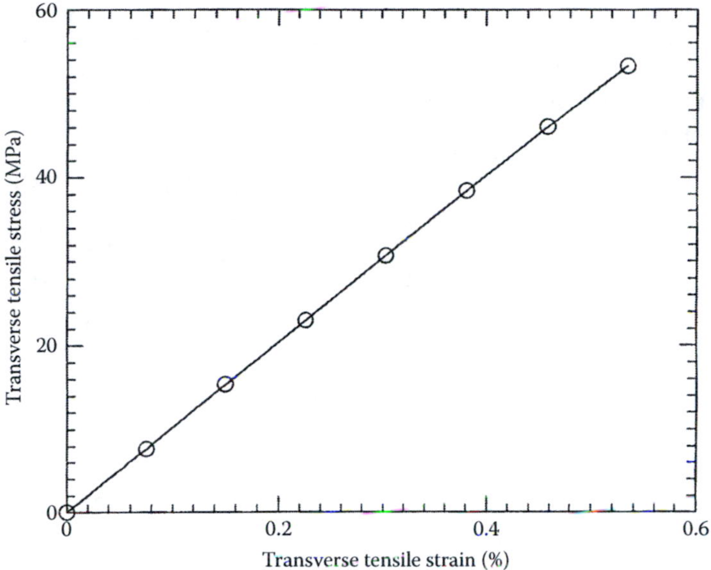

FIGURE 3.34
Stress–strain curve for a $[90]_{16}$ graphite/epoxy laminate under a transverse tensile load. (Data courtesy of Dr. R.Y. Kim, University of Dayton Research Institute, Dayton, OH.)

of residual stresses due to thermal expansion mismatch between the fiber and matrix. Possible modes of failure under transverse tensile stress include matrix tensile failure accompanied by fiber matrix debonding and/or fiber splitting.

3.4.4 Transverse Compressive Strength

Equation (3.182), which was developed for evaluating transverse tensile strength, can be used to find the transverse compressive strengths of a lamina. The actual compressive strength is again lower due to imperfect fiber/matrix interfacial bond and longitudinal fiber splitting. Using compressive parameters in Equation (3.182),

$$(\sigma_2^C)_{ult} = E_2(\varepsilon_2^C)_{ult}, \tag{3.183}$$

where

$$(\varepsilon_2^C)_{ult} = \left[\frac{d}{s}\frac{E_m}{E_f} + \left(1 - \frac{d}{s}\right)\right](\varepsilon_m^C)_{ult}, \tag{3.184}$$

$(\varepsilon_m^C)_{ult}$ = ultimate compressive failure strain of matrix.

Example 3.16

Find the ultimate transverse compressive strength of a glass/epoxy lamina with 70% fiber volume fraction. Use the properties of glass and epoxy from Table 3.1 and Table 3.2, respectively. Assume that the fibers are circular and packed in a square array.

Solution

From Table 3.1, the Young's modulus of the fiber is $E_f = 85$ GPa.

From Table 3.2, the Young's modulus of the matrix is $E_m = 3.4$ GPa and the ultimate compressive strength of the matrix is

$$(\sigma_m^C)_{ult} = 102 \ MPa.$$

From Example 3.4, the transverse Young's modulus is $E_2 = 10.37$ GPa.

From Example 3.14, the fiber diameter to fiber spacing ratio is

$$\frac{d}{s} = 0.9441.$$

The ultimate compressive strain of the matrix is

$$(\varepsilon_m^C) = \frac{102 \times 10^6}{3.4 \times 10^9}$$
$$= 0.03.$$

From Equation (3.184), the ultimate transverse compressive strain of the lamina is

$$(\varepsilon_2^C) = \left[0.9441 \frac{3.4 \times 10^9}{85 \times 10^9} + (1 - 0.9441)\right](0.03)$$
$$= 0.2810 \times 10^{-2},$$

and from Equation (3.183), the ultimate transverse compressive strength is

$$(\sigma_2^C)_{ult} = (10.37 \times 10^9)(0.2810 \times 10^{-2}) = 29.14 \ MPa.$$

Experimental evaluation: The procedure for finding the transverse compressive strength is the same as that for finding the longitudinal compressive strength. The only difference is in the specimen dimensions. The width of the specimen is 12.7 mm (1/2 in.) and 30 to 40 plies are used in the test.

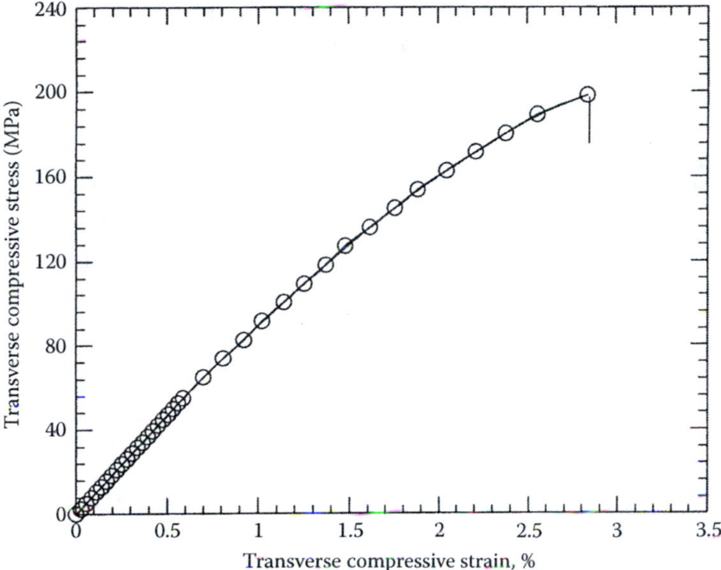

FIGURE 3.35
Stress–strain curve for a $[90]_{40}$ graphite/epoxy laminate under a transverse compressive load perpendicular to the fibers. (Data reprinted with permission from *Experimental Characterization of Advanced Composites*, Carlsson, L.A. and Pipes, R.B., Technomic Publishing Co., Inc., 1987, p. 79.)

Figure 3.35 shows the typical stress–strain curve for a 90° graphite/epoxy laminate. From Figure 3.35, the following data are obtained[26]:

$$E_2^c = 93\ GPa,$$

$$(\sigma_2^c)_{ult} = 198\ MPa,$$

$$(\varepsilon_2^c)_{ult} = 2.7\%.$$

Discussion: Methods for predicting transverse compressive strength are also not yet satisfactory. Several modes of failure possible under a transverse compressive stress include matrix compressive failure, matrix shear failure, and matrix shear failure with fiber–matrix debonding and/or fiber crushing.

3.4.5 In-Plane Shear Strength

The procedure for finding the ultimate shear strength for a unidirectional lamina using a mechanics of materials approach follows that described in Section 3.4.3. Assume that one is applying a shear stress of magnitude τ_{12}

and then that the shearing deformation in the representative element is given by the sum of the deformations in the fiber and matrix,

$$\Delta_c = \Delta_f + \Delta_m. \tag{3.185}$$

By definition of shearing strain,

$$\Delta_c = s(\gamma_{12})_c, \tag{3.186a}$$

$$\Delta_f = d(\gamma_{12})_f, \tag{3.186b}$$

and

$$\Delta_m = (s-d)(\gamma_{12})_m, \tag{3.186c}$$

where $(\gamma_{12})_{c,f,m}$ = the in-plane shearing strains in the composite, fiber, and matrix, respectively.

Substituting the Equation (3.186a-c) in Equation (3.185),

$$(\gamma_{12})_c = \frac{d}{s}(\gamma_{12})_f + \left(1 - \frac{d}{s}\right)(\gamma_{12})_m. \tag{3.187}$$

Now, under shearing stress loading, one assumes that the shear stress in the fiber and matrix are equal (see derivation of shear modulus in Section 3.3.1.4). Then, the shearing strains in the fiber and matrix are related as

$$(\gamma_{12})_m G_m = (\gamma_{12})_f G_f. \tag{3.188}$$

Substituting the expression for $(\gamma_{12})_f$ from Equation (3.188) in Equation (3.187),

$$(\gamma_{12})_c = \left[\frac{d}{s}\frac{G_m}{G_f} + \left(1 - \frac{d}{s}\right)\right](\gamma_{12})_m. \tag{3.189}$$

If one assumes that the shear failure is due to failure of the matrix, then

$$(\gamma_{12})_{ult} = \left[\frac{d}{s}\frac{G_m}{G_f} + \left(1 - \frac{d}{s}\right)\right](\gamma_{12})_{mult}, \tag{3.190}$$

where $(\gamma_{12})_{m\,ult}$ = ultimate shearing strain of the matrix.

The ultimate shear strength is then given by

$$(\tau_{12})_{ult} = G_{12}(\gamma_{12})_{ult}$$

$$= G_{12}\left[\frac{d}{s}\frac{G_m}{G_f} + \left(1 - \frac{d}{s}\right)\right](\gamma_{12})_{mult}. \tag{3.191}$$

Example 3.17

Find the ultimate shear strength for a glass/epoxy lamina with 70% fiber volume fraction. Use properties for glass and epoxy from Table 3.1 and Table 3.2, respectively. Assume that the fibers are circular and arranged in a square array.

Solution

From Example 3.6, the shear modulus of the fiber is

$$G_f = 35.42\ GPa,$$

the shear modulus of the matrix is

$$G_m = 1.308\ GPa,$$

and the in-plane shear modulus of the lamina is

$$G_{12} = 4.014\ GPa.$$

From Example 3.14, the fiber diameter to fiber spacing ratio is

$$\frac{d}{s} = 0.9441.$$

From Table 3.2, the ultimate shear strength of the matrix is

$$(\tau_{12})_{mult} = 34\ MPa.$$

Then, the ultimate shearing strain of the matrix is

$$(\gamma_{12})_{mult} = \frac{34 \times 10^6}{1.308 \times 10^9}$$

$$= 0.2599 \times 10^{-1}.$$

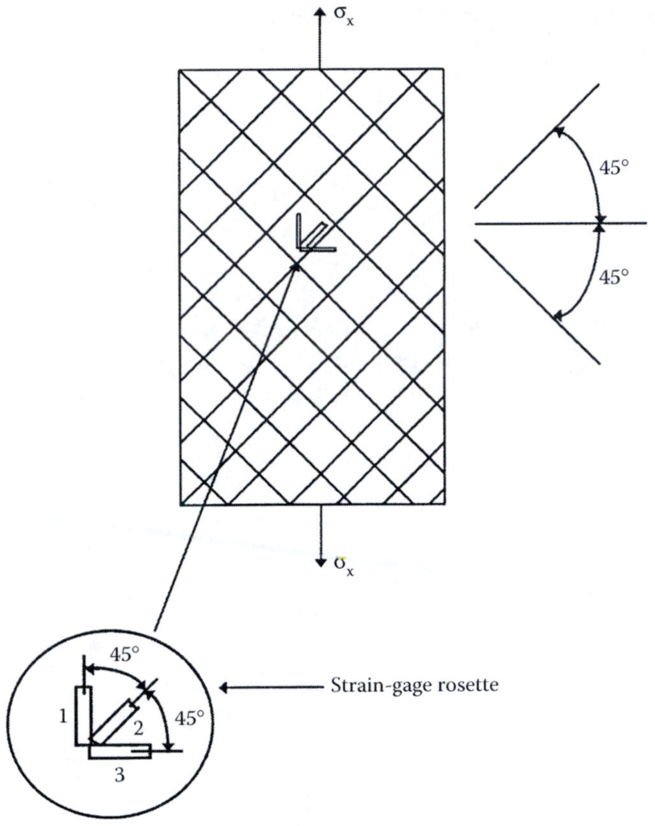

FIGURE 3.36
Schematic of a $[\pm 45]_{2S}$ laminate shear test.

Using Equation (3.191), the ultimate in-plane shear strength of the unidirectional lamina is

$$(\tau_{12})_{ult} = (4.014 \times 10^9)\left[0.9441\frac{1.308 \times 10^9}{35.42 \times 10^9} + (1 - 0.9441)\right](0.2599 \times 10^{-1})$$

$$= 9.469\ MPa.$$

Experimental determination: One of the most recommended methods[27] for calculating the in-plane shear strength is the $[\pm 45]_{2S}$ laminated tensile coupon* (Figure 3.36). A $[\pm 45]_{2S}$ laminate is an eight-ply laminate with [+45/–45/+45/–45/–45/+45/–45/+45] distribution of plies on top of each other.

* See Section 4.2 of Chapter 4 for an explanation on laminate codes.

An axial stress σ_x is applied to the eight-ply laminate; the axial strain ε_x and transverse strain ε_y are measured. If the laminate fails at a load of $(\sigma_x)_{ult}$, the ultimate shear strength of a unidirectional lamina is given by

$$(\tau_{12})_{ult} = \frac{(\sigma_x)_{ult}}{2}, \tag{3.192}$$

and the ultimate shear strain of a unidirectional lamina is

$$(\gamma_{12})_{ult} = \left(|\varepsilon_x|\right)_{ult} + \left(|\varepsilon_y|\right)_{ult}. \tag{3.193}$$

An eight-ply $[\pm 45]_{2S}$ laminate is used for several reasons. First, according to maximum stress and strain failure theories of Chapter 2, each lamina fails in the shear mode and at the same load. The stress at which it fails is simply twice the shear strength of a unidirectional lamina and is independent of the other mechanical properties of the lamina, as reflected in Equation (3.192). Second, the shear strain is measured simply by strain gages in two perpendicular directions and does not require the values of elastic constants of the lamina.

Equation (3.192) and Equation (3.193) can be derived using concepts from Chapter 4 and Chapter 5. The in-plane shear strength is simply half of the maximum uniaxial stress that can be applied to the laminate. The initial slope of the τ_{12} vs. γ_{12} curve gives the shear modulus, G_{12}. A total of 40 to 50 points are taken for the stress and strains until the specimen fails. From Figure 3.37, the following values are obtained for a typical graphite/epoxy lamina:

$$G_{12} = 5.566\ GPa,$$

$$(\tau_{12}) = 87.57\ MPa,$$

$$(\gamma_{12})_{ult} = 2.619\%.$$

Discussion: The prediction of the ultimate shear strength is complex. Similar parameters, such as weak interfaces, the presence of voids, and Poisson's ratio mismatch, make modeling quite complex.

Theoretical methods for obtaining the strength parameters also include statistical and advanced methods. Statistical methods include accounting for variations in fiber strength, fiber–matrix adhesion, voids, fiber spacing, fiber diameter, alignment of fibers, etc. Advanced methods use elasticity, finite element methods, boundary element methods, finite difference methods, etc.

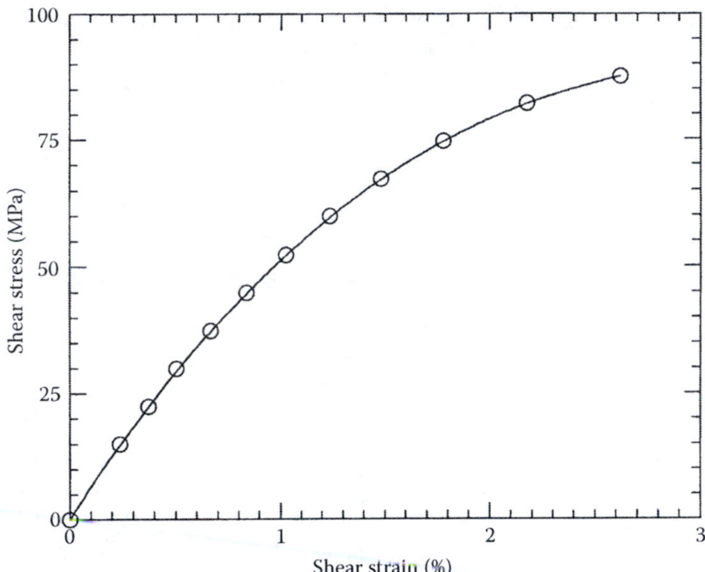

FIGURE 3.37
Shear stress–shear strain curve obtained from a $[\pm 45]_{2S}$ graphite/epoxy laminate under a tensile load. (Data courtesy of Dr. R.Y. Kim, University of Dayton Research Institute, Dayton, OH.)

3.5 Coefficients of Thermal Expansion

When a body undergoes a temperature change, its dimensions relative to its original dimensions change in proportion to the temperature change. The coefficient of thermal expansion is defined as the change in the linear dimension of a body per unit length per unit change of temperature.

For a unidirectional lamina, the dimensions changes differ in the two directions 1 and 2. Thus, the two coefficients of thermal expansion are defined as

α_1 = linear coefficient of thermal expansion in direction 1, m/m/°C (in./in./°F)

α_2 = linear coefficient of thermal expansion in direction 2, m/m/°C (in./in./°F)

The following are the expressions developed for the two thermal expansion coefficients using the thermoelastic extremum principle[28]:

$$\alpha_1 = \frac{1}{E_1}(\alpha_f E_f V_f + \alpha_m E_m V_m), \tag{3.194}$$

$$\alpha_2 = (1+\nu_f)\alpha_f V_f + (1+\nu_m)\alpha_m V_m - \alpha_1\nu_{12}, \tag{3.195}$$

where α_f and α_m are the coefficients of thermal expansion for the fiber and the matrix, respectively.

3.5.1 Longitudinal Thermal Expansion Coefficient

As an example, Equation (3.194) can be derived using the mechanics of materials approach.[29] Consider the expansion of a unidirectional lamina in the longitudinal direction under a temperature change of ΔT. If only the temperature ΔT is applied, the unidirectional lamina has zero overall load, F_1, in the longitudinal direction. Then

$$F_1 = \sigma_1 A_c = 0 = \sigma_f A_f + \sigma_m A_m \tag{3.196}$$

$$\sigma_f V_f + \sigma_m V_m = 0\,, \tag{3.197}$$

where

$A_{c,f,m}$ = the cross-sectional area of composite, fiber, and matrix, respectively

$\sigma_{1,f,m}$ = the stress in composite, fiber, and matrix, respectively

Although the overall load in the longitudinal direction 1 is zero, stresses are caused in the fiber and the matrix by the thermal expansion mismatch between the fiber and the matrix. These stresses are

$$\sigma_f = E_f(\varepsilon_f - \alpha_f \Delta T), \tag{3.198a}$$

and

$$\sigma_m = E_m(\varepsilon_m - \alpha_m \Delta T)\,. \tag{3.198b}$$

Substituting Equation (3.198a) and Equation (3.198b) in Equation (3.197) and realizing that the strains in the fiber and matrix are equal ($\varepsilon_f = \varepsilon_m = \varepsilon_1$),

$$\varepsilon_f = \frac{\alpha_f E_f V_f + \alpha_m E_m V_m}{E_f V_f + E_m V_m}\Delta T. \tag{3.199}$$

For free expansion in the composite in the longitudinal direction 1, the longitudinal strain is

$$\varepsilon_1 = \alpha_1 \Delta T. \tag{3.200}$$

Because the strains in the fiber and composite are also equal ($\varepsilon_1 = \varepsilon_f$), from Equation (3.199) and Equation (3.200),

$$\alpha_1 = \frac{\alpha_f E_f V_f + \alpha_m E_m V_m}{E_f V_f + E_m V_m}.$$

Using Equation (3.34) for the definition of longitudinal Young's modulus,

$$\alpha_1 = \frac{1}{E_1}(\alpha_f E_f V_f + \alpha_m E_m V_m) . \tag{3.201}$$

The longitudinal coefficient of thermal expansion can be rewritten as

$$\alpha_1 = \left(\frac{\alpha_f E_f}{E_1}\right) V_f + \left(\frac{\alpha_m E_m}{E_1}\right) V_m , \tag{3.202}$$

which shows that it also follows the rule of mixtures based on the weighted mean of $\alpha E/E_1$ of the constituents.

3.5.2 Transverse Thermal Expansion Coefficient

Due to temperature change, ΔT, assume that the compatibility condition that the strain in the fiber and matrix is equal in direction 1 — that is,

$$\varepsilon_m = \varepsilon_f = \varepsilon_1 . \tag{3.203}$$

Now, the stress in the fiber in the longitudinal direction 1 is

$$\begin{aligned}(\sigma_f)_1 &= E_f(\varepsilon_f)_1 \\ &= E_f \varepsilon_1 \\ &= E_f(\alpha_1 - \alpha_f)\Delta T\end{aligned} \tag{3.204}$$

and the stress in the matrix in longitudinal direction 1 is

$$\begin{aligned}(\sigma_m)_1 &= E_m(\varepsilon_m)_1 \\ &= E_m \varepsilon_1 \\ &= -E_m(\alpha_m - \alpha_1)\Delta T.\end{aligned} \tag{3.205}$$

The strains in the fiber and matrix in the transverse direction 2 are given by using Hooke's law:

$$\left(\varepsilon_f\right)_2 = \alpha_f \Delta T - \frac{\nu_f\left(\sigma_f\right)_1}{E_f} \tag{3.206}$$

$$\left(\varepsilon_m\right)_2 = \alpha_m \Delta T - \frac{\nu_m\left(\sigma_m\right)_1}{E_m} . \tag{3.207}$$

The transverse strain of the composite is given by the rule of mixtures as

$$\varepsilon_2 = \left(\varepsilon_f\right)_2 V_f + \left(\varepsilon_m\right)_2 V_m . \tag{3.208}$$

Substituting Equation (3.206) and Equation (3.207) in Equation (3.208),

$$\varepsilon_2 = \left[\alpha_f \Delta T - \frac{\nu_f E_f\left(\alpha_1 - \alpha_f\right)\Delta T}{E_f}\right] V_f + \left[\alpha_m \Delta T + \frac{\nu_m E_m\left(\alpha_m - \alpha_1\right)\Delta T}{E_m}\right] V_m \tag{3.209}$$

and, because

$$\varepsilon_2 = \alpha_2 \Delta T , \tag{3.210}$$

we get

$$\alpha_2 = \left[\alpha_f - \nu_f\left(\alpha_1 - \alpha_f\right)\right] V_f + \left[\alpha_m + \nu_m\left(\alpha_m - \alpha_1\right)\right] V_m . \tag{3.211}$$

Substituting

$$\nu_{12} = \nu_f V_f + \nu_m V_m \tag{3.212}$$

in the preceding equation, it can be rewritten as

$$\alpha_2 = \left(1 + \nu_f\right)\alpha_f V_f + \left(1 + \nu_m\right)\alpha_m V_m - \alpha_1 \nu_{12} . \tag{3.213}$$

Example 3.18

Find the coefficients of thermal expansion for a glass/epoxy lamina with 70% fiber volume fraction. Use the properties of glass and epoxy from Table 3.1 and Table 3.2, respectively.

Solution

From Table 3.1, the Young's modulus of the fiber is

$$E_f = 85\ GPa$$

and the Poisson's ratio of the fiber is

$$\nu_f = 0.2.$$

The coefficient of thermal expansion of the fiber is

$$\alpha_f = 5 \times 10^{-6}\ m/m/^\circ C.$$

From Table 3.2, the Young's modulus of the matrix is

$$E_m = 3.4\ GPa,$$

the Poisson's ratio of the matrix is

$$\nu_m = 0.3,$$

and the coefficient of thermal expansion of the matrix is

$$\alpha_m = 63 \times 10^{-6}\ m/m/^\circ C.$$

From Example 3.3, the longitudinal Young's modulus of the unidirectional lamina is

$$E_1 = 60.52\ GPa.$$

From Example 3.5, the major Poisson's ratio of the unidirectional lamina is

$$\nu_{12} = 0.2300.$$

Now, substituting the preceding values in Equation (3.194) and Equation (3.195), the coefficients of thermal expansion are

$$\alpha_1 = \frac{1}{60.52\times 10^9}\left[(5\times 10^{-6})(85\times 10^9)(0.7) + (63\times 10^{-6})(3.4\times 10^9)(0.3)\right]$$

$$= 5.978\times 10^{-6}\ m/m/°C,$$

$$\alpha_2 = (1+0.2)(5.0\times 10^{-6})(0.7) + (1+0.3)(63\times 10^{-6})(0.3) - (5.978\times 10^{-6})(0.23)$$

$$= 27.40\times 10^{-6}\ m/m/°C$$

In Figure 3.38, the two coefficients of thermal expansion of glass/epoxy are plotted as a function of fiber volume fraction.

It should be noted that the longitudinal thermal expansion coefficient is lower than the transverse thermal expansion coefficient in polymeric matrix composites. Also, in some cases, the thermal expansion coefficient of the fibers is negative, and it is thus possible for a lamina to have zero thermal expansions in the fiber directions. This property is widely used in the manufacturing of antennas, doors, etc., when dimensional stability in the presence of wide temperature fluctuations is desired.

Experimental determinations: The linear coefficients of thermal expansion are determined experimentally by measuring the dimensional changes in a

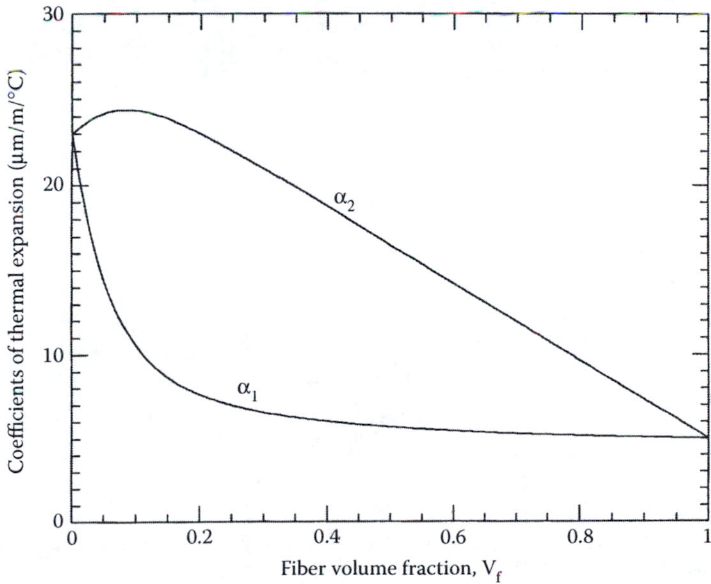

FIGURE 3.38
Longitudinal and transverse coefficients of thermal expansion as a function of fiber volume fraction for a glass/epoxy unidirectional lamina. (Properties of glass and epoxy from Table 3.1 and Table 3.2.)

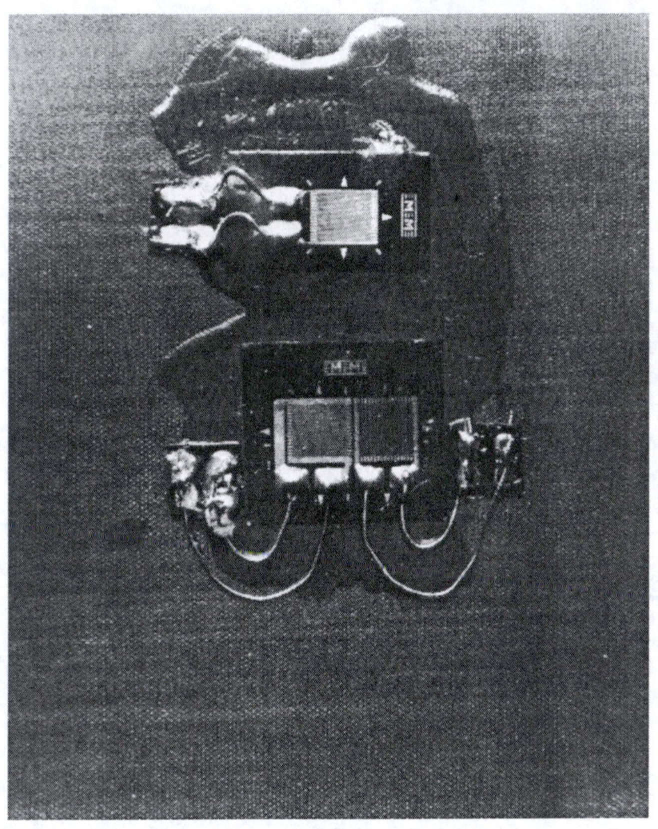

FIGURE 3.39
Unidirectional graphite/epoxy specimen with strain gages and temperature sensors for finding coefficients of thermal expansion. (Reprinted with permission from *Experimental Characterization of Advanced Composites*, Carlsson, L.A. and Pipes, R.B., Technomic Publishing Co., Inc., 1987, p. 98. Copyright CRC Press, Boca Raton, FL.)

lamina that is free of external stresses. A test specimen is made of a 50 × 50 mm (2 in. × 2 in.), eight-ply laminated unidirectional composite (Figure 3.39). Two strain gages are placed perpendicular to each other on the specimen. A temperature sensor is also placed. The specimen is put in an oven and the temperature is slowly increased. Strain and temperature measurements are taken and plotted as a function of each other as given in Figure 3.40. The data are reduced using linear regression. The slope of the two strain-temperature curves directly gives the coefficient of thermal expansion.

From Figure 3.40, the following values are obtained for a typical graphite/epoxy laminate[26]:

$$\alpha_1 = -1.3 \times 10^{-6}\ m/m/°C$$

$$\alpha_2 = 33.9 \times 10^{-6}\ m/m/°C.$$

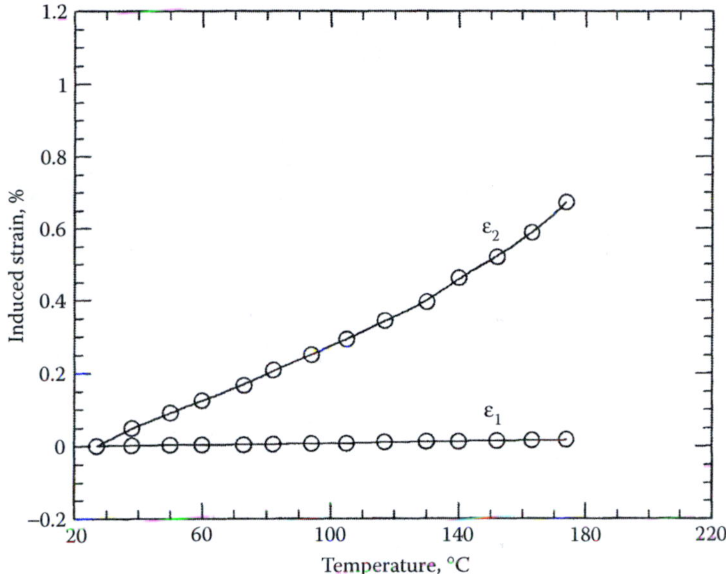

FIGURE 3.40
Induced strain as a function of temperature to find the coefficients of thermal expansion of a unidirectional graphite/epoxy laminate. (Reprinted with permission from *Experimental Characterization of Advanced Composites*, Carlsson, L.A. and Pipes, R.B., Technomic Publishing Co., Inc., 1987, p. 102. Copyright CRC Press, Boca Raton, FL.)

3.6 Coefficients of Moisture Expansion

When a body absorbs water, as is the case for resins in polymeric matrix composites, it expands. The change in dimensions of the body are measured by the coefficient of moisture expansion defined as the change in the linear dimension of a body per unit length per unit change in weight of moisture content per unit weight of the body. Similar to the coefficients of thermal expansion, there are two coefficients of moisture expansion: one in the longitudinal direction 1 and the other in the transverse direction 2:

β_1 = linear coefficient of moisture expansion in direction 1, m/m/kg/kg (in./in./lb/lb)

β_2 = linear coefficient of moisture expansion in direction 2, m/m/kg/kg (in./in./lb/lb)

The following are the expressions for the two coefficients of moisture expansion[30]:

$$\beta_1 = \frac{\beta_f \Delta C_f V_f E_f + \beta_m \Delta C_m V_m E_m}{E_1(\Delta C_f \rho_f V_f + \Delta C_m \rho_m V_m)} \rho_c, \tag{3.214}$$

$$\beta_2 = \frac{V_f(1+\nu_f)\Delta C_f \beta_f + V_m(1+\nu_m)\Delta C_m \beta_m}{(V_m \rho_m \Delta C_m + V_f \rho_f \Delta C_f)} \rho_c - \beta_1 \nu_{12}, \tag{3.215}$$

where

ΔC_f = the moisture concentration in the fiber, kg/kg (lb/lb)
ΔC_m = the moisture concentration in the matrix, kg/kg (lb/lb)
β_f = the coefficient of moisture expansion of the fiber, m/m/kg/kg (in./in./lb/lb)
β_m = the coefficient of moisture expansion of the matrix, m/m/kg/kg (in./in./lb/lb)

Note that, unlike the coefficients of thermal expansion, the content of moisture enters into the formula because the moisture absorption capacity in each constituent can be different. However, in most polymeric matrix composites, fibers do not absorb or deabsorb moisture, so the expressions for coefficients of moisture expansion do become independent of moisture contents. Substituting $\Delta C_f = 0$ in Equation (3.214) and Equation (3.215),

$$\beta_1 = \frac{E_m}{E_1} \frac{\rho_c}{\rho_m} \beta_m, \tag{3.216}$$

$$\beta_2 = (1+\nu_m)\frac{\rho_c}{\rho_m}\beta_m - \beta_1 \nu_{12}. \tag{3.217}$$

Further simplification for composites such as graphite/epoxy with high fiber-to-matrix moduli ratio (E_f/E_m) and no moisture absorption by fibers leads to

$$\beta_1 = 0, \text{ and} \tag{3.218}$$

$$\beta_2 = (1+\nu_m)\frac{\rho_c}{\rho_m}\beta_m. \tag{3.219}$$

Similar to the derivation for the longitudinal coefficient of thermal expansion in Section 3.5, Equation (3.214) can be derived using the mechanics of materials approach. Consider the expansion of a unidirectional lamina in the longitudinal direction because of change in moisture content in the composite. The overall load in the composite, F_1, is zero — that is,

$$F_1 = \sigma_1 A_c = 0 = \sigma_f A_f + \sigma_m A_m, \text{ and}$$

$$\sigma_f V_f + \sigma_m V_m = 0 , \tag{3.220}$$

where

$A_{c,f,m}$ = the cross-sectional areas of the fiber, matrix, and composite, respectively

$\sigma_{1,f,m}$ = the stresses in the fiber, matrix, and composite, respectively

The stresses in the fiber and matrix caused by moisture are

$$\sigma_f = E_f(\varepsilon_f - \beta_f \Delta C_f), \tag{3.221}$$

$$\sigma_m = E_m(\varepsilon_m - \beta_m \Delta C_m) . \tag{3.222}$$

Substituting Equation (3.221) and (3.222) in Equation (3.220) and knowing that the strains in the fiber and matrix are equal ($\varepsilon_f = \varepsilon_m$),

$$\varepsilon_f = \frac{\beta_f \Delta C_f V_f E_f + \beta_m \Delta C_m V_m E_m}{E_f V_f + E_m V_m} . \tag{3.223}$$

For free expansion of the composite in the longitudinal direction, the longitudinal strain is

$$\varepsilon_1 = \beta_1 \Delta C_c , \tag{3.224}$$

where ΔC_c = the moisture concentration in composite.

Because the strains in the fiber and the matrix are equal,

$$\beta_1 = \frac{\beta_f \Delta C_f V_f E_f + \beta_m \Delta C_m V_m E_m}{(E_f V_f + E_m V_m)\Delta C_c} . \tag{3.225}$$

Equation (3.225) can be simplified by relating the moisture concentration in the composite (ΔC_c) to the moisture concentration in the fiber (ΔC_f) and the matrix (ΔC_m).

The moisture content in the composite is the sum of the moisture contents in the fiber and the matrix,

$$\Delta C_c w_c = \Delta C_f w_f + \Delta C_m w_m , \tag{3.226}$$

where $w_{c,f,m}$ = the mass of composite, fiber, and matrix, respectively. Thus,

$$\Delta C_c = \Delta C_f W_f + \Delta C_m W_m \,, \tag{3.227}$$

where $W_{f,m}$ = the mass fractions of the fiber and matrix, respectively.
Substituting Equation (3.227) in Equation (3.225),

$$\beta_1 = \frac{\beta_f \Delta C_f V_f E_f + \beta_m \Delta C_m V_m E_m}{(E_f V_f + E_m V_m)(\Delta C_f W_f + \Delta C_m W_m)}. \tag{3.228}$$

Using Equation (3.4) and Equation (3.34), one can rewrite Equation (3.228) in terms of fiber volume fractions and the longitudinal Young's modulus as

$$\beta_1 = \frac{\beta_f \Delta C_f V_f E_f + \beta_m \Delta C_m V_m E_m}{E_1(\Delta C_f \rho_f V_f + \Delta C_m \rho_m V_m)} \rho_c. \tag{3.229}$$

Example 3.19

Find the two coefficients of moisture expansion for a glass/epoxy lamina with 70% fiber volume fraction. Use properties for glass and epoxy from Table 3.1 and Table 3.2, respectively. Assume that glass does not absorb moisture.

Solution

From Table 3.1, the density of the fiber is

$$\rho_f = 2500 \; kg/m^3.$$

From Table 3.2, the density of the matrix is

$$\rho_m = 1200 \; kg/m^3,$$

the swelling coefficient of the matrix is

$$\beta_m = 0.33 \; m/m/kg/kg,$$

and the Young's modulus of the matrix is

$$E_m = 3.4 \; GPa.$$

The Poisson's ratio of the matrix is

$$\nu_m = 0.3.$$

From Example 3.1, the density of the composite is

$$\rho_c = 2110 \; kg/m^3.$$

From Example 3.3, the longitudinal Young's modulus of the lamina is

$$E_1 = 60.52 \; GPa.$$

From Example 3.5, the major Poisson's ratio is

$$\nu_{12} = 0.230.$$

Thus, the longitudinal coefficient of moisture expansion from Equation (3.216) is

$$\beta_1 = \frac{3.4 \times 10^9}{60.52 \times 10^9} \frac{2110}{1200} (0.33)$$
$$= 0.3260 \times 10^{-1} \; m/m/kg/kg,$$

and the transverse coefficient of moisture expansion from Equation (3.217) is

$$\beta_2 = (1 + 0.3) \frac{2110}{1200} (0.33) - (0.3260 \times 10^{-1})(0.230)$$
$$= 0.7468 \; m/m/kg/kg.$$

Experimental determination: A specimen is placed in water and the moisture expansion strain is measured in the longitudinal and transverse directions. Because moisture attacks strain gage adhesives, micrometers are used to find the swelling strains.

3.7 Summary

After developing the concepts of fiber volume and weight fractions, we developed equations for density and void content. We found the four elastic moduli constants of a unidirectional lamina using three analytical approaches: strength of materials, Halphin–Tsai, and elasticity. Analytical models and experimental techniques for the five strength parameters, the two coefficients of thermal expansion, and the two coefficients of moisture expansion for a unidirectional lamina were discussed.

Key Terms

Volume fraction
Weight (mass) fraction
Density
Void volume fraction
Void content
Elastic moduli
Array packing
Halphin–Tsai equations
Elasticity models
Transversely isotropic fibers
Strength
ASTM standards
Failure modes
IITRI compression test
Shear test
Coefficient of thermal expansion
Coefficient of moisture expansion

Exercise Set

3.1 The weight fraction of glass in a glass/epoxy composite is 0.8. If the specific gravity of glass and epoxy is 2.5 and 1.2, respectively, find the

1. Fiber and matrix volume fractions
2. Density of the composite

3.2 A hybrid lamina uses glass and graphite fibers in a matrix of epoxy for its construction. The fiber volume fractions of glass and graphite are 40 and 20%, respectively. The specific gravity of glass, graphite, and epoxy is 2.6, 1.8, and 1.2, respectively. Find

1. Mass fractions
2. Density of the composite

3.3 The acid digestion test left 2.595 g of fiber from a composite specimen weighing 3.697 g. The composite specimen weighs 1.636 g in water. If the specific gravity of the fiber and matrix is 2.5 and 1.2, respectively, find the

1. Theoretical volume fraction of fiber and matrix
2. Theoretical density of composite

3. Experimental density
4. Weight fraction of fiber and matrix
5. Void fraction

3.4 A resin hybrid lamina is made by reinforcing graphite fibers in two matrices: resin A and resin B. The fiber weight fraction is 40%; for resin A and resin B, the weight fraction is 30% each. If the specific gravity of graphite, resin A, and resin B is 1.2, 2.6, and 1.7, respectively, find

1. Fiber volume fraction
2. Density of composite

3.5 Find the elastic moduli of a glass/epoxy unidirectional lamina with 40% fiber volume fraction. Use the properties of glass and epoxy from Table 3.3 and Table 3.4, respectively.

3.6 Show that

$$G_{12} = \frac{G_m}{1 - V_f}$$

if the fibers are much stiffer than the matrix — that is, $G_f >> G_m$.

3.7 Assume that fibers in a unidirectional lamina are circularly shaped and in a square array. Calculate the ratio of fiber diameter to fiber center-to-center spacing ratio in terms of the fiber volume fraction.

3.8 Circular graphite fibers of 10 μm diameter are packed in a hexagonal array in an epoxy matrix. The fiber weight fraction is 50%. Find the fiber-to-fiber spacing between the centers of the fibers. The density of graphite fibers is 1800 kg/m^3 and epoxy is 1200 kg/m^3.

3.9 Find the elastic moduli for problem 3.5 using Halphin–Tsai equations. Assume that the fibers are circularly shaped and are in a square array. Compare your results with those of problem 3.5.

3.10 A unidirectional glass/epoxy lamina with a fiber volume fraction of 70% is replaced by a graphite/epoxy lamina with the same longitudinal Young's modulus. Find the fiber volume fraction required in the graphite/epoxy lamina. Use properties of glass, graphite, and epoxy from Table 3.1 and Table 3.2.

3.11 Sometimes, the properties of a fiber are determined from the measured properties of a unidirectional lamina. As an example, find the experimentally determined value of the Poisson's ratio of an isotropic fiber from the following measured properties of a unidirectional lamina:

1. Major Poisson's ratio of composite = 0.27
2. Poisson's ratio of the matrix = 0.35
3. Fiber volume fraction = 0.65

3.12 Using elasticity model equations, find the elastic moduli of a glass/epoxy unidirectional lamina with 40% fiber volume fraction. Use the properties of glass and epoxy from Table 3.3 and Table 3.4, respectively. Compare your results with those obtained by using the strength of materials approach and the Halphin–Tsai approach. Assume that the fibers are circularly shaped and are in a square array for the Halphin–Tsai approach.

3.13 A measure of degree of orthotropy of a material is given by the ratio of the longitudinal to transverse Young's modulus. Given the properties of glass, graphite, and epoxy from Table 3.1 and Table 3.2 and using the mechanics of materials approach to find the longitudinal and transverse Young's modulus, find the fiber volume fraction at which the degree of orthotropy is maximum for graphite/epoxy and glass/epoxy unidirectional laminae.

3.14 What are three common modes of failure of a unidirectional composite subjected to longitudinal tensile load?

3.15 Do high fiber volume fractions increase the transverse strength of a unidirectional lamina?

3.16 Find the five strength parameters of a unidirectional glass/epoxy lamina with 40% fiber volume fraction. Use the properties of glass and epoxy from Table 3.3 and Table 3.4.

3.17 A rod is designed to carry a uniaxial tensile load of 1400 N with a factor of safety of two. The designer has two options for the materials: steel or 66% fiber volume fraction graphite/epoxy. Use the properties of graphite and epoxy from Table 3.1 and Table 3.2. Assume the following properties for steel:

- Young's modulus of steel = 210 GPa
- Poisson's ratio of steel = 0.3
- Tensile strength of steel = 450 MPa
- Specific gravity of steel = 7.8

The cost of graphite/epoxy is five times that of steel by weight. List your material of choice if the criterion depends on just

1. Mass
2. Cost

3.18 Find the coefficients of thermal expansion for a 60% unidirectional glass/epoxy lamina with a 60% fiber volume fraction. Use properties of glass and epoxy from Table 3.3 and Table 3.4, respectively.

3.19 If one plots the transverse coefficient of thermal expansion, α_2, as a function of fiber volume fraction, V_f, for a unidirectional glass/epoxy lamina, $\alpha_2 > \alpha_m$ for a certain fiber volume fraction. Find this range of fiber volume fraction. Use properties of glass and epoxy from Table 3.1 and Table 3.2, respectively.

3.20 Find the fiber volume fraction for which the unidirectional glass/epoxy lamina transverse thermal expansion coefficient is a maximum. Use properties of glass and epoxy from Table 3.1 and Table 3.2, respectively.

3.21 Prove[31] that it is possible to have the transverse coefficient of thermal expansion of a unidirectional lamina greater than the coefficient of thermal expansion of the matrix ($\alpha_2 > \alpha_m$) only if

$$\frac{E_f}{E_m} > \frac{1+\nu_f}{\nu_m} \text{ or } \frac{E_f}{E_m} < \frac{1+\nu_f}{1+\nu_m}$$

3.22 The coefficient of thermal expansion perpendicular to the fibers of a unidirectional glass/epoxy lamina is given as 28 μm/m/°C. Use the properties of glass and epoxy from Table 3.3 and Table 3.4 to find the coefficient of thermal expansion of the unidirectional glass/epoxy lamina in the direction parallel to the fibers.

3.23 There are large excursions of temperature in space and thus composites with zero or near zero thermal expansion coefficients are attractive. Find the volume fraction of the graphite fibers for which the thermal expansion coefficient is zero in the longitudinal direction of a graphite/epoxy unidirectional lamina. Use all the properties of graphite and epoxy from Table 3.1 and Table 3.2, respectively, but assume that the longitudinal coefficient of thermal expansion of graphite fiber is -1.3×10^{-6} m/m/°C.

3.24 Find the coefficients of moisture expansion of a glass/epoxy lamina with 40% fiber volume fraction. Use the properties of glass and epoxy from Table 3.1 and Table 3.2, respectively.

3.25 Assume a 60% fiber volume fraction glass/epoxy lamina of cuboid dimensions 25 cm (along the fibers) × 10 cm × 0.125 mm. Epoxy absorbs water as much as 8% of its weight. Use the properties of glass and epoxy from Table 3.1 and Table 3.2, respectively, and find

1. Maximum mass of water that the specimen can absorb
2. Change in volume of the lamina if the maximum possible water is absorbed

Assume that the coefficient of moisture expansion through the thickness is the same as the coefficient of moisture expansion in the transverse direction and that the glass fibers absorb no moisture.

References

1. Judd, N.C.W. and Wright, W.W., Voids and their effects on the mechanical properties of composites — an appraisal, *SAMPE J.*, 10, 14, 1978.

2. Geier, M.H., *Quality Handbook for Composite Materials*, Chapman & Hall, London, 1994.
3. Adams, R.D., Damping properties analysis of composites, in *Engineered Materials Handbook*, vol. 1, ASM International, Metals Park, OH, 1987.
4. Hashin, Z., Theory of fiber reinforced materials, NASA tech. rep. contract no: NAS1-8818, November 1970.
5. Chamis, L.C. and Sendeckyj, G.P., Critique on theories predicting thermoelastic properties of fibrous composites, *J. Composite Mater.*, 2, 332, 1968.
6. Halphin, J.C. and Tsai, S.W., Effect of environment factors on composite materials, Air Force tech. rep. AFML-TR-67-423, June 1969.
7. Foye, R.L., An evaluation of various engineering estimates of the transverse properties of unidirectional composite, *SAMPE*, 10, 31, 1966.
8. Hewitt, R.L. and Malherbe, M.D.De, An approximation for the longitudinal shear modulus of continuous fiber composites, *J. Composite Mater.*, 4, 280, 1970.
9. Hashin, Z. and Rosen, B.W., 1964, The elastic moduli of fiber reinforced materials, *ASME J. Appl. Mech.*, 31, 223, 1964.
10. Hashin, Z., Analysis of composite materials — a survey, *ASME J. Appl. Mech.*, 50, 481, 1983.
11. Knott, T.W. and Herakovich, C.T., Effect of fiber orthotropy on effective composite properties, *J. Composite Mater.*, 25, 732, 1991.
12. Christensen, R.M., Solutions for effective shear properties in three phase sphere and cylinder models, *J. Mech. Phys. Solids*, 27, 315, 1979.
13. Timoshenko, S.P. and Goodier, J.N., *Theory of Elasticity*, McGraw–Hill, New York, 1970.
14. Maple 9.0, Advancing mathematics. See http://www.maplesoft.com.
15. Hashin, Z., Analysis of properties of fiber composites with anisotropic constituents, *ASME J. Appl. Mech.*, 46, 543, 1979.
16. Hyer, M.W., *Stress Analysis of Fiber-Reinforced Materials,* WCB McGraw–Hill, New York, 1998.
17. Hashin, Z., *Theory of Fiber Reinforced Materials*, NASA CR-1974, 1972.
18. Whitney, J.M. and Riley, M.B., Elastic properties of fiber reinforced composite materials, *AIAA J.*, 4, 1537, 1966.
19. Hill, R., Theory of mechanical properties of fiber-strengthened materials — I. Elastic behavior, *J. Mech. Phys. Solids*, 12, 199, 1964.
20. Agrawal, B.D. and Broutman, L.J., *Analysis and Performance of Fiber Composites*, John Wiley & Sons, New York, 1990.
21. Dow, N.F. and Rosen, B.W., Evaluations of filament reinforced composites for aerospace structural applications, NASA CR-207, April 1965.
22. Schuerch, H., Prediction of compressive strength in uniaxial boron fiber metal matrix composites, *AIAA J.*, 4, 102, 1966.
23. Hull, D., *An Introduction to Composite Materials*, Cambridge University Press, 1981.
24. Hofer, K.E., Rao, N., and Larsen, D., Development of engineering data on mechanical properties of advanced composite materials, Air Force tech. rep. AFML-TR-72-205, part 1, September 1972.
25. Kies, J.A., Maximum strains in the resin of fiber-glass composites, NRL rep. No. 5752, AD-274560, 1962.
26. Carlsson, L.A. and Pipes, R.B., *Experimental Characterization of Advanced Composite Materials*, Technomic Publishing Company, Inc., Lancaster, PA, 1996.

27. Rosen, B.W., A simple procedure for experimental determination of the longitudinal shear modulus of unidirectional composites, *J. Composite Mater.*, 21, 552, 1972.
28. Shapery, R.A., Thermal expansion coefficients of composite materials based on energy principles, *J. Composite Mater.*, 2, 380, 1968.
29. Greszak, L.B., Thermoelastic properties of filamentary composites, presented at AIAA 6th Structures and Materials Conference, April 1965.
30. Tsai, S.W. and Hahn, H.T., *Introduction to Composite Materials*, Technomic Publishing Company, Inc., Lancaster, PA, 1980.
31. Kaw, A.K., On using a symbolic manipulator in mechanics of composites, *ASEE Computers Educ. J.*, 3, 61, 1993.

4

Macromechanical Analysis of Laminates

Chapter Objectives

- Understand the code for laminate stacking sequence.
- Develop relationships of mechanical and hygrothermal loads applied to a laminate to strains and stresses in each lamina.
- Find the elastic moduli of laminate based on the elastic moduli of individual laminae and the stacking sequence.
- Find the coefficients of thermal and moisture expansion of a laminate based on elastic moduli, coefficients of thermal and moisture expansion of individual laminae, and stacking sequence.

4.1 Introduction

In Chapter 2, stress–strain equations were developed for a single lamina. A real structure, however, will not consist of a single lamina but a laminate consisting of more than one lamina bonded together through their thickness. Why? First, lamina thicknesses are on the order of 0.005 in. (0.125 mm), implying that several laminae will be required to take realistic loads (a typical glass/epoxy lamina will fail at about only 750 lb per inch [131,350 N/m] width of a normal load along the fibers). Second, the mechanical properties of a typical unidirectional lamina are severely limited in the transverse direction. If one stacks several unidirectional layers, this may be an optimum laminate for unidirectional loads. However, for complex loading and stiffness requirements, this would not be desirable. This problem can be overcome by making a laminate with layers stacked at different angles for given loading and stiffness requirements. This approach increases the cost and weight of the laminate and thus it is necessary to optimize the ply angles. Moreover, layers of different composite material systems may be used to develop a more optimum laminate.

Similar to what was done in Chapter 2, the macromechanical analysis will be developed for a laminate. Based on applied in-plane loads of extension, shear, bending, and torsion, stresses and strains will be found in the local and global axes of each ply. Stiffnesses of whole laminates will also be calculated. Because laminates can also be subjected to hygrothermal loads of temperature change and moisture absorption during processing and servicing, stresses and strains in each ply will also be calculated due to these loads. Intuitively, one can see that the strengths, stiffnesses, and hygrothermal properties of a laminate will depend on

- Elastic moduli
- Stacking position
- Thickness
- Angle of orientation
- Coefficients of thermal expansion
- Coefficients of moisture expansion

of each lamina.

4.2 Laminate Code

A laminate is made of a group of single layers bonded to each other. Each layer can be identified by its location in the laminate, its material, and its angle of orientation with a reference axis (Figure 4.1). Each lamina is represented by the angle of ply and separated from other plies by a slash sign. The first ply is the top ply of the laminate. Special notations are used for

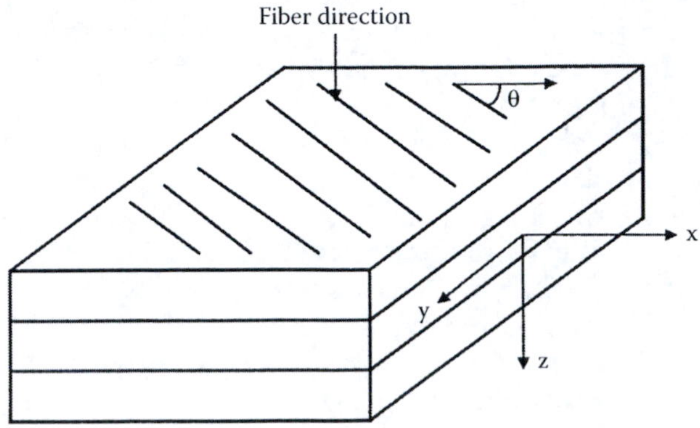

FIGURE 4.1
Schematic of a laminate.

symmetric laminates, laminates with adjacent lamina of the same orientation or of opposite angles, and hybrid laminates. The following examples illustrate the laminate code.

0
–45
90
60
30

[0/–45/90/60/30] denotes the code for the above laminate. It consists of five plies, each of which has a different angle to the reference x-axis. A slash separates each lamina. The code also implies that each ply is made of the same material and is of the same thickness. Sometimes, $[0/-45/90/60/30]_T$ may also denote this laminate, where the subscript T stands for a total laminate.

0
–45
90
90
60
0

$[0/-45/90_2/60/0]$ denotes the laminate above, which consists of six plies. Because two 90° plies are adjacent to each other, 90_2 denotes them, where the subscript 2 is the number of adjacent plies of the same angle.

0
–45
60
60
–45
0

$[0/-45/60]_s$ denotes the laminate above consisting of six plies. The plies above the midplane are of the same orientation, material, and thickness as the plies below the midplane, so this is a symmetric laminate. The top three plies are written in the code, and the subscript s outside the brackets represents that the three plies are repeated in the reverse order.

0
–45
60
–45
0

$[0/-45/\overline{60}]_s$ denotes this laminate, which consists of five plies. The number of plies is odd and symmetry exists at the midsurface; therefore, the 60° ply is denoted with a bar on the top.

Graphite/epoxy	0
Boron/epoxy	45
Boron/epoxy	–45
Boron/epoxy	–45
Boron/epoxy	45
Graphite/epoxy	0

$[0^{Gr}/\pm 45^{B}]_s$ denotes the above laminate. It consists of six plies; the 0° plies are made of graphite/epoxy and the ±45° angle plies are made of boron/epoxy. Note the symmetry of the laminate. Also, the ±45° notation indicates that the 0° ply should be followed by a +45° angle ply and then by a –45° angle ply. A notation of ±45° would indicate the –45° angle ply is followed by a +45° angle ply.

4.3 Stress–Strain Relations for a Laminate

4.3.1 One–Dimensional Isotropic Beam Stress–Strain Relation

Consider a prismatic beam of cross-section A (Figure 4.2a) under a simple load P; the normal stress at any cross-section is given by

$$\sigma_x = \frac{P}{A}. \tag{4.1}$$

The corresponding normal strain for a linearly elastic isotropic beam is

$$\varepsilon_x = \frac{P}{AE}, \tag{4.2}$$

where E is the Young's modulus of the beam. Note the assumption that the normal stress and strain are uniform and constant in the beam and are dependent on the load P being applied at the centroid of the cross section.

Now consider the same prismatic beam in a pure bending moment M (Figure 4.2b). The beam is assumed to be initially straight and the applied loads pass through a plane of symmetry to avoid twisting. Based on the elementary strength of material assumptions,

- The transverse shear is neglected
- Cross-sections retain their original shape
- The yz-plane before and after bending stays the same and normal to the x-axis.

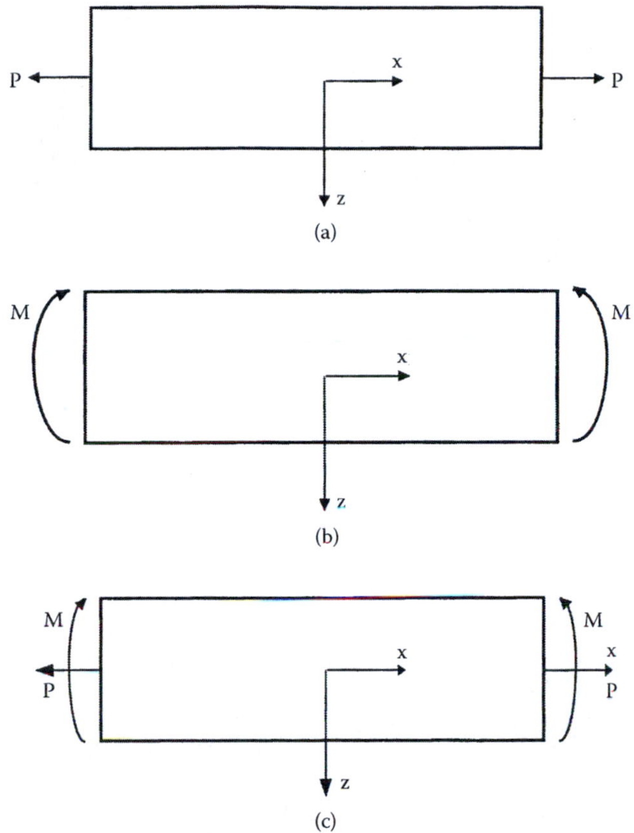

FIGURE 4.2
A beam under (a) axial load, (b) bending moment, and (c) combined axial and bending moment.

Then, at a distance, z, from the centroidal line,

$$\varepsilon_{xx} = \frac{z}{\rho}, \tag{4.3}$$

where ρ is the radius of curvature of the beam.

If the material is linearly elastic and isotropic,

$$\sigma_{xx} = \frac{Ez}{\rho} \tag{4.4}$$

and

$$\sigma_{xx} = \frac{Mz}{I}, \tag{4.5}$$

where $I = \int_A z^2 dA$ and is defined as the second moment of area, and M is the overall bending moment.

Now, if the same beam is under the influence of axial load P and bending moment M (Figure 4.2c), then

$$\varepsilon_{xx} = \left(\frac{1}{AE}\right)P + \left(\frac{z}{EI}\right)M \tag{4.6}$$

$$\varepsilon_{xx} = \varepsilon_0 + z\left(\frac{1}{\rho}\right)$$

$$\varepsilon_{xx} = \varepsilon_0 + z\kappa, \tag{4.7}$$

where ε_0 is the strain at $z = 0$ that is the centroid line of the beam, and κ = curvature of the beam. This shows that, under a combined uniaxial and bending load, the strain varies linearly through the thickness of the beam. Introducing the stress–strain relations of a laminate in this manner has been important because it forms a clear basis for developing similar relationships for a laminate in the next section. There, the strain-displacement equations, similar to Equation (4.7), will be developed in two dimensions.

4.3.2 Strain-Displacement Equations

In the previous section, the axial strain in a beam was related to the midplane strain and curvature of the beam under a uniaxial load and bending. In this section, similar relationships will be developed for a plate under in-plane loads such as shear and axial forces, and bending and twisting moments (Figure 4.3). The classical lamination theory is used to develop these relationships. The following assumptions are made in the classical lamination theory to develop the relationships[1]:

- Each lamina is orthotropic.
- Each lamina is homogeneous.
- A line straight and perpendicular to the middle surface remains straight and perpendicular to the middle surface during deformation ($\gamma_{xz} = \gamma_{yz} = 0$).

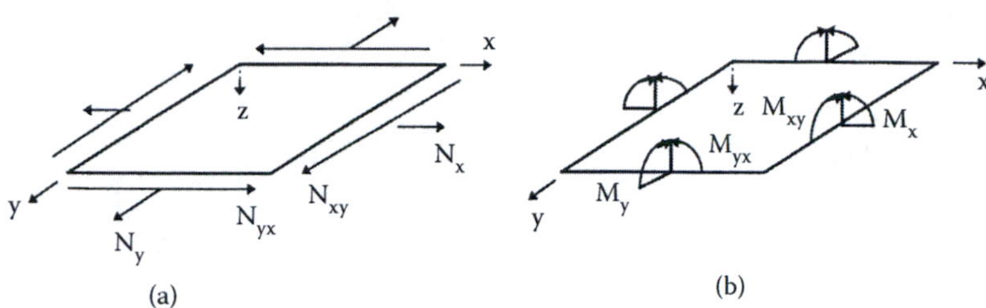

FIGURE 4.3
Resultant forces and moments on a laminate.

- The laminate is thin and is loaded only in its plane (plane stress) ($\sigma_z = \tau_{xz} = \tau_{yz} = 0$).
- Displacements are continuous and small throughout the laminate ($|u|, |v|, |w| << |h|$), where h is the laminate thickness.
- Each lamina is elastic.
- No slip occurs between the lamina interfaces.

Consider a side view of a plate in the Cartesian x–y–z coordinate system as shown in Figure 4.4. The origin of the plate is at the midplane of the plate, that is, $z = 0$. Assume u_0, v_0, and w_0 to be displacements in the x, y, and z directions, respectively, at the midplane and u, v, and w are the displacements at any point in the x, y, and z directions, respectively. At any point other than the midplane, the two displacements in the x–y plane will depend on the axial location of the point and the slope of the laminate midplane with the x and y directions. For example, as shown in Figure 4.4,

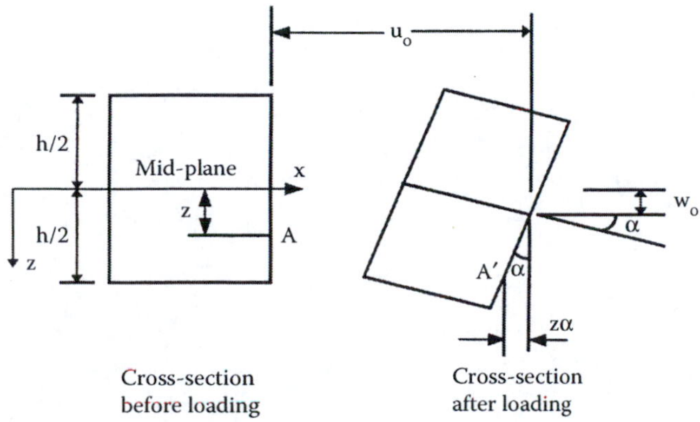

FIGURE 4.4
Figure showing the relationship between displacements through the thickness of a plate to midplane displacements and curvatures.

$$u = u_0 - z\alpha, \tag{4.8}$$

where

$$\alpha = \frac{\partial w_0}{\partial x}. \tag{4.9}$$

Thus, the displacement u in the x-direction is

$$u = u_0 - z\frac{\partial w_0}{\partial x}. \tag{4.10}$$

Similarly, taking a cross-section in the y–z plane would give the displacement in the y-direction as

$$v = v_0 - z\frac{\partial w_o}{\partial y}. \tag{4.11}$$

Now, from the definitions of the three strains (Equation 2.16) in the x–y plane and Equation (4.10) and Equation (4.11),

$$\begin{aligned}\varepsilon_x &= \frac{\partial u}{\partial x}\\ &= \frac{\partial u_o}{\partial x} - z\frac{\partial^2 w_0}{\partial x^2},\end{aligned} \tag{4.12a}$$

$$\begin{aligned}\varepsilon_y &= \frac{\partial v}{\partial y}\\ &= \frac{\partial v_o}{\partial y} - z\frac{\partial^2 w_0}{\partial y^2},\end{aligned} \tag{4.12b}$$

and

$$\begin{aligned}\gamma_{xy} &= \frac{\partial u}{\partial y} + \frac{\partial v}{\partial x}\\ &= \frac{\partial u_0}{\partial y} + \frac{\partial v_0}{\partial x} - 2z\frac{\partial^2 w_0}{\partial x \partial y}.\end{aligned} \tag{4.12c}$$

The strain-displacement equations (4.12a to 4.12c) can be written in matrix form as

$$\begin{Bmatrix} \varepsilon_x \\ \varepsilon_y \\ \gamma_{xy} \end{Bmatrix} = \begin{Bmatrix} \dfrac{\partial u_0}{\partial x} \\ \dfrac{\partial v_0}{\partial y} \\ \dfrac{\partial u_0}{\partial y} + \dfrac{\partial v_0}{\partial x} \end{Bmatrix} + z \begin{Bmatrix} -\dfrac{\partial^2 w_0}{\partial x^2} \\ -\dfrac{\partial^2 w_0}{\partial y^2} \\ -2\dfrac{\partial^2 w_0}{\partial x \partial y} \end{Bmatrix}. \tag{4.13}$$

The two arrays on the right-hand sides of Equation (4.13) are the definitions of the midplane strains:

$$\begin{Bmatrix} \varepsilon_x^0 \\ \varepsilon_y^0 \\ \gamma_{xy}^0 \end{Bmatrix} = \begin{Bmatrix} \dfrac{\partial u_0}{\partial x} \\ \dfrac{\partial v_0}{\partial y} \\ \dfrac{\partial u_0}{\partial y} + \dfrac{\partial v_0}{\partial x} \end{Bmatrix} \tag{4.14}$$

and the midplane curvatures

$$\begin{Bmatrix} \kappa_x \\ \kappa_y \\ \kappa_{xy} \end{Bmatrix} = \begin{Bmatrix} -\dfrac{\partial^2 w_0}{\partial x^2} \\ -\dfrac{\partial^2 w_0}{\partial y^2} \\ -2\dfrac{\partial^2 w_0}{\partial x \partial y} \end{Bmatrix}, \tag{4.15}$$

respectively.

Therefore, the laminate strains can be written as

$$\begin{Bmatrix} \varepsilon_x \\ \varepsilon_y \\ \gamma_{xy} \end{Bmatrix} = \begin{Bmatrix} \varepsilon_x^0 \\ \varepsilon_y^0 \\ \gamma_{xy}^0 \end{Bmatrix} + z \begin{Bmatrix} \kappa_x \\ \kappa_y \\ \kappa_{xy} \end{Bmatrix}. \tag{4.16}$$

Equation (4.16) shows the linear relationship of the strains in a laminate to the curvatures of the laminate. It also indicates that the strains are independent of the x and y coordinates. Also, note the similarity between Equation (4.16) and Equation (4.7), which was developed for the one-dimensional beam.

Example 4.1

A 0.010 in. thick laminate is subjected to in-plane loads. The midplane strains and curvatures are given as follows:

$$\begin{Bmatrix} \varepsilon_x^0 \\ \varepsilon_y^0 \\ \gamma_{xy}^0 \end{Bmatrix} = \begin{Bmatrix} 2751 \\ -1331 \\ -1125 \end{Bmatrix} \mu \; in/in$$

and

$$\begin{Bmatrix} \kappa_x \\ \kappa_y \\ \kappa_{xy} \end{Bmatrix} = \begin{Bmatrix} 1.965 \\ 0.2385 \\ -1.773 \end{Bmatrix} \frac{in}{in}.$$

Find the global strains at the top surface of the laminate.

Solution

Because the value of z in Equation (4.16) is measured from the midplane, $z = -0.005$ in. at the top surface of the laminate. The global strains at the top surface from Equation (4.16) are

$$\begin{Bmatrix} \varepsilon_x \\ \varepsilon_y \\ \gamma_{xy} \end{Bmatrix} = \begin{Bmatrix} 2751 \\ -1331 \\ -1125 \end{Bmatrix} (10^{-6}) - 0.005 \begin{Bmatrix} 1.965 \\ 0.2385 \\ -1.773 \end{Bmatrix}.$$

$$= \begin{Bmatrix} -7074 \\ -2524 \\ 7740 \end{Bmatrix} \mu \; in/in$$

4.3.3 Strain and Stress in a Laminate

If the strains are known at any point along the thickness of the laminate, the stress–strain Equation (2.103) calculates the global stresses in each lamina:

$$\begin{bmatrix} \sigma_x \\ \sigma_y \\ \tau_{xy} \end{bmatrix} = \begin{bmatrix} \bar{Q}_{11} & \bar{Q}_{12} & \bar{Q}_{16} \\ \bar{Q}_{12} & \bar{Q}_{22} & \bar{Q}_{26} \\ \bar{Q}_{16} & \bar{Q}_{26} & \bar{Q}_{66} \end{bmatrix} \begin{bmatrix} \varepsilon_x \\ \varepsilon_y \\ \gamma_{xy} \end{bmatrix}. \tag{4.17}$$

The reduced transformed stiffness matrix, $[\bar{Q}]$, corresponds to that of the ply located at the point along the thickness of the laminate.

Substituting Equation (4.16) into Equation (4.17),

$$\begin{bmatrix} \sigma_x \\ \sigma_y \\ \tau_{xy} \end{bmatrix} = \begin{bmatrix} \bar{Q}_{11} & \bar{Q}_{12} & \bar{Q}_{16} \\ \bar{Q}_{12} & \bar{Q}_{22} & \bar{Q}_{26} \\ \bar{Q}_{16} & \bar{Q}_{26} & \bar{Q}_{66} \end{bmatrix} \begin{bmatrix} \varepsilon_x^0 \\ \varepsilon_y^0 \\ \gamma_{xy}^0 \end{bmatrix} + z \begin{bmatrix} \bar{Q}_{11} & \bar{Q}_{12} & \bar{Q}_{16} \\ \bar{Q}_{12} & \bar{Q}_{22} & \bar{Q}_{26} \\ \bar{Q}_{16} & \bar{Q}_{26} & \bar{Q}_{66} \end{bmatrix} \begin{bmatrix} \kappa_x \\ \kappa_y \\ \kappa_{xy} \end{bmatrix}. \tag{4.18}$$

From Equation (4.18), the stresses vary linearly only through the thickness of each lamina (Figure 4.5). The stresses, however, may jump from lamina to lamina because the transformed reduced-stiffness matrix $[\bar{Q}]$ changes from ply to ply because $[\bar{Q}]$ depends on the material and orientation of the ply. These global stresses can then be transformed to local stresses through the transformation Equation (2.94). Local strains can be transformed to global strains through Equation (2.99). The local stresses and strains can then be used in the failure criteria, discussed in Chapter 2, to find when a laminate fails. The only question remaining in the macromechanical analysis of a laminate now is how to find the midplane strains and curvatures if the

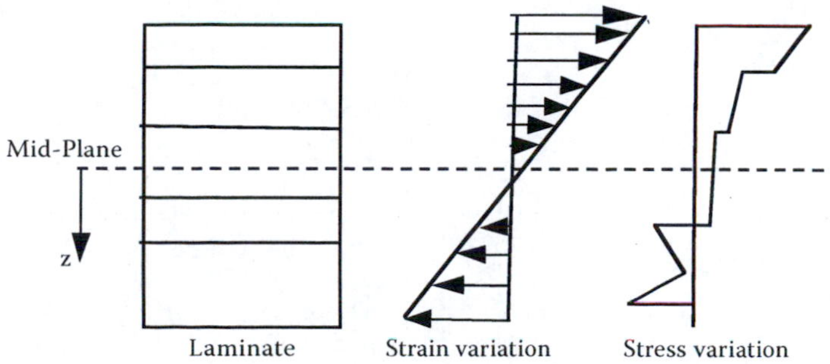

FIGURE 4.5
Strain and stress variation through the thickness of the laminate.

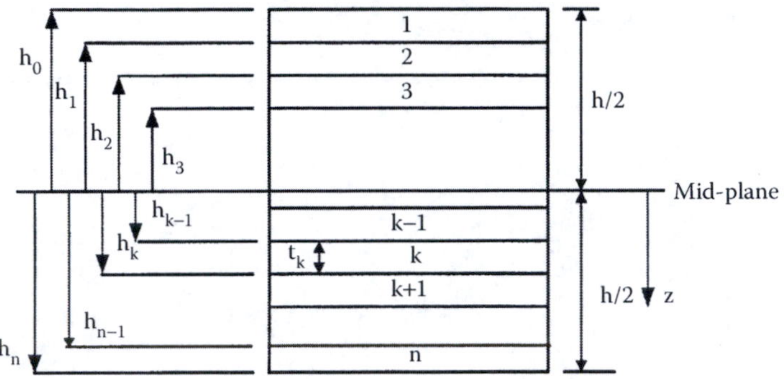

FIGURE 4.6
Coordinate locations of plies in a laminate.

loads applied to the laminate are known. This question is answered in the next section.

4.3.4 Force and Moment Resultants Related to Midplane Strains and Curvatures

The midplane strains and plate curvatures in Equation (4.16) are the unknowns for finding the lamina strains and stresses. However, Equation (4.18) gives the stresses in each lamina in terms of these unknowns. The stresses in each lamina can be integrated through the laminate thickness to give resultant forces and moments (or applied forces and moments). The forces and moments applied to a laminate will be known, so the midplane strains and plate curvatures can then be found. This relationship between the applied loads and the midplane strains and curvatures is developed in this section.

Consider a laminate made of n plies shown in Figure 4.6. Each ply has a thickness of t_k. Then the thickness of the laminate h is

$$h = \sum_{k=1}^{n} t_k. \tag{4.19}$$

Then, the location of the midplane is $h/2$ from the top or the bottom surface of the laminate. The z-coordinate of each ply k surface (top and bottom) is given by

Ply 1:

$$h_0 = -\frac{h}{2} \ (top\ surface),$$

$$h_1 = -\frac{h}{2} + t_1 \text{ (bottom surface)}.$$

Ply k: (k = 2, 3,…n – 2, n – 1):

$$h_{k-1} = -\frac{h}{2} + \sum_{\ell=1}^{k-1} t_\ell \text{ (top surface)}$$

$$h_k = -\frac{h}{2} + \sum_{\ell=1}^{k} t_\ell \text{ (bottom surface)}.$$

Ply n:

$$h_{n-1} = \frac{h}{2} - t_n \text{ (top surface)}$$

$$h_n = \frac{h}{2} \text{ (bottom surface)}. \tag{4.20}$$

Integrating the global stresses in each lamina gives the resultant forces per unit length in the x–y plane through the laminate thickness as

$$N_x = \int_{-h/2}^{h/2} \sigma_x dz, \tag{4.21a}$$

$$N_y = \int_{-h/2}^{h/2} \sigma_y dz, \tag{4.21b}$$

$$N_{xy} = \int_{-h/2}^{h/2} \tau_{xy} dz,, \tag{4.21c}$$

where $h/2$ is the half thickness of the laminate.

Similarly, integrating the global stresses in each lamina gives the resulting moments per unit length in the x–y plane through the laminate thickness as

$$M_x = \int_{-h/2}^{h/2} \sigma_x z dz, \tag{4.22a}$$

$$M_y = \int_{-h/2}^{h/2} \sigma_y z dz, \tag{4.22b}$$

$$M_{xy} = \int_{-h/2}^{h/2} \tau_{xy} z dz, \tag{4.22c}$$

where

N_x, N_y = normal force per unit length
N_{xy} = shear force per unit length
M_x, M_y = bending moments per unit length
M_{xy} = twisting moments per unit length

The resulting force and moment in the laminate are written in matrix form per Equation (4.21) and Equation (4.22) as

$$\begin{bmatrix} N_x \\ N_y \\ N_{xy} \end{bmatrix} = \int_{-h/2}^{h/2} \begin{bmatrix} \sigma_x \\ \sigma_y \\ \tau_{xy} \end{bmatrix} dz, \tag{4.23a}$$

$$\begin{bmatrix} M_x \\ M_y \\ M_{xy} \end{bmatrix} = \int_{-h/2}^{h/2} \begin{bmatrix} \sigma_x \\ \sigma_y \\ \tau_{xy} \end{bmatrix} z dz, \tag{4.23b}$$

which gives

$$\begin{bmatrix} N_x \\ N_y \\ N_{xy} \end{bmatrix} = \sum_{k=1}^{n} \int_{h_{k-1}}^{h_k} \begin{bmatrix} \sigma_x \\ \sigma_y \\ \tau_{xy} \end{bmatrix}_k dz, \tag{4.24a}$$

$$\begin{bmatrix} M_x \\ M_y \\ M_{xy} \end{bmatrix} = \sum_{k=1}^{n} \int_{h_{k-1}}^{h_k} \begin{bmatrix} \sigma_x \\ \sigma_y \\ \tau_{xy} \end{bmatrix}_k z dz. \tag{4.24b}$$

Substituting Equation (4.18) in Equation (4.24), the resultant forces and moments can be written in terms of the midplane strains and curvatures:

$$\begin{bmatrix} N_x \\ N_y \\ N_{xy} \end{bmatrix} = \sum_{k=1}^{n} \int_{h_{k-1}}^{h_k} \begin{bmatrix} \bar{Q}_{11} & \bar{Q}_{12} & \bar{Q}_{16} \\ \bar{Q}_{12} & \bar{Q}_{22} & \bar{Q}_{26} \\ \bar{Q}_{16} & \bar{Q}_{26} & \bar{Q}_{66} \end{bmatrix}_k \begin{bmatrix} \varepsilon_x^0 \\ \varepsilon_y^0 \\ \gamma_{xy}^0 \end{bmatrix} dz$$

$$+ \sum_{k=1}^{n} \int_{h_{k-1}}^{h_k} \begin{bmatrix} \bar{Q}_{11} & \bar{Q}_{12} & \bar{Q}_{16} \\ \bar{Q}_{12} & \bar{Q}_{22} & \bar{Q}_{26} \\ \bar{Q}_{16} & \bar{Q}_{26} & \bar{Q}_{66} \end{bmatrix}_k \begin{bmatrix} \kappa_x \\ \kappa_y \\ \kappa_{xy} \end{bmatrix} z dz \tag{4.25a}$$

$$\begin{bmatrix} M_x \\ M_y \\ M_{xy} \end{bmatrix} = \sum_{k=1}^{n} \int_{h_{k-1}}^{h_k} \begin{bmatrix} \bar{Q}_{11} & \bar{Q}_{12} & \bar{Q}_{16} \\ \bar{Q}_{12} & \bar{Q}_{22} & \bar{Q}_{26} \\ \bar{Q}_{16} & \bar{Q}_{26} & \bar{Q}_{66} \end{bmatrix}_k \begin{bmatrix} \varepsilon_x^0 \\ \varepsilon_y^0 \\ \gamma_{xy}^0 \end{bmatrix} z dz$$

$$+ \sum_{k=1}^{n} \int_{h_{k-1}}^{h_k} \begin{bmatrix} \bar{Q}_{11} & \bar{Q}_{12} & \bar{Q}_{16} \\ \bar{Q}_{12} & \bar{Q}_{22} & \bar{Q}_{26} \\ \bar{Q}_{16} & \bar{Q}_{26} & \bar{Q}_{66} \end{bmatrix}_k \begin{bmatrix} \kappa_x \\ \kappa_y \\ \kappa_{xy} \end{bmatrix} z^2 dz\,. \tag{4.25b}$$

In Equation (4.25a) and Equation (4.25b), the midplane strains and plate curvatures are independent of the z-coordinate. Also, the transformed reduced stiffness matrix, $[\bar{Q}]_k$,is constant for each ply. Therefore, Equation (4.25) can be rewritten as

$$\begin{bmatrix} N_x \\ N_y \\ N_{xy} \end{bmatrix} = \left\{ \sum_{k=1}^{n} \begin{bmatrix} \bar{Q}_{11} & \bar{Q}_{12} & \bar{Q}_{16} \\ \bar{Q}_{12} & \bar{Q}_{22} & \bar{Q}_{26} \\ \bar{Q}_{16} & \bar{Q}_{26} & \bar{Q}_{66} \end{bmatrix}_k \int_{h_{k-1}}^{h_k} dz \right\} \begin{bmatrix} \varepsilon_x^0 \\ \varepsilon_y^0 \\ \gamma_{xy}^0 \end{bmatrix}$$

$$+ \left\{ \sum_{k=1}^{n} \begin{bmatrix} \bar{Q}_{11} & \bar{Q}_{12} & \bar{Q}_{16} \\ \bar{Q}_{12} & \bar{Q}_{22} & \bar{Q}_{26} \\ \bar{Q}_{16} & \bar{Q}_{26} & \bar{Q}_{66} \end{bmatrix}_k \int_{h_{k-1}}^{h_k} z\,dz \right\} \begin{bmatrix} \kappa_x \\ \kappa_y \\ \kappa_{xy} \end{bmatrix} \tag{4.26a}$$

$$\begin{bmatrix} M_x \\ M_y \\ M_{xy} \end{bmatrix} = \left\{ \sum_{k=1}^{n} \begin{bmatrix} \bar{Q}_{11} & \bar{Q}_{12} & \bar{Q}_{16} \\ \bar{Q}_{12} & \bar{Q}_{22} & \bar{Q}_{26} \\ \bar{Q}_{16} & \bar{Q}_{26} & \bar{Q}_{66} \end{bmatrix}_k \int_{h_{k-1}}^{h_k} z\,dz \right\} \begin{bmatrix} \varepsilon_x^0 \\ \varepsilon_y^0 \\ \gamma_{xy}^0 \end{bmatrix}$$

$$+ \left\{ \sum_{k=1}^{n} \begin{bmatrix} \bar{Q}_{11} & \bar{Q}_{12} & \bar{Q}_{16} \\ \bar{Q}_{12} & \bar{Q}_{22} & \bar{Q}_{26} \\ \bar{Q}_{16} & \bar{Q}_{26} & \bar{Q}_{66} \end{bmatrix}_k \int_{h_{k-1}}^{h_k} z^2\,dz \right\} \begin{bmatrix} \kappa_x \\ \kappa_y \\ \kappa_{xy} \end{bmatrix}. \tag{4.26b}$$

Knowing that

$$\int_{h_{k-1}}^{h_k} dz = (h_k - h_{k-1}),$$

$$\int_{h_{k-1}}^{h_k} z\,dz = \frac{1}{2}(h_k^2 - h_{k-1}^2),$$

$$\int_{h_{k-1}}^{h_k} z^2\,dz = \frac{1}{3}(h_k^3 - h_{k-1}^3),$$

and substituting in Equation (4.26) gives

$$\begin{bmatrix} N_x \\ N_y \\ N_{xy} \end{bmatrix} = \begin{bmatrix} A_{11} & A_{12} & A_{16} \\ A_{12} & A_{22} & A_{26} \\ A_{16} & A_{26} & A_{66} \end{bmatrix} \begin{bmatrix} \varepsilon_x^0 \\ \varepsilon_y^0 \\ \gamma_{xy}^0 \end{bmatrix} + \begin{bmatrix} B_{11} & B_{12} & B_{16} \\ B_{12} & B_{22} & B_{26} \\ B_{16} & B_{26} & B_{66} \end{bmatrix} \begin{bmatrix} \kappa_x \\ \kappa_y \\ \kappa_{xy} \end{bmatrix}, \quad (4.27a)$$

$$\begin{bmatrix} M_x \\ M_y \\ M_{xy} \end{bmatrix} = \begin{bmatrix} B_{11} & B_{12} & B_{16} \\ B_{12} & B_{22} & B_{26} \\ B_{16} & B_{26} & B_{66} \end{bmatrix} \begin{bmatrix} \varepsilon_x^0 \\ \varepsilon_y^0 \\ \gamma_{xy}^0 \end{bmatrix} + \begin{bmatrix} D_{11} & D_{12} & D_{16} \\ D_{12} & D_{22} & D_{26} \\ D_{16} & D_{26} & D_{66} \end{bmatrix} \begin{bmatrix} \kappa_x \\ \kappa_y \\ \kappa_{xy} \end{bmatrix}, \quad (4.27b)$$

where

$$A_{ij} = \sum_{k=1}^{n} [(\bar{Q}_{ij})]_k (h_k - h_{k-1}), \quad i = 1,2,6; \quad j = 1,2,6, \quad (4.28a)$$

$$B_{ij} = \frac{1}{2} \sum_{k=1}^{n} [(\bar{Q}_{ij})]_k (h_k^2 - h_{k-1}^2), \quad i = 1,2,6; \quad j = 1,2,6, \quad (4.28b)$$

$$D_{ij} = \frac{1}{3} \sum_{k=1}^{n} [(\bar{Q}_{ij})]_k (h_k^3 - h_{k-1}^3), \quad i = 1,2,6; \quad j = 1,2,6. \quad (4.28c)$$

The $[A]$, $[B]$, and $[D]$ matrices are called the extensional, coupling, and bending stiffness matrices, respectively. Combining Equation (4.27a) and Equation (4.27b) gives six simultaneous linear equations and six unknowns as:

$$\begin{bmatrix} N_x \\ N_y \\ N_{xy} \\ M_x \\ M_y \\ M_{xy} \end{bmatrix} = \begin{bmatrix} A_{11} & A_{12} & A_{16} & B_{11} & B_{12} & B_{16} \\ A_{12} & A_{22} & A_{26} & B_{12} & B_{22} & B_{26} \\ A_{16} & A_{26} & A_{66} & B_{16} & B_{26} & B_{66} \\ B_{11} & B_{12} & B_{16} & D_{11} & D_{12} & D_{16} \\ B_{12} & B_{22} & B_{26} & D_{12} & D_{22} & D_{26} \\ B_{16} & B_{26} & B_{66} & D_{16} & D_{26} & D_{66} \end{bmatrix} \begin{bmatrix} \varepsilon_x^0 \\ \varepsilon_y^0 \\ \gamma_{xy}^0 \\ \kappa_x \\ \kappa_y \\ \kappa_{xy} \end{bmatrix}. \quad (4.29)$$

The extensional stiffness matrix $[A]$ relates the resultant in-plane forces to the in-plane strains, and the bending stiffness matrix $[D]$ relates the resultant bending moments to the plate curvatures. The coupling stiffness matrix $[B]$ couples the force and moment terms to the midplane strains and midplane curvatures.

The following are the steps for analyzing a laminated composite subjected to the applied forces and moments:

1. Find the value of the reduced stiffness matrix $[Q]$ for each ply using its four elastic moduli, E_1, E_2, ν_{12}, and G_{12} in Equation (2.93).
2. Find the value of the transformed reduced stiffness matrix $[\bar{Q}]$ for each ply using the $[Q]$ matrix calculated in step 1 and the angle of the ply in Equation (2.104) or in Equation (2.137) and Equation (2.138).
3. Knowing the thickness, t_k, of each ply, find the coordinate of the top and bottom surface, h_i, $i = 1\ldots, n$, of each ply, using Equation (4.20).
4. Use the $[\bar{Q}]$ matrices from step 2 and the location of each ply from step 3 to find the three stiffness matrices $[A]$, $[B]$, and $[D]$ from Equation (4.28).
5. Substitute the stiffness matrix values found in step 4 and the applied forces and moments in Equation (4.29).
6. Solve the six simultaneous equations (4.29) to find the midplane strains and curvatures.
7. Now that the location of each ply is known, find the global strains in each ply using Equation (4.16).
8. For finding the global stresses, use the stress–strain Equation (2.103).
9. For finding the local strains, use the transformation Equation (2.99).
10. For finding the local stresses, use the transformation Equation (2.94).

Example 4.2

Find the three stiffness matrices $[A]$, $[B]$, and $[D]$ for a three-ply [0/30/–45] graphite/epoxy laminate as shown in Figure 4.7. Use the unidirectional

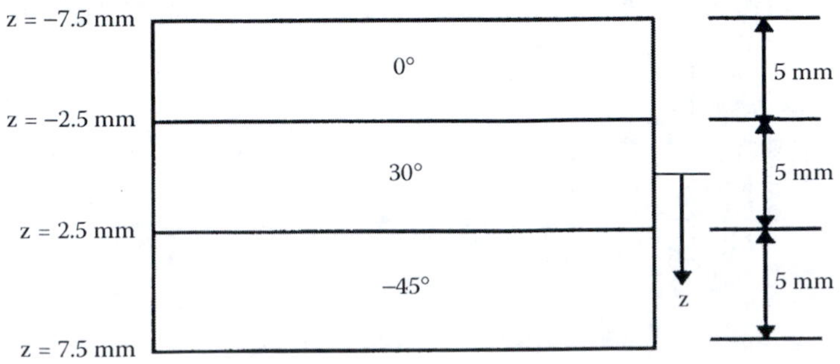

FIGURE 4.7
Thickness and coordinate locations of the three-ply laminate in Example 4.2 and Example 4.3.

properties from Table 2.1 of graphite/epoxy. Assume that each lamina has a thickness of 5 mm.

Solution

From Example 2.6, the reduced stiffness matrix for the 0° graphite/epoxy ply is

$$[Q] = \begin{bmatrix} 181.8 & 2.897 & 0 \\ 2.897 & 10.35 & 0 \\ 0 & 0 & 7.17 \end{bmatrix} (10^9)\ Pa\ .$$

From Equation (2.104), the transformed reduced stiffness matrix $[\bar{Q}]$ for each of the three plies is

$$[\bar{Q}]_0 = \begin{bmatrix} 181.8 & 2.897 & 0 \\ 2.897 & 10.35 & 0 \\ 0 & 0 & 7.17 \end{bmatrix} (10^9)\ Pa$$

$$[\bar{Q}]_{30} = \begin{bmatrix} 109.4 & 32.46 & 54.19 \\ 32.46 & 23.65 & 20.05 \\ 54.19 & 20.05 & 36.74 \end{bmatrix} (10^9)\ Pa$$

$$[\bar{Q}]_{-45} = \begin{bmatrix} 56.66 & 42.32 & -42.87 \\ 42.32 & 56.66 & -42.87 \\ -42.87 & -42.87 & 46.59 \end{bmatrix} (10^9)\ Pa\ .$$

The total thickness of the laminate is $h = (0.005)(3) = 0.015$ m.

The midplane is 0.0075 m from the top and the bottom of the laminate. Thus, using Equation (4.20), the locations of the ply surfaces are

$h_0 = -0.0075$ m

$h_1 = -0.0025$ m

$h_2 = 0.0025$ m

$h_3 = 0.0075$ m

From Equation (4.28a), the extensional stiffness matrix $[A]$ is

$$A_{ij} = \sum_{k=1}^{3} [\bar{Q}_{ij}]_k (h_k - h_{k-1})$$

$$[A] = \begin{bmatrix} 181.8 & 2.897 & 0 \\ 2.897 & 10.35 & 0 \\ 0 & 0 & 7.17 \end{bmatrix} (10^9)[(-0.0025) - (-0.0075)]$$

$$+ \begin{bmatrix} 109.4 & 32.46 & 54.19 \\ 32.46 & 23.65 & 20.05 \\ 54.19 & 20.05 & 36.74 \end{bmatrix} (10^9)[0.0025 - (-0.0025)]$$

$$+ \begin{bmatrix} 56.66 & 42.32 & -42.87 \\ 42.32 & 56.66 & -42.87 \\ -42.87 & -42.87 & 46.59 \end{bmatrix} (10^9)[0.0075 - 0.0025]$$

$$[A] = \begin{bmatrix} 1.739 \times 10^9 & 3.884 \times 10^8 & 5.663 \times 10^7 \\ 3.884 \times 10^8 & 4.533 \times 10^8 & -1.141 \times 10^8 \\ 5.663 \times 10^7 & -1.141 \times 10^8 & 4.525 \times 10^8 \end{bmatrix} Pa\text{-}m\,.$$

From Equation (4.28b), the coupling stiffness matrix $[B]$ is

$$B_{ij} = \frac{1}{2} \sum_{k=1}^{3} [\bar{Q}_{ij}]_k (h_k^2 - h_{k-1}^2)$$

$$[B] = \frac{1}{2} \begin{bmatrix} 181.8 & 2.897 & 0 \\ 2.897 & 10.35 & 0 \\ 0 & 0 & 7.17 \end{bmatrix} (10^9)[(-0.0025)^2 - (-0.0075)^2]$$

$$+ \frac{1}{2} \begin{bmatrix} 109.4 & 32.46 & 54.19 \\ 32.46 & 23.65 & 20.05 \\ 54.19 & 20.05 & 36.74 \end{bmatrix} (10^9)[(0.0025)^2 - (-0.0025)^2]$$

$$+\frac{1}{2}\begin{bmatrix} 56.66 & 42.32 & -42.87 \\ 42.32 & 56.66 & -42.87 \\ -42.87 & -42.87 & 46.59 \end{bmatrix}(10^9)[(0.0075)^2-(0.0025)^2]$$

$$[B]=\begin{bmatrix} -3.129\times10^6 & 9.855\times10^5 & -1.972\times10^6 \\ 9.855\times10^5 & 1.158\times10^6 & -1.972\times10^6 \\ -1.072\times10^6 & -1.072\times10^6 & 9.855\times10^5 \end{bmatrix} Pa\text{-}m^2 .$$

From Equation (4.28c), the bending stiffness matrix $[D]$ is

$$D_{ij}=\frac{1}{3}\sum_{k=1}^{3}[\bar{Q}_{ij}]_k(h_k^3-h_{k-1}^3)$$

$$[D]=\frac{1}{3}\begin{bmatrix} 181.8 & 2.897 & 0 \\ 2.897 & 10.35 & 0 \\ 0 & 0 & 7.17 \end{bmatrix}(10^9)[(-0.0025)^3-(-0.0075)^3]$$

$$+\frac{1}{3}\begin{bmatrix} 109.4 & 32.46 & 54.19 \\ 32.46 & 23.65 & 20.05 \\ 54.19 & 20.05 & 36.74 \end{bmatrix}(10^9)[(0.0025)^3-(-0.0025)^3]$$

$$+\frac{1}{3}\begin{bmatrix} 56.66 & 42.32 & -42.87 \\ 42.32 & 56.66 & -42.87 \\ -42.87 & -42.87 & 46.59 \end{bmatrix}(10^9)[(0.0075)^3-(0.0025)^3]$$

$$[D]=\begin{bmatrix} 3.343\times10^4 & 6.461\times10^3 & -5.240\times10^3 \\ 6.461\times10^3 & 9.320\times10^3 & -5.596\times10^3 \\ -5.240\times10^3 & -5.596\times10^3 & 7.663\times10^3 \end{bmatrix} Pa\text{-}m^3 .$$

Example 4.3

A [0/30/–45] graphite/epoxy laminate is subjected to a load of $N_x = N_y =$ 1000 N/m. Using the properties of unidirectional graphite/epoxy from Table 2.1 and assuming that each lamina is 5 mm thick, find

1. Midplane strains and curvatures
2. Global and local stresses on top surface of 30° ply
3. Percentage of load, N_x, taken by each ply

Solution

1. From Example 4.2, the three stiffness matrices [A], [B], and [D] are

$$[A]=\begin{bmatrix} 1.739\times10^9 & 3.884\times10^8 & 5.663\times10^7 \\ 3.884\times10^8 & 4.533\times10^8 & -1.141\times10^8 \\ 5.663\times10^7 & -1.141\times10^8 & 4.525\times10^8 \end{bmatrix} Pa\text{-}m\ ,$$

$$[B]=\begin{bmatrix} -3.129\times10^6 & 9.855\times10^5 & -1.072\times10^6 \\ 9.855\times10^5 & 1.158\times10^6 & -1.072\times10^6 \\ -1.072\times10^6 & -1.072\times10^6 & 9.855\times10^5 \end{bmatrix} Pa\text{-}m^2\ ,$$

$$[D]=\begin{bmatrix} 3.343\times10^4 & 6.461\times10^3 & -5.240\times10^3 \\ 6.461\times10^3 & 9.320\times10^3 & -5.596\times10^3 \\ -5.240\times10^3 & -5.596\times10^3 & 7.663\times10^3 \end{bmatrix} Pa\text{-}m^3\ .$$

Because the applied load is $N_x = N_y = 1000$ N/m, the midplane strains and curvatures can be found by solving the following set of six simultaneous linear equations (Equation 4.29):

$$\begin{bmatrix} N_x \\ N_y \\ N_{xy} \\ M_x \\ M_y \\ M_{xy} \end{bmatrix} =$$

$$\begin{bmatrix} 1.739\times10^9 & 3.884\times10^8 & 5.663\times10^7 & -3.129\times10^6 & 9.855\times10^5 & -1.072\times10^6 \\ 3.884\times10^8 & 4.533\times10^8 & -1.141\times10^8 & 9.855\times10^5 & 1.158\times10^6 & -1.072\times10^6 \\ 5.663\times10^7 & -1.141\times10^8 & 4.525\times10^8 & -1.072\times10^6 & -1.072\times10^6 & 9.855\times10^5 \\ -3.129\times10^6 & 9.855\times10^5 & -1.072\times10^6 & 3.343\times10^4 & 6.461\times10^3 & -5.240\times10^3 \\ 9.855\times10^5 & 1.158\times10^6 & -1.072\times10^6 & 6.461\times10^3 & 9.320\times10^3 & -5.596\times10^3 \\ -1.072\times10^6 & -1.072\times10^6 & 9.855\times10^5 & -5.240\times10^3 & -5.596\times10^3 & 7.663\times10^3 \end{bmatrix} \begin{bmatrix} \varepsilon_x^0 \\ \varepsilon_y^0 \\ \gamma_{xy}^0 \\ \kappa_x \\ \kappa_y \\ \kappa_{xy} \end{bmatrix}$$

This gives

$$\begin{bmatrix} \varepsilon_x^0 \\ \varepsilon_y^0 \\ \gamma_{xy}^0 \\ \hline \kappa_x \\ \kappa_y \\ \kappa_{xy} \end{bmatrix} = \begin{bmatrix} 3.123\times10^{-7} \\ 3.492\times10^{-6} \\ -7.598\times10^{-7} \\ \hline 2.971\times10^{-5} \\ -3.285\times10^{-4} \\ 4.101\times10^{-4} \end{bmatrix} \begin{matrix} m/m \\ \\ 1/m \end{matrix}.$$

2. The strains and stresses at the top surface of the 30° ply are found as follows. First, the top surface of the 30° ply is located at $z = h_1 = -0.0025$ m. From Equation (4.16),

$$\begin{bmatrix} \varepsilon_x \\ \varepsilon_y \\ \gamma_{xy} \end{bmatrix}_{30^\circ, top} = \begin{bmatrix} 3.123\times10^{-7} \\ 3.492\times10^{-6} \\ -7.598\times10^{-7} \end{bmatrix} + (-0.0025) \begin{bmatrix} 2.971\times10^{-5} \\ -3.285\times10^{-4} \\ 4.101\times10^{-4} \end{bmatrix}$$

$$= \begin{bmatrix} 2.380\times10^{-7} \\ 4.313\times10^{-6} \\ -1.785\times10^{-6} \end{bmatrix} m/m\ .$$

Using the stress–strain Equation (2.103) for an angle ply,

$$\begin{bmatrix} \sigma_x \\ \sigma_y \\ \tau_{xy} \end{bmatrix}_{30^\circ, top} = \begin{bmatrix} 109.4 & 32.46 & 54.19 \\ 32.46 & 23.65 & 20.05 \\ 54.19 & 20.05 & 36.74 \end{bmatrix} (10^9) \begin{bmatrix} 2.380\times10^{-7} \\ 4.313\times10^{-6} \\ -1.785\times10^{-6} \end{bmatrix}$$

$$= \begin{bmatrix} 6.930\times10^{4} \\ 7.391\times10^{4} \\ 3.381\times10^{4} \end{bmatrix} Pa\ .$$

The local strains and local stress as in the 30° ply at the top surface are found using transformation Equation (2.99) as

$$\begin{bmatrix} \varepsilon_1 \\ \varepsilon_2 \\ \gamma_{12}/2 \end{bmatrix} = \begin{bmatrix} 0.7500 & 0.2500 & 0.8660 \\ 0.2500 & 0.7500 & -0.8660 \\ -4.330 & 0.4330 & 0.5000 \end{bmatrix} \begin{bmatrix} 2.380 \times 10^{-7} \\ 4.313 \times 10^{-6} \\ -1.785 \times 10^{-6}/2 \end{bmatrix}$$

$$\begin{bmatrix} \varepsilon_1 \\ \varepsilon_2 \\ \gamma_{12} \end{bmatrix} = \begin{bmatrix} 4.837 \times 10^{-7} \\ 4.067 \times 10^{-6} \\ 2.636 \times 10^{-6} \end{bmatrix} m/m$$

and transformation Equation (2.94) as

$$\begin{bmatrix} \sigma_1 \\ \sigma_2 \\ \tau_{12} \end{bmatrix} = \begin{bmatrix} 0.7500 & 0.2500 & 0.8660 \\ 0.2500 & 0.7500 & -0.8660 \\ -0.4330 & 0.4330 & 0.5000 \end{bmatrix} \begin{bmatrix} 6.930 \times 10^{4} \\ 7.391 \times 10^{4} \\ 3.381 \times 10^{4} \end{bmatrix}$$

$$= \begin{bmatrix} 9.973 \times 10^{4} \\ 4.348 \times 10^{4} \\ 1.890 \times 10^{4} \end{bmatrix} Pa .$$

The values of global and local strains and stresses at the top, middle, and bottom surfaces of each ply are shown in Table 4.1 to Table 4.4.

3. The portion of the load, N_x, taken by each ply can be calculated by integrating the stress, σ_{xx}, through the thickness of each ply. However, because the stress varies linearly through each ply, the portion of the load, N_x, taken is simply the product of the stress, σ_{xx}, at the middle of each ply (see Table 4.2) and the thickness of the ply.

TABLE 4.1

Global Strains (m/m) in Example 4.3

Ply no.	Position	ε_x	ε_y	γ_{xy}
1 (0°)	Top	8.944×10^{-8}	5.955×10^{-6}	-3.836×10^{-6}
	Middle	1.637×10^{-7}	5.134×10^{-6}	-2.811×10^{-6}
	Bottom	2.380×10^{-7}	4.313×10^{-6}	-1.785×10^{-6}
2 (30°)	Top	2.380×10^{-7}	4.313×10^{-6}	-1.785×10^{-6}
	Middle	3.123×10^{-7}	3.492×10^{-6}	-7.598×10^{-7}
	Bottom	3.866×10^{-7}	2.670×10^{-6}	2.655×10^{-7}
3 (–45)	Top	3.866×10^{-7}	2.670×10^{-6}	2.655×10^{-7}
	Middle	4.609×10^{-7}	1.849×10^{-6}	1.291×10^{-6}
	Bottom	5.352×10^{-7}	1.028×10^{-6}	2.316×10^{-6}

TABLE 4.2

Global Stresses (Pa) in Example 4.3

Ply no.	Position	σ_x	σ_y	τ_{xy}
1 (0°)	Top	3.351×10^4	6.188×10^4	-2.750×10^4
	Middle	4.464×10^4	5.359×10^4	-2.015×10^4
	Bottom	5.577×10^4	4.531×10^4	-1.280×10^4
2 (30°)	Top	6.930×10^4	7.391×10^4	3.381×10^4
	Middle	1.063×10^5	7.747×10^4	5.903×10^4
	Bottom	1.434×10^5	8.102×10^4	8.426×10^4
3 (–45°)	Top	1.235×10^5	1.563×10^5	-1.187×10^5
	Middle	4.903×10^4	6.894×10^4	-3.888×10^4
	Bottom	-2.547×10^4	-1.840×10^4	4.091×10^4

TABLE 4.3

Local Strains (m/m) in Example 4.3

Ply no.	Position	ε_1	ε_2	γ_{12}
1 (0°)	Top	8.944×10^{-8}	5.955×10^{-6}	-3.836×10^{-6}
	Middle	1.637×10^{-7}	5.134×10^{-6}	-2.811×10^{-6}
	Bottom	2.380×10^{-7}	4.313×10^{-6}	-1.785×10^{-6}
2 (30°)	Top	4.837×10^{-7}	4.067×10^{-6}	2.636×10^{-6}
	Middle	7.781×10^{-7}	3.026×10^{-6}	2.374×10^{-6}
	Bottom	1.073×10^{-6}	1.985×10^{-6}	2.111×10^{-6}
3 (–45°)	Top	1.396×10^{-6}	1.661×10^{-6}	-2.284×10^{-6}
	Middle	5.096×10^{-7}	1.800×10^{-6}	-1.388×10^{-6}
	Bottom	-3.766×10^{-7}	1.940×10^{-6}	-4.928×10^{-7}

TABLE 4.4

Local Stresses (Pa) in Example 4.3

Ply no.	Position	σ_1	σ_2	σ_{12}
1 (0°)	Top	3.351×10^4	6.188×10^4	-2.750×10^4
	Middle	4.464×10^4	5.359×10^4	-2.015×10^4
	Bottom	5.577×10^4	4.531×10^4	-1.280×10^4
2 (30°)	Top	9.973×10^4	4.348×10^4	1.890×10^4
	Middle	1.502×10^5	3.356×10^4	1.702×10^4
	Bottom	2.007×10^5	2.364×10^4	1.513×0^4
3 (–45°)	Top	2.586×10^5	2.123×10^4	-1.638×10^4
	Middle	9.786×10^4	2.010×10^4	-9.954×10^3
	Bottom	-6.285×10^4	1.898×10^4	-3.533×10^3

$$\text{Portion of load } N_x \text{ taken by } 0° \text{ ply} = (4.464 \times 10^4)(5 \times 10^{-3}) = 223.2 \text{ N/m}$$

$$\text{Portion of load } N_x \text{ taken by } 30° \text{ ply} = (1.063 \times 10^5)(5 \times 10^{-3}) = 531.5 \text{ N/m}$$

$$\text{Portion of load } N_x \text{ taken by } -45° \text{ ply} = (4.903 \times 10^4)(5 \times 10^{-3}) = 245.2 \text{ N/m}.$$

The sum total of the loads shared by each ply is 1000 N/m, (223.2 + 531.5 + 245.2), which is the applied load in the x-direction, N_x.

$$\text{Percentage of load } N_x \text{ taken by } 0^\circ \text{ ply} = \frac{223.2}{1000} \times 100 = 22.32\%$$

$$\text{Percentage of load } N_x \text{ taken by } 30^\circ \text{ ply} = \frac{531.5}{1000} \times 100 = 53.15\%$$

$$\text{Percentage of load } N_x \text{ taken by } -45^\circ \text{ ply} = \frac{245.2}{1000} \times 100 = 24.52\%.$$

4.4 In-Plane and Flexural Modulus of a Laminate

Laminate engineering constants are another way of defining laminate stiffnesses. Showing Equation (4.29), in short notation,

$$\left[\frac{N}{M}\right] = \left[\begin{array}{c|c} A & B \\ \hline B & D \end{array}\right] \left[\frac{\varepsilon^0}{\kappa}\right], \tag{4.30}$$

where

$$[N] = \begin{bmatrix} N_x \\ N_y \\ N_{xy} \end{bmatrix},$$

$$[M] = \begin{bmatrix} M_x \\ M_y \\ M_{xy} \end{bmatrix},$$

$$[\varepsilon^0] = \begin{bmatrix} \varepsilon_x^0 \\ \varepsilon_y^0 \\ \gamma_{xy}^0 \end{bmatrix},$$

$$[\kappa] = \begin{bmatrix} \kappa_x \\ \kappa_y \\ \kappa_{xy} \end{bmatrix}.$$

Inverting Equation (4.30) gives

$$\left[\frac{\varepsilon^0}{\kappa}\right] = \left[\begin{array}{c|c} A^* & B^* \\ \hline C^* & D^* \end{array}\right]\left[\frac{N}{M}\right], \tag{4.31}$$

where

$$\left[\begin{array}{c|c} A^* & B^* \\ \hline C^* & D^* \end{array}\right] = \left[\begin{array}{c|c} A & B \\ \hline B & D \end{array}\right]^{-1} \tag{4.32a}$$

and

$$[C^*] = [B^*]^T . \tag{4.32b}$$

The $[A^*]$, $[B^*]$, and $[D^*]$ matrices are called the extensional compliance matrix, coupling compliance matrix, and bending compliance matrix, respectively.

4.4.1 In-Plane Engineering Constants of a Laminate

For a symmetric laminate, $[B] = 0$ and it can be shown that $[A^*] = [A]^{-1}$ and $[D^*] = [D]^{-1}$. Then, from Equation (4.31)

$$\begin{bmatrix} \varepsilon_x^0 \\ \varepsilon_y^0 \\ \gamma_{xy}^0 \end{bmatrix} = \begin{bmatrix} A_{11}^* & A_{12}^* & A_{16}^* \\ A_{12}^* & A_{22}^* & A_{26}^* \\ A_{16}^* & A_{26}^* & A_{66}^* \end{bmatrix}\begin{bmatrix} N_x \\ N_y \\ N_{xy} \end{bmatrix}. \tag{4.33}$$

The preceding equations allow us to define effective in-plane moduli in terms of the extensional compliance matrix $[A^*]$ as follows[2]:

Effective in-plane longitudinal modulus, E_x:

Apply the load $N_x \neq 0$, $N_y = 0$, $N_{xy} = 0$ and then substitute in Equation (4.33) as

$$\begin{bmatrix} \varepsilon_x^0 \\ \varepsilon_y^0 \\ \gamma_{xy}^0 \end{bmatrix} = \begin{bmatrix} A_{11}^* & A_{12}^* & A_{16}^* \\ A_{12}^* & A_{22}^* & A_{26}^* \\ A_{16}^* & A_{26}^* & A_{66}^* \end{bmatrix} \begin{bmatrix} N_x \\ 0 \\ 0 \end{bmatrix}, \tag{4.34}$$

which gives

$$\varepsilon_x^0 = A_{11}^* N_x.$$

Now the effective in-plane longitudinal modulus E_x is

$$E_x \equiv \frac{\sigma_x}{\varepsilon_x^0} = \frac{N_x/h}{A_{11}^* N_x} = \frac{1}{hA_{11}^*}. \tag{4.35}$$

Effective in-plane transverse modulus, E_y:

Apply the load $N_x = 0$, $N_y \neq 0$, $N_{xy} = 0$ and then substitute in Equation (4.33) as

$$\begin{bmatrix} \varepsilon_x^0 \\ \varepsilon_y^0 \\ \gamma_{xy}^0 \end{bmatrix} = \begin{bmatrix} A_{11}^* & A_{12}^* & A_{16}^* \\ A_{12}^* & A_{22}^* & A_{26}^* \\ A_{16}^* & A_{26}^* & A_{66}^* \end{bmatrix} \begin{bmatrix} 0 \\ N_y \\ 0 \end{bmatrix}, \tag{4.36}$$

which gives

$$\varepsilon_y^0 = A_{22}^* N_y.$$

The effective in-plane transverse modulus E_y is

$$E_y \equiv \frac{\sigma_y}{\varepsilon_y^0} = \frac{N_y/h}{A_{22}^* N_y} = \frac{1}{hA_{22}^*}. \tag{4.37}$$

Effective in-plane shear modulus G_{xy}:

Apply $N_x = 0$, $N_y = 0$, $N_{xy} \neq 0$ and then substitute in Equation (4.33) as

$$\begin{bmatrix} \varepsilon_x^0 \\ \varepsilon_y^0 \\ \gamma_{xy}^0 \end{bmatrix} = \begin{bmatrix} A_{11}^* & A_{12}^* & A_{16}^* \\ A_{12}^* & A_{22}^* & A_{26}^* \\ A_{16}^* & A_{26}^* & A_{66}^* \end{bmatrix} \begin{bmatrix} 0 \\ 0 \\ N_{xy} \end{bmatrix}, \tag{4.38}$$

which gives

$$\gamma_{xy}^0 = A_{66}^* N_{xy} \, .$$

The effective in-plane shear modulus G_{xy} is

$$G_{xy} \equiv \frac{\tau_{xy}}{\gamma_{xy}^0} = \frac{N_{xy}/h}{A_{66}^* N_{xy}} = \frac{1}{hA_{66}^*} \tag{4.39}$$

Effective in-plane Poisson's ratio, ν_{xy}:

From the derivation for the effective longitudinal Young's modulus, E_x, where the load applied is $N_x \neq 0$, $N_y = 0$, $N_{xy} = 0$, from Equation (4.34), we have

$$\varepsilon_y^0 = A_{12}^* N_x \tag{4.40}$$

$$\varepsilon_x^0 = A_{11}^* N_x \, . \tag{4.41}$$

The effective Poisson's ratio, ν_{xy}, is then defined as

$$\nu_{xy} \equiv -\frac{\varepsilon_y^0}{\varepsilon_x^0} = -\frac{A_{12}^* N_x}{A_{11}^* N_x} = -\frac{A_{12}^*}{A_{11}^*}. \tag{4.42}$$

Effective in-plane Poisson's ratio ν_{yx}:

From the derivation for the effective transverse Young's modulus, E_y, where the load applied is $N_x = 0$, $N_y \neq 0$, $N_{xy} = 0$, from Equation (4.36), we have

$$\varepsilon_x^0 = A_{12}^* N_y \tag{4.43}$$

$$\varepsilon_y^0 = A_{22}^* N_y \tag{4.44}$$

The effective Poisson's ratio, ν_{yx}, is then defined as

$$\nu_{yx} \equiv -\frac{\varepsilon_x^0}{\varepsilon_y^0}$$

$$= -\frac{A_{12}^*}{A_{22}^*}\frac{N_y}{N_y} \tag{4.45}$$

$$= -\frac{A_{12}^*}{A_{22}^*}.$$

Note here that a reciprocal relationship exists between the two effective Poisson's ratios ν_{xy} and ν_{yx}. From Equation (4.35) and Equation (4.42),

$$\frac{\nu_{xy}}{E_x} = \left(-\frac{A_{12}^*}{A_{11}^*}\right) hA_{11}^*$$
$$= -A_{12}^* h. \tag{4.46a}$$

From Equation (4.37) and Equation (4.45),

$$\frac{\nu_{yx}}{E_y} = \left(-\frac{A_{12}^*}{A_{22}^*}\right) hA_{22}^*$$
$$= -A_{12}^* h. \tag{4.46b}$$

From Equation (4.46a) and Equation (4.46b),

$$\frac{\nu_{xy}}{E_x} = \frac{\nu_{yx}}{E_y}. \tag{4.47}$$

4.4.2 Flexural Engineering Constants of a Laminate

Also, for a symmetric laminate, the coupling matrix $[B] = 0$; then, from Equation (4.31),

$$\begin{bmatrix} \kappa_x \\ \kappa_y \\ \kappa_{xy} \end{bmatrix} = \begin{bmatrix} D_{11}^* & D_{12}^* & D_{16}^* \\ D_{12}^* & D_{22}^* & D_{26}^* \\ D_{16}^* & D_{26}^* & D_{66}^* \end{bmatrix} \begin{bmatrix} M_x \\ M_y \\ M_{xy} \end{bmatrix}. \tag{4.48}$$

Equation (4.48) allows us to define effective flexural moduli in terms of the bending compliance matrix $[D^*]$ as follows[2]:

Apply $M_x \neq 0$, $M_y = 0$, $M_{xy} = 0$ and then substitute in Equation (4.48) as

$$\begin{bmatrix} \kappa_x \\ \kappa_y \\ \kappa_{xy} \end{bmatrix} = \begin{bmatrix} D_{11}^* & D_{12}^* & D_{16}^* \\ D_{12}^* & D_{22}^* & D_{26}^* \\ D_{16}^* & D_{26}^* & D_{66}^* \end{bmatrix} \begin{bmatrix} M_x \\ 0 \\ 0 \end{bmatrix}, \tag{4.49}$$

which gives

$$\kappa_x = D_{11}^* M_x . \tag{4.50}$$

The effective flexural longitudinal modulus, E_x^f, is

$$E_x^f \equiv \frac{12 M_x}{\kappa_x h^3} = \frac{12}{h^3 D_{11}^*}. \tag{4.51}$$

Similarly, one can show that the other flexural elastic moduli are given by

$$E_y^f = \frac{12}{h^3 D_{22}^*}, \tag{4.52}$$

$$G_{xy}^f = \frac{12}{h^3 D_{66}^*}, \tag{4.53}$$

$$\nu_{xy}^f = -\frac{D_{12}^*}{D_{11}^*}, \tag{4.54}$$

$$\nu_{yx}^f = -\frac{D_{12}^*}{D_{22}^*} . \tag{4.55}$$

Flexural Poisson's ratios ν_{xy}^f and ν_{yx}^f also have a reciprocal relationship, as given by

$$\frac{\nu_{xy}^f}{E_x^f} = \frac{\nu_{yx}^f}{E_y^f}. \tag{4.56}$$

In unsymmetric laminates, the stress–strain relationships in Equation (4.29) are not uncoupled between force and moment terms. Therefore, in those cases, the effective in-plane stiffness constants and flexural stiffness constants are not meaningful.

Example 4.4

Find the in-plane and flexural stiffness constants for a three-ply $[0/\overline{90}]_s$ graphite/epoxy laminate. Use the unidirectional properties of graphite/epoxy from Table 2.1. Each lamina is 5 mm thick.

Solution

From Example 4.2, the transformed reduced stiffness matrix is

$$[\bar{Q}]_0 = \begin{bmatrix} 181.8 & 2.897 & 0 \\ 2.897 & 10.35 & 0 \\ 0 & 0 & 7.17 \end{bmatrix} (10^9)\ Pa\ .$$

Then, from Equation (2.104), the transformed reduced stiffness matrix is

$$[\bar{Q}]_{90} = \begin{bmatrix} 10.35 & 2.897 & 0 \\ 2.897 & 181.8 & 0 \\ 0 & 0 & 7.17 \end{bmatrix} (10^9)\ Pa\ .$$

The total thickness of the laminate is $h = 0.005 \times 3 = 0.015$ m.

The midplane is 0.0075 m from the top and bottom surfaces of the laminate. Thus,

$$h_0 = -0.0075 \text{ m}$$

$$h_1 = -0.0025 \text{ m}$$

$$h_2 = 0.0025 \text{ m}$$

$$h_3 = 0.0075 \text{ m}.$$

From Equation (4.28a),

$$A_{ij} = \sum_{k=1}^{3} [\bar{Q}_{ij}]_k (h_k - h_{k-1})$$

$$[A] = \begin{bmatrix} 181.8 & 2.897 & 0 \\ 2.897 & 10.35 & 0 \\ 0 & 0 & 7.17 \end{bmatrix} (10^9)[-0.0025 - (-0.0075)]$$

$$+ \begin{bmatrix} 10.35 & 2.897 & 0 \\ 2.897 & 181.8 & 0 \\ 0 & 0 & 7.17 \end{bmatrix} (10^9)[0.0025 - (-0.0025)]$$

$$+ \begin{bmatrix} 181.8 & 2.897 & 0 \\ 2.897 & 10.35 & 0 \\ 0 & 0 & 7.17 \end{bmatrix} (10^9)[0.0075 - 0.0025] \, ,$$

which gives

$$[A] = \begin{bmatrix} 1.870 \times 10^9 & 4.345 \times 10^7 & 0 \\ 4.345 \times 10^7 & 1.013 \times 10^9 & 0 \\ 0 & 0 & 1.076 \times 10^8 \end{bmatrix} Pa\text{-}m \, .$$

Inverting the extensional stiffness matrix $[A]$, we get the extensional compliance matrix as

$$[A^*] = \begin{bmatrix} 5.353 \times 10^{-10} & -2.297 \times 10^{-11} & 0 \\ -2.297 \times 10^{-11} & 9.886 \times 10^{-10} & 0 \\ 0 & 0 & 9.298 \times 10^{-9} \end{bmatrix} \frac{1}{Pa\text{-}m} .$$

The in-plane engineering constants are found as follows:
From Equation (4.35),

$$E_x = \frac{1}{hA_{11}^*} = \frac{1}{(0.015)(5.353 \times 10^{-10})} = 124.5 \; GPa \; ;$$

from Equation (4.37),

$$E_y = \frac{1}{hA_{22}^*} = \frac{1}{(0.015)(9.886 \times 10^{-10})} = 67.43 \; GPa \; .$$

From Equation (4.39),

$$G_{xy} = \frac{1}{hA^*_{66}} = \frac{1}{(0.015)(9.289\times 10^{-9})} = 7.17\ GPa\ ,$$

from Equation (4.42),

$$\nu_{xy} = -\frac{A^*_{12}}{A^*_{11}} = -\frac{-2.297\times 10^{-11}}{5.353\times 10^{-10}} = 0.04292\ ,$$

and from Equation (4.45),

$$\nu_{yx} = -\frac{A^*_{12}}{A^*_{22}} = -\frac{-2.297\times 10^{-11}}{9.886\times 10^{-10}} = 0.02323\ .$$

Note that the reciprocal relationship of the Poisson's ratios:

$$\frac{\nu_{xy}}{E_x} = \frac{\nu_{yx}}{E_y}$$

can be verified with the preceding values.

From Equation (4.28c),

$$D_{ij} = \frac{1}{3}\sum_{k=1}^{3}[\bar{Q}_{ij}]_k(h_k^3 - h_{k-1}^3)$$

$$[D] = \frac{1}{3}\begin{bmatrix} 181.8 & 2.897 & 0 \\ 2.897 & 10.35 & 0 \\ 0 & 0 & 7.17 \end{bmatrix}(10^9)[(-0.0025)^3 - (-0.0075)^3]$$

$$+\frac{1}{3}\begin{bmatrix} 10.35 & 2.897 & 0 \\ 2.897 & 181.8 & 0 \\ 0 & 0 & 7.17 \end{bmatrix}(10^9)[(0.0025)^3 - (-0.0025)^3]$$

$$+\frac{1}{3}\begin{bmatrix} 181.8 & 2.897 & 0 \\ 2.897 & 10.35 & 0 \\ 0 & 0 & 7.17 \end{bmatrix}(10^9)[(0.0075)^3 - (0.0025)^3]\ ,$$

which gives

$$[D] = \begin{bmatrix} 4.925 \times 10^4 & 8.148 \times 10^2 & 0 \\ 8.148 \times 10^2 & 4.696 \times 10^3 & 0 \\ 0 & 0 & 2.017 \times 10^3 \end{bmatrix} Pa\text{-}m^3 .$$

Inverting the bending stiffness matrix $[D]$, we get

$$[D^*] = \begin{bmatrix} 2.032 \times 10^{-5} & -3.526 \times 10^{-6} & 0 \\ -3.526 \times 10^{-6} & 2.136 \times 10^{-4} & 0 \\ 0 & 0 & 4.959 \times 10^{-4} \end{bmatrix} \frac{1}{Pa\text{-}m^3}$$

The flexural engineering constants are found as follows:
From Equation (4.51),

$$E_x^f = \frac{12}{h^3 D_{11}^*} = \frac{12}{(0.015)^3 (2.032 \times 10^{-5})} = 175.0 \ GPa ,$$

and from Equation (4.52),

$$E_y^f = \frac{12}{h^3 D_{22}^*} = \frac{12}{(0.015)^3 (2.136 \times 10^{-4})} = 16.65 \ GPa .$$

From Equation (4.53),

$$G_{xy}^f = \frac{12}{h^3 D_{66}^*} = \frac{12}{(0.015)^3 (4.959 \times 10^{-4})} = 7.17 \ GPa ,$$

from Equation (4.54),

$$\nu_{xy}^f = -\frac{D_{12}^*}{D_{11}^*} = -\frac{-3.526 \times 10^{-6}}{(2.032 \times 10^{-5})} = 0.1735 ,$$

and from Equation (4.55),

$$\nu_{yx}^f = -\frac{D_{12}^*}{D_{22}^*} = -\frac{-3.526 \times 10^{-6}}{2.136 \times 10^{-4}} = 0.01651.$$

The reciprocal relationship of the Poisson's ratios:

$$\frac{\nu_{xy}^f}{E_x^f} = \frac{\nu_{yx}^f}{E_y^f}$$

can be verified with the preceding values. Also, note that in the preceding example of a cross-ply laminate, the in-plane shear moduli and the flexural shear moduli are the same.

4.5 Hygrothermal Effects in a Laminate

In Section 2.9, the hygrothermal strains were calculated for an angle and unidirectional lamina subjected to a temperature change, ΔT, and moisture content change, ΔC. As mentioned, if the lamina is free to expand, no residual mechanical stresses would develop in the lamina at the macromechanical level. However, in a laminate with various plies of different angles or materials, each individual lamina is not free to deform. This results in residual stresses in the laminate.[3]

4.5.1 Hygrothermal Stresses and Strains

Sources of hygrothermal loads include cooling down from processing temperatures, operating temperatures different from processing temperatures, and humid environments such as in aircraft flying at high altitudes. Each ply in a laminate gets stressed by the deformation differences of adjacent lamina. Only the strains in excess of or less than the hygrothermal strains in the unrestricted lamina produce the residual stresses. These strain differences are called mechanical strains and the stresses caused by them are called mechanical stresses.

The mechanical strains induced by hygrothermal loads alone are

$$\begin{bmatrix} \varepsilon_x^M \\ \varepsilon_y^M \\ \gamma_{xy}^M \end{bmatrix} = \begin{bmatrix} \varepsilon_x \\ \varepsilon_y \\ \gamma_{xy} \end{bmatrix} - \begin{bmatrix} \varepsilon_x^T \\ \varepsilon_y^T \\ \gamma_{xy}^T \end{bmatrix} - \begin{bmatrix} \varepsilon_x^C \\ \varepsilon_x^C \\ \gamma_{xy}^C \end{bmatrix}, \tag{4.57}$$

where the superscript M represents the mechanical strains, T stands for the free expansion thermal strain, and C refers to the free expansion moisture strains.

Using stress–strain Equation (2.103), the hygrothermal stresses in a lamina are then given by

$$\begin{bmatrix} \sigma_x^{TC} \\ \sigma_y^{TC} \\ \gamma_{xy}^{TC} \end{bmatrix} = \begin{bmatrix} \bar{Q}_{11} & \bar{Q}_{12} & \bar{Q}_{16} \\ \bar{Q}_{16} & \bar{Q}_{26} & \bar{Q}_{26} \\ \bar{Q}_{16} & \bar{Q}_{26} & \bar{Q}_{66} \end{bmatrix} \begin{bmatrix} \varepsilon_x^{M} \\ \varepsilon_y^{M} \\ \gamma_{xy}^{M} \end{bmatrix}, \tag{4.58}$$

where *TC* stands for combined thermal and moisture effects. Hygrothermal stresses induce zero resultant forces and moments in the laminate and thus in the *n*-ply laminate shown in Figure 4.6,

$$\int_{-h/2}^{h/2} \begin{bmatrix} \sigma_x^{TC} \\ \sigma_y^{TC} \\ \tau_{xy}^{TC} \end{bmatrix} dz = 0 = \sum_{k=1}^{n} \int_{h_{k-1}}^{h_k} \begin{bmatrix} \sigma_x^{TC} \\ \sigma_y^{TC} \\ \tau_{xy}^{TC} \end{bmatrix}_k dz\ , \tag{4.59}$$

$$\int_{-h/2}^{h/2} \begin{bmatrix} \sigma_x^{TC} \\ \sigma_y^{TC} \\ \tau_{xy}^{TC} \end{bmatrix} zdz = 0 = \sum_{k=1}^{n} \int_{h_{k-1}}^{h_k} \begin{bmatrix} \sigma_x^{TC} \\ \sigma_y^{TC} \\ \tau_{xy}^{TC} \end{bmatrix}_k zdz\ . \tag{4.60}$$

From Equation (4.58) to Equation (4.60),

$$\sum_{k=1}^{n} \int_{h_{k-1}}^{h_k} \begin{bmatrix} \bar{Q}_{11} & \bar{Q}_{12} & \bar{Q}_{16} \\ \bar{Q}_{12} & \bar{Q}_{22} & \bar{Q}_{26} \\ \bar{Q}_{16} & \bar{Q}_{26} & \bar{Q}_{66} \end{bmatrix} \begin{bmatrix} \varepsilon_x^{M} \\ \varepsilon_y^{M} \\ \gamma_{xy}^{M} \end{bmatrix} dz = 0\ , \tag{4.61a}$$

and

$$\sum_{k=1}^{n} \int_{h_{k-1}}^{h_k} \begin{bmatrix} \bar{Q}_{11} & \bar{Q}_{12} & \bar{Q}_{16} \\ \bar{Q}_{12} & \bar{Q}_{22} & \bar{Q}_{26} \\ \bar{Q}_{16} & \bar{Q}_{26} & \bar{Q}_{66} \end{bmatrix} \begin{bmatrix} \varepsilon_x^{M} \\ \varepsilon_y^{M} \\ \gamma_{xy}^{M} \end{bmatrix} zdz = 0\ . \tag{4.61b}$$

Substituting Equation (4.57) and Equation (4.16) in Equation (4.61) gives

$$\begin{bmatrix} A_{11} & A_{12} & A_{16} \\ A_{12} & A_{22} & A_{26} \\ A_{16} & A_{26} & A_{66} \end{bmatrix} \begin{bmatrix} \varepsilon_x^o \\ \varepsilon_y^0 \\ \gamma_{xy}^0 \end{bmatrix} + \begin{bmatrix} B_{11} & B_{12} & B_{16} \\ B_{12} & B_{22} & B_{26} \\ B_{16} & B_{26} & B_{66} \end{bmatrix} \begin{bmatrix} \kappa_x \\ \kappa_y \\ \kappa_{xy} \end{bmatrix} = \begin{bmatrix} N_x^T \\ N_y^T \\ N_{xy}^T \end{bmatrix} + \begin{bmatrix} N_x^C \\ N_y^C \\ N_{xy}^C \end{bmatrix}, \quad (4.62)$$

$$\begin{bmatrix} B_{11} & B_{12} & B_{16} \\ B_{12} & B_{22} & B_{26} \\ B_{16} & B_{26} & B_{66} \end{bmatrix} \begin{bmatrix} \varepsilon_x^o \\ \varepsilon_y^o \\ \gamma_{xy}^o \end{bmatrix} + \begin{bmatrix} D_{11} & D_{12} & D_{16} \\ D_{12} & D_{22} & D_{26} \\ D_{16} & D_{26} & D_{66} \end{bmatrix} \begin{bmatrix} \kappa_x \\ \kappa_y \\ \kappa_{xy} \end{bmatrix} = \begin{bmatrix} M_x^T \\ M_y^T \\ M_{xy}^T \end{bmatrix} + \begin{bmatrix} M_x^C \\ M_y^C \\ M_{xy}^T \end{bmatrix} \quad (4.63)$$

The four arrays on the right-hand sides of Equation (4.62) and Equation (4.63) are given by

$$[N^T] = \begin{bmatrix} N_x^T \\ N_y^T \\ N_{xy}^T \end{bmatrix} = \Delta T \sum_{k=1}^{n} \begin{bmatrix} \bar{Q}_{11} & \bar{Q}_{12} & \bar{Q}_{16} \\ \bar{Q}_{12} & \bar{Q}_{22} & \bar{Q}_{26} \\ \bar{Q}_{16} & \bar{Q}_{26} & \bar{Q}_{66} \end{bmatrix}_k \begin{bmatrix} \alpha_x \\ \alpha_y \\ \alpha_{xy} \end{bmatrix}_k (h_k - h_{k-1}), \quad (4.64)$$

$$[M^T] = \begin{bmatrix} M_x^T \\ M_y^T \\ M_{xy}^T \end{bmatrix} = \frac{1}{2} \Delta T \sum_{k=1}^{n} \begin{bmatrix} \bar{Q}_{11} & \bar{Q}_{12} & \bar{Q}_{16} \\ \bar{Q}_{12} & \bar{Q}_{22} & \bar{Q}_{26} \\ \bar{Q}_{16} & \bar{Q}_{26} & \bar{Q}_{66} \end{bmatrix}_k \begin{bmatrix} \alpha_x \\ \alpha_y \\ \alpha_{xy} \end{bmatrix}_k (h_k^2 - h_{k-1}^2), \quad (4.65)$$

$$[N^C] = \begin{bmatrix} N_x^C \\ N_y^C \\ N_{xy}^C \end{bmatrix} = \Delta C \sum_{k=1}^{n} \begin{bmatrix} \bar{Q}_{11} & \bar{Q}_{12} & \bar{Q}_{16} \\ \bar{Q}_{12} & \bar{Q}_{22} & \bar{Q}_{26} \\ \bar{Q}_{16} & \bar{Q}_{26} & \bar{Q}_{66} \end{bmatrix}_k \begin{bmatrix} \beta_x \\ \beta_y \\ \beta_{xy} \end{bmatrix}_k (h_k - h_{k-1}), \quad (4.66)$$

$$[M^C] = \begin{bmatrix} M_x^C \\ M_y^C \\ M_{xy}^C \end{bmatrix} = \frac{1}{2} \Delta C \sum_{k=1}^{n} \begin{bmatrix} \bar{Q}_{11} & \bar{Q}_{12} & \bar{Q}_{16} \\ \bar{Q}_{12} & \bar{Q}_{22} & \bar{Q}_{26} \\ \bar{Q}_{16} & \bar{Q}_{26} & \bar{Q}_{66} \end{bmatrix}_k \begin{bmatrix} \beta_x \\ \beta_y \\ \beta_{xy} \end{bmatrix}_k (h_k^2 - h_{k-1}^2). \quad (4.67)$$

The loads in Equation (4.64) to Equation (4.67) are called fictitious hygrothermal loads and are known. One can calculate the midplane strains and curvatures by combining Equation (4.62) and Equation (4.63):

$$\begin{bmatrix} N^T \\ \hline M^T \end{bmatrix} + \begin{bmatrix} N^C \\ \hline M^C \end{bmatrix} = \left[\begin{array}{c|c} A & B \\ \hline B & D \end{array}\right] \begin{bmatrix} \varepsilon^0 \\ \hline \kappa \end{bmatrix}. \tag{4.68}$$

Using Equation (4.16),

$$\begin{bmatrix} \varepsilon_x \\ \varepsilon_y \\ \gamma_{xy} \end{bmatrix} = \begin{bmatrix} \varepsilon_x^0 \\ \varepsilon_y^0 \\ \gamma_{xy}^0 \end{bmatrix} + z \begin{bmatrix} \kappa_x \\ \kappa_y \\ \kappa_{xy} \end{bmatrix}, \tag{4.69}$$

one can calculate the global strains in any ply of the laminate. These global strains are the actual strains in the laminate. However, it is the difference between the actual strains and the free expansion strains, which results in mechanical stresses. The mechanical strains in the kth ply are given by Equation (4.57) as

$$\begin{bmatrix} \varepsilon_x^M \\ \varepsilon_y^M \\ \gamma_{xy}^M \end{bmatrix}_k = \begin{bmatrix} \varepsilon_x \\ \varepsilon_y \\ \gamma_{xy} \end{bmatrix}_k - \begin{bmatrix} \varepsilon_x^T \\ \varepsilon_y^T \\ \gamma_{xy}^T \end{bmatrix} - \begin{bmatrix} \varepsilon_x^C \\ \varepsilon_y^C \\ \gamma_{xy}^C \end{bmatrix}_k. \tag{4.70}$$

The mechanical stresses in the kth ply are then calculated by

$$\begin{bmatrix} \sigma_x \\ \sigma_y \\ \tau_{xy} \end{bmatrix}_k = \begin{bmatrix} \bar{Q}_{11} & \bar{Q}_{12} & \bar{Q}_{16} \\ \bar{Q}_{12} & \bar{Q}_{22} & \bar{Q}_{26} \\ \bar{Q}_{16} & \bar{Q}_{26} & \bar{Q}_{66} \end{bmatrix}_k \begin{bmatrix} \varepsilon_x^M \\ \varepsilon_y^M \\ \gamma_{xy}^M \end{bmatrix}_k. \tag{4.71}$$

The fictitious hygrothermal loads represent the loads in Equation (4.64) to Equation (4.67), which one can apply mechanically to induce the same stresses and strains as by the hygrothermal load. Thus, if both mechanical and hygrothermal loads are applied, one can add the mechanical loads to the fictitious hygrothermal loads to find the ply-by-ply stresses and strains in the laminate or separately apply the mechanical and hygrothermal loads and then add the resulting stresses and strains from the solution of the two problems.

Example 4.5

Calculate the residual stresses at the bottom surface of the 90° ply in a two-ply [0/90] graphite/epoxy laminate subjected to a temperature change of

–75°C. Use the unidirectional properties of a graphite/epoxy lamina from Table 2.1. Each lamina is 5 mm thick.

Solution

From Table 2.1, the coefficients of thermal expansion for a 0° graphite/epoxy ply are

$$\begin{bmatrix} \alpha_1 \\ \alpha_2 \\ \alpha_{12} \end{bmatrix} = \begin{bmatrix} 0.200 \times 10^{-7} \\ 0.225 \times 10^{-4} \\ 0 \end{bmatrix} m/m/°C .$$

From Equation (2.175), the transformed coefficients of thermal expansion are

$$\begin{bmatrix} \alpha_x \\ \alpha_y \\ \alpha_{xy} \end{bmatrix}_{0°} = \begin{bmatrix} 0.200 \times 10^{-7} \\ 0.225 \times 10^{-4} \\ 0 \end{bmatrix} m/m/°C$$

and

$$\begin{bmatrix} \alpha_x \\ \alpha_y \\ \alpha_{xy} \end{bmatrix}_{90°} = \begin{bmatrix} 0.225 \times 10^{-4} \\ 0.200 \times 10^{-7} \\ 0 \end{bmatrix} m/m/°C .$$

From Example 4.4, the reduced transformed stiffness matrices are

$$[\bar{Q}]_0 = \begin{bmatrix} 181.8 & 2.897 & 0 \\ 2.897 & 10.35 & 0 \\ 0 & 0 & 7.17 \end{bmatrix} GPa$$

$$[\bar{Q}]_{90} = \begin{bmatrix} 10.35 & 2.897 & 0 \\ 2.897 & 181.8 & 0 \\ 0 & 0 & 7.17 \end{bmatrix} GPa .$$

According to Equation (4.64), the fictitious thermal forces are

$$\begin{bmatrix} N_x^T \\ N_y^T \\ N_{xy}^T \end{bmatrix} = +(-75)\begin{bmatrix} 181.8 & 2.897 & 0 \\ 2.897 & 10.35 & 0 \\ 0 & 0 & 7.17 \end{bmatrix}(10^9)\begin{bmatrix} 0.200(10^{-7}) \\ 0.225(10^{-4}) \\ 0 \end{bmatrix}[0.000-(-0.005)]$$

$$+(-75)\begin{bmatrix} 10.35 & 2.897 & 0 \\ 2.897 & 181.8 & 0 \\ 0 & 0 & 7.17 \end{bmatrix}(10^9)\begin{bmatrix} 0.225(10^4) \\ 0.200(10^7) \\ 0 \end{bmatrix}[0.005-0.000]$$

$$= \begin{bmatrix} -1.131\times 10^5 \\ -1.131\times 10^5 \\ 0 \end{bmatrix} Pa\text{-}m.$$

According to Equation (4.65), the fictitious thermal moments are

$$\begin{bmatrix} M_x^T \\ M_y^T \\ M_{xy}^T \end{bmatrix} = \frac{1}{2}(-75)\begin{bmatrix} 181.8 & 2.897 & 0 \\ 2.897 & 10.35 & 0 \\ 0 & 0 & 7.17 \end{bmatrix}(10^9)\begin{bmatrix} 0.200\times 10^{-7} \\ 0.225\times 10^{-4} \\ 0 \end{bmatrix}[(0.000)^2-(-0.005)^2]$$

$$+\frac{1}{2}(-75)\begin{bmatrix} 10.35 & 2.897 & 0 \\ 2.897 & 10.35 & 0 \\ 0 & 0 & 7.17 \end{bmatrix}(10^9)\begin{bmatrix} 0.225\times 10^{-4} \\ 0.200\times 10^{-7} \\ 0 \end{bmatrix}[(0.005)^2-(0.000)^2]$$

$$= \begin{bmatrix} -1.538\times 10^2 \\ 1.538\times 10^2 \\ 0 \end{bmatrix} Pa\text{-}m.$$

Using Equation (4.28), the stiffness matrices $[A]$, $[B]$, and $[D]$ are calculated as

$$[A] = \begin{bmatrix} 9.608\times 10^8 & 2.897\times 10^7 & 0 \\ 2.897\times 10^7 & 9.608\times 10^8 & 0 \\ 0 & 0 & 7.170\times 10^7 \end{bmatrix} Pa\text{-}m,$$

$$[B] = \begin{bmatrix} -2.143\times10^{6} & 0 & 0 \\ 0 & 2.143\times10^{6} & 0 \\ 0 & 0 & 0 \end{bmatrix} Pa\text{-}m^{2},$$

$$[D] = \begin{bmatrix} 8.007\times10^{3} & 2.414\times10^{2} & 0 \\ 2.414\times10^{2} & 8.007\times10^{3} & 0 \\ 0 & 0 & 5.975\times10^{2} \end{bmatrix} Pa\text{-}m^{3}.$$

These stiffness matrices, $[A]$, $[B]$, and $[D]$, are to be used in Equation (4.68) to give the midplane strains and curvatures:

$$\left[\frac{N^T}{M^T}\right] = \left[\begin{array}{c|c} A & B \\ \hline B & D \end{array}\right]\left[\frac{\varepsilon^0}{\kappa}\right]$$

$$\begin{bmatrix} -1.131\times10^{5} \\ -1.131\times10^{5} \\ 0 \\ -1.538\times10^{2} \\ 1.538\times10^{2} \\ 0 \end{bmatrix} =$$

$$\begin{bmatrix} 9.608\times10^{8} & 2.897\times10^{7} & 0 & -2.143\times10^{6} & 0 & 0 \\ 2.897\times10^{7} & 9.608\times10^{8} & 0 & 0 & 2.143\times10^{6} & 0 \\ 0 & 0 & 7.170\times10^{7} & 0 & 0 & 0 \\ -2.143\times10^{6} & 0 & 0 & 8.007\times10^{3} & 2.414\times10^{2} & 0 \\ 0 & 2.143\times10^{6} & 0 & 2.414\times10^{2} & 8.007\times10^{3} & 0 \\ 0 & 0 & 0 & 0 & 0 & 5.975\times10^{2} \end{bmatrix} \begin{bmatrix} \varepsilon_x^0 \\ \varepsilon_y^0 \\ \gamma_{xy}^0 \\ \kappa_x \\ \kappa_y \\ \kappa_{xy} \end{bmatrix}.$$

This gives

$$\begin{bmatrix} \varepsilon_x^0 \\ \varepsilon_y^0 \\ \gamma_{xy}^0 \\ \hline \kappa_x \\ \kappa_y \\ \kappa_{xy} \end{bmatrix} = \begin{bmatrix} -3.907\times10^{-4} \\ -3.907\times10^{-4} \\ 0 \\ \hline -1.276\times10^{-1} \\ 1.276\times10^{-1} \\ 0 \end{bmatrix} \begin{matrix} m/m \\ \\ 1/m \end{matrix}$$

Equation (4.16) gives the actual strains at the bottom surface ($h_2 = 0.005$) of the 90° ply as

$$\begin{bmatrix} \varepsilon_x \\ \varepsilon_y \\ \gamma_{xy} \end{bmatrix}_{90^\circ} = \begin{bmatrix} -3.907 \times 10^{-4} \\ -3.907 \times 10^{-4} \\ 0 \end{bmatrix} + (0.005) \begin{bmatrix} -1.276 \times 10^{-1} \\ 1.276 \times 10^{-1} \\ 0 \end{bmatrix}$$

$$= \begin{bmatrix} -1.029 \times 10^{-3} \\ 2.475 \times 10^{-4} \\ 0 \end{bmatrix} m/m \; .$$

The mechanical strains result in the residual stresses. Thus, if one subtracts the strains that would have been caused by free expansion from the actual strains, one can calculate the mechanical strains. Equation (2.179) gives the free expansion thermal strains in the 90° ply:

$$\begin{bmatrix} \varepsilon_x^T \\ \varepsilon_y^T \\ \gamma_{xy}^T \end{bmatrix} = \begin{bmatrix} 0.225 \times 10^{-4} \\ 0.200 \times 10^{-7} \\ 0 \end{bmatrix} (-75)$$

$$= \begin{bmatrix} -0.16875 \times 10^{-2} \\ -0.1500 \times 10^{-5} \\ 0 \end{bmatrix} m/m \; .$$

From Equation (4.57), the mechanical strains at the bottom of the 90° ply are thus:

$$\begin{bmatrix} \varepsilon_x^M \\ \varepsilon_y^M \\ \gamma_{xy}^M \end{bmatrix} = \begin{bmatrix} -1.029 \times 10^{-3} \\ 2.475 \times 10^{-4} \\ 0 \end{bmatrix} - \begin{bmatrix} -0.16875 \times 10^{-2} \\ -0.1500 \times 10^{-5} \\ 0 \end{bmatrix}$$

$$= \begin{bmatrix} 0.6585 \times 10^{-3} \\ 0.2490 \times 10^{-3} \\ 0 \end{bmatrix} .$$

TABLE 4.5

Global Strains for Example 4.5

Ply no.	Position	ε_x	ε_y	γ_{xy}
1 (0°)	Top	2.475×10^{-4}	-1.029×10^{-3}	0.0
	Middle	-7.160×10^{-5}	-7.098×10^{-4}	0.0
	Bottom	-3.907×10^{-4}	-3.907×10^{-4}	0.0
2 (90°)	Top	-3.907×10^{-4}	-3.907×10^{-4}	0.0
	Middle	-7.098×10^{-4}	-7.160×10^{-5}	0.0
	Bottom	-1.029×10^{-3}	2.475×10^{-4}	0.0

TABLE 4.6

Global Stresses (Pa) for Example 4.5

Ply no.	Position	σ_y	σ_y	τ_{xy}
1 (0°)	Top	4.718×10^{7}	7.535×10^{6}	0.0
	Middle	-9.912×10^{6}	9.912×10^{6}	0.0
	Bottom	-6.701×10^{7}	1.229×10^{7}	0.0
2 (90°)	Top	1.229×10^{7}	-6.701×10^{7}	0.0
	Middle	9.912×10^{6}	-9.912×10^{6}	0.0
	Bottom	7.535×10^{6}	4.718×10^{7}	0.0

The stress–strain Equation (4.71) gives the mechanical stresses at the bottom surface of the 90° ply as

$$\begin{bmatrix} \sigma_x \\ \sigma_y \\ \tau_{xy} \end{bmatrix}_{90^\circ} = \begin{bmatrix} 10.35 & 2.897 & 0 \\ 2.897 & 181.8 & 0 \\ 0 & 0 & 7.17 \end{bmatrix} (10^9) \begin{bmatrix} 0.6585 \times 10^{-3} \\ 0.2490 \times 10^{-3} \\ 0 \end{bmatrix}$$

$$= \begin{bmatrix} 7.535 \times 10^{6} \\ 4.718 \times 10^{7} \\ 0 \end{bmatrix} Pa.$$

The global strains and stresses in all the plies of the laminate are given in Table 4.5 and Table 4.6, respectively.

4.5.2 Coefficients of Thermal and Moisture Expansion of Laminates

The concept of finding coefficients of thermal and moisture expansion of laminates is again well suited only for symmetric laminates because, in this case, the coupling stiffness matrix $[B] = 0$ and no bending occurs under hygrothermal loads.

The coefficients of thermal expansion are defined as the change in length per unit length per unit change of temperature. Three coefficients of thermal expansion, one in direction x (α_x) and the others in direction y (α_y) and in the plane xy (α_{xy}), are defined for a laminate.

Assuming $\Delta T = 1$ and $C = 0$,

$$\begin{bmatrix} \alpha_x \\ \alpha_y \\ \alpha_{xy} \end{bmatrix} \equiv \begin{bmatrix} \varepsilon_x^0 \\ \varepsilon_y^0 \\ \gamma_{xy}^0 \end{bmatrix}_{\substack{\Delta C=0 \\ \Delta T=1}} = \begin{bmatrix} A_{11}^* & A_{12}^* & A_{16}^* \\ A_{12}^* & A_{22}^* & A_{26}^* \\ A_{16}^* & A_{26}^* & A_{66}^* \end{bmatrix} \begin{bmatrix} N_x^T \\ N_y^T \\ N_{xy}^T \end{bmatrix}, \tag{4.72}$$

where $[N^T]$ is the resultant thermal force given by Equation (4.64) corresponding to $\Delta T = 1$ and $\Delta C = 0$.

Similarly, assuming $\Delta T = 0$ and $\Delta C = 1$, the moisture expansion coefficients can be defined as

$$\begin{bmatrix} \beta_x \\ \beta_y \\ \beta_{xy} \end{bmatrix} \equiv \begin{bmatrix} \varepsilon_x^0 \\ \varepsilon_y^0 \\ \gamma_{xy}^0 \end{bmatrix}_{\substack{\Delta T=0 \\ \Delta C=1}} = \begin{bmatrix} A_{11}^* & A_{12}^* & A_{16}^* \\ A_{12}^* & A_{22}^* & A_{26}^* \\ A_{16}^* & A_{26}^* & A_{66}^* \end{bmatrix} \begin{bmatrix} N_x^C \\ N_y^C \\ N_{xy}^C \end{bmatrix}, \tag{4.73}$$

where $[N^C]$ is the resultant moisture force given by Equation (4.66) corresponding to $\Delta T = 0$ and $\Delta C = 1$.

Example 4.6

Find the coefficients of thermal and moisture expansion of a $[0/\overline{90}]_s$ graphite/epoxy laminate. Use the properties of a unidirectional graphite/epoxy lamina from Table 2.1.

Solution

From Example 4.4, the extensional compliance matrix is

$$[A^*] = \begin{bmatrix} 5.353\times10^{-10} & -2.297\times10^{-11} & 0 \\ -2.297\times10^{-11} & 9.886\times10^{-10} & 0 \\ 0 & 0 & 9.298\times10^{-9} \end{bmatrix} \frac{1}{Pa\text{-}m}.$$

Corresponding to a temperature change of $\Delta T = 1°C$, the fictitious thermal forces are

$$\begin{bmatrix} N_x^T \\ N_y^T \\ N_{xy}^T \end{bmatrix} = \Delta T \sum_{k=1}^{3} \begin{bmatrix} \bar{Q}_{11} & \bar{Q}_{12} & \bar{Q}_{16} \\ \bar{Q}_{12} & \bar{Q}_{22} & \bar{Q}_{26} \\ \bar{Q}_{16} & \bar{Q}_{26} & \bar{Q}_{66} \end{bmatrix}_k \begin{bmatrix} \alpha_x \\ \alpha_y \\ \alpha_{xy} \end{bmatrix}_k (h_k - h_{k-1})$$

$$= (1)\begin{bmatrix} 181.8 & 2.897 & 0 \\ 2.897 & 10.35 & 0 \\ 0 & 0 & 7.17 \end{bmatrix}(10^9)\begin{bmatrix} 0.200\times 10^{-7} \\ 0.225\times 10^{-4} \\ 0 \end{bmatrix}[-0.0025-(-0.0075)]$$

$$+ (1)\begin{bmatrix} 10.35 & 2.897 & 0 \\ 2.897 & 181.35 & 0 \\ 0 & 0 & 7.17 \end{bmatrix}(10^9)\begin{bmatrix} 0.225\times 10^{-4} \\ 0.200\times 10^{-7} \\ 0 \end{bmatrix}[0.0025-(-0.0025)]$$

$$+ (1)\begin{bmatrix} 181.8 & 2.897 & 0 \\ 2.897 & 10.35 & 0 \\ 0 & 0 & 7.17 \end{bmatrix}(10^9)\begin{bmatrix} 0.200\times 10^{-7} \\ 0.225\times 10^{-4} \\ 0 \end{bmatrix}[0.0075-0.0025]$$

$$= \begin{bmatrix} 1.852\times 10^{3} \\ 2.673\times 10^{3} \\ 0 \end{bmatrix} Pa\text{-}m\ .$$

Equation (4.72) and the extensional compliance matrix from Example 4.4 then give the coefficients of thermal expansion for the laminate:

$$\begin{bmatrix} \alpha_x \\ \alpha_y \\ \alpha_{xy} \end{bmatrix} = \begin{bmatrix} 5.353\times 10^{-10} & -2.297\times 10^{-11} & 0 \\ -2.297\times 10^{-11} & 9.886\times 10^{-10} & 0 \\ 0 & 0 & 9.298\times 10^{-9} \end{bmatrix}\begin{bmatrix} 1.852\times 10^{3} \\ 2.673\times 10^{3} \\ 0 \end{bmatrix}$$

$$= \begin{bmatrix} 9.303\times 10^{-7} \\ 2.600\times 10^{-6} \\ 0 \end{bmatrix} m/m/^{\circ}C.$$

Corresponding to a moisture content of $\Delta C = 1$ kg/kg, the fictitious moisture forces are

$$\begin{bmatrix} N_x^C \\ N_y^C \\ N_{xy}^C \end{bmatrix} = \Delta C \sum_{k=1}^{n} \begin{bmatrix} \bar{Q}_{11} & \bar{Q}_{12} & \bar{Q}_{16} \\ \bar{Q}_{12} & \bar{Q}_{22} & \bar{Q}_{26} \\ \bar{Q}_{16} & \bar{Q}_{26} & \bar{Q}_{66} \end{bmatrix}_k \begin{bmatrix} \beta_x \\ \beta_y \\ \beta_{xy} \end{bmatrix} (h_k - h_{k-1})$$

$$= (1)\begin{bmatrix} 181.8 & 2.897 & 0 \\ 2.897 & 10.35 & 0 \\ 0 & 0 & 7.17 \end{bmatrix}(10^9)\begin{bmatrix} 0 \\ 0.6 \\ 0 \end{bmatrix}[-0.0025-(-0.0075)]$$

$$+ (1)\begin{bmatrix} 10.35 & 2.897 & 0 \\ 2.897 & 181.35 & 0 \\ 0 & 0 & 7.17 \end{bmatrix}(10^9)\begin{bmatrix} 0.6 \\ 0 \\ 0 \end{bmatrix}[0.0025-(-0.0025)]$$

$$+ (1)\begin{bmatrix} 181.8 & 2.897 & 0 \\ 2.897 & 10.35 & 0 \\ 0 & 0 & 7.17 \end{bmatrix}(10^9)\begin{bmatrix} 0 \\ 0.6 \\ 0 \end{bmatrix}[0.0075-0.0025]$$

$$= \begin{bmatrix} 4.842 \times 10^7 \\ 7.077 \times 10^7 \\ 0 \end{bmatrix} Pa\text{-}m.$$

Equation (4.73) gives the coefficients of moisture expansion for the laminate as

$$\begin{bmatrix} \beta_x \\ \beta_y \\ \beta_{xy} \end{bmatrix} = \begin{bmatrix} 5.353 \times 10^{-10} & -2.297 \times 10^{-11} & 0 \\ -2.297 \times 10^{-11} & 9.886 \times 10^{-10} & 0 \\ 0 & 0 & 9.298 \times 10^{-9} \end{bmatrix} \begin{bmatrix} 4.842 \times 10^7 \\ 7.077 \times 10^7 \\ 0 \end{bmatrix}$$

$$= \begin{bmatrix} 2.430 \times 10^{-2} \\ 6.885 \times 10^{-2} \\ 0 \end{bmatrix} m/m/kg/kg\ .$$

4.5.3 Warpage of Laminates

In laminates that are not symmetric, a temperature difference results in out-of-plane deformations. This deformation is also called warpage[4] and is calculated by integrating the curvature-displacement Equation (4.15):

$$\kappa_x = -\frac{\partial^2 w}{\partial x^2}, \tag{4.74a}$$

$$\kappa_y = -\frac{\partial^2 w}{\partial y^2}, \tag{4.74b}$$

$$\kappa_{xy} = -2\frac{\partial^2 w}{\partial x \partial y}. \tag{4.74c}$$

From the integration of Equation (4.74), the out-of-plane deflection, w, can be derived. Integrating Equation (4.74a),

$$w = -\kappa_x \frac{x^2}{2} + f_1(y)x + f_2(y), \tag{4.75}$$

where $f_1(y)$ and $f_2(y)$ are unknown functions. Substituting Equation (4.75) in Equation (4.74c),

$$\kappa_{xy} = -2\frac{\partial^2 w}{\partial x \partial y} = -2\frac{df_1(y)}{dy}. \tag{4.76}$$

This gives

$$f_1(y) = -\kappa_{xy}\frac{y}{2} + C_1, \tag{4.77}$$

where C_1 is an unknown constant of integration. From Equation (4.75) and Equation (4.77),

$$w = -\kappa_x \frac{x^2}{2} - \kappa_{xy}\frac{xy}{2} + C_1 x + f_2(y). \tag{4.78}$$

Substituting Equation (4.78) in Equation (4.74b),

$$\kappa_y = -\frac{\partial^2 w}{\partial y^2} = -\frac{d^2 f_2(y)}{dy^2}. \tag{4.79}$$

This gives

$$f_2(y) = -\kappa_y \frac{y^2}{2} + C_2 y + C_3 . \tag{4.80}$$

Substituting Equation (4.80) in Equation (4.78),

$$w = -\frac{1}{2}(\kappa_x x^2 + \kappa_y y^2 + \kappa_{xy} xy) + (C_1 x + C_2 y + C_3). \tag{4.81}$$

The terms $(C_1 x + C_2 y + C_3)$ are simply rigid body motion terms and one can relate the warpage to be

$$w = -\frac{1}{2}(\kappa_x x^2 + \kappa_y y^2 + \kappa_{xy} xy). \tag{4.82}$$

Example 4.7

Find the warpage in [0/90] graphite/epoxy laminate under a temperature change of –75°C. Use the properties of graphite/epoxy from Table 2.1.

Solution

From Example 4.5, the midplane curvatures of the laminate are given by

$$\begin{bmatrix} \kappa_x \\ \kappa_y \\ \kappa_{xy} \end{bmatrix} = \begin{bmatrix} -1.276 \times 10^{-1} \\ 1.276 \times 10^{-1} \\ 0 \end{bmatrix} 1/m .$$

Thus, the warpage at any point (x,y) on the plane from Equation (4.82) is $w = 0.6383 \times 10^{-1}\ x^2 - 0.6383 \times 10^{-1}\ y^2$. Note that this warpage is calculated relative to the point $(x,y) = (0,0)$.

4.6 Summary

In this chapter, we introduced the laminate code for laminate stacking sequence. Then, we developed the theory for the elastic response of a laminate subjected to mechanical loads such as in-plane loads and bending moments, and thermal and hygrothermal loads. This theory allowed us to calculate ply-by-ply global as well as local stresses and strains in each ply, and also to calculate the effective in-plane and bending flexural moduli for a laminate.

Key Terms

Laminate code
Classical lamination theory
Midplane strains and stresses
Ply-by-ply strains and stresses
In-plane and flexural modulus
Hygrothermal stresses in a laminate
Warpage

Exercise Set

4.1 Condense the following expanded laminate codes:

1. [0/45/–45/90]
2. [0/45/–45/–45/45/0]
3. [0/90/60/60/90/0]
4. [0/45/60/45/0]
5. [45/–45/45/–45/–45/45/–45/45]

4.2 Expand the following laminate codes:

1. $[45/-45]_S$
2. $[45/-45_2/90]_S$
3. $[45/0]_{3S}$
4. $[45/\pm 30]_2$
5. $[45/\pm 30]_2$

4.3 A laminate of 0.015 in. thickness under a complex load gives the following midplane strains and curvatures:

$$\begin{bmatrix} \epsilon_x^0 \\ \epsilon_y^0 \\ \gamma_{xy}^0 \\ \hline \kappa_x \\ \kappa_y \\ \kappa_{xy} \end{bmatrix} = \begin{bmatrix} 2\times 10^{-6} \\ 3\times 10^{-6} \\ 4\times 10^{-6} \\ \hline 1.2\times 10^{-4} \\ 1.5\times 10^{-4} \\ 2.6\times 10^{-4} \end{bmatrix} \begin{matrix} in./in. \\ \\ \\ 1/in. \end{matrix}.$$

Find the global strains at the top, middle and bottom surface of the laminate.

4.4 Do global strains vary linearly through the thickness of a laminate?

4.5 Do global stresses vary linearly through the thickness of a laminate?

4.6 The global strains at the top surface of a $[0/45/60]_s$ laminate are given as

$$\begin{bmatrix} \epsilon_x \\ \epsilon_y \\ \gamma_{xy} \end{bmatrix} = \begin{bmatrix} 1.686 \times 10^{-8} \\ -6.500 \times 10^{-8} \\ -2.143 \times 10^{-7} \end{bmatrix}$$

and the midplane strains in this laminate are given as

$$\begin{bmatrix} \varepsilon_x^0 \\ \varepsilon_y^0 \\ \gamma_{xy}^0 \end{bmatrix} = \begin{bmatrix} 8.388 \times 10^{-6} \\ 4.762 \times 10^{-4} \\ -3.129 \times 10^{-3} \end{bmatrix}.$$

What are the midplane curvatures in this laminate, if each ply is 0.005 in. thick?

4.7 The global stresses in a three-ply laminate are given at the top and bottom surface of each ply. Each ply is 0.005 in. thick. Find the resultant forces and moments on the laminate if it has a top cross-section of 4 in. × 4 in.

	σ_{xx} (psi)	
Ply no.	**Top**	**Bottom**
1	-3.547×10^4	-2.983×10^3
2	-9.267×10^3	1.658×10^4
3	7.201×10^3	2.435×10^4

	σ_{yy} (psi)	
Ply no.	**Top**	**Bottom**
1	-2.425×10^4	-7.087×10^3
2	-1.638×10^4	9.432×10^3
3	3.155×10^3	3.553×10^4

	τ_{xy} (psi)	
Ply no.	**Top**	**Bottom**
1	-2.946×10^4	-5.564×10^3
2	-1.299×10^4	1.317×10^4
3	5.703×10^3	2.954×10^4

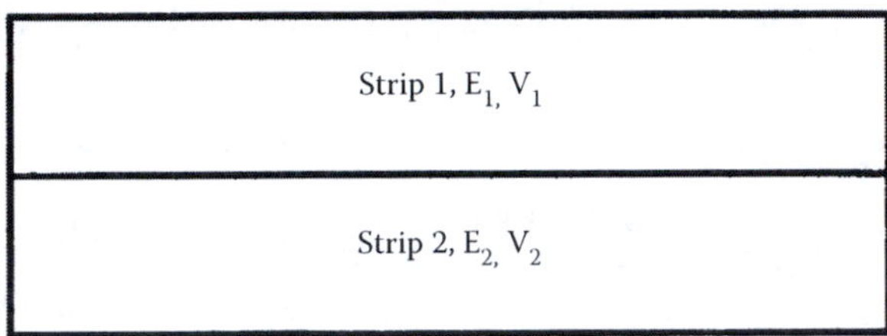

FIGURE 4.8
Laminate made of two isotropic plies.

4.8 Find the three stiffness matrices [*A*], [*B*], and [*D*] for a [0/60/–60] glass/epoxy laminate. Use the properties of glass/epoxy unidirectional lamina from Table 2.2 and assume the lamina thickness to be 0.005 in. Also, find the mass of the laminate if the top surface area of the laminate is 5 in. × 7 in. Use densities of glass and epoxy from Table 3.3 and Table 3.4, respectively.

4.9 Give expressions for the stiffness matrices [*A*], [*B*], and [*D*] for an isotropic material in terms of its thickness, *t*, Young's modulus, *E*, and Poisson's ratio, ν.

4.10 Show that, for a symmetric laminate, the coupling stiffness matrix is equal to zero.

4.11 A beam is made of two bonded isotropic strips as shown in the Figure 4.8. The two strips are of equal thickness. Find the stiffness matrices [*A*], [*B*], and [*D*].

4.12 Rewrite the expressions for the stiffness matrices [*A*], [*B*], and [*D*] in terms of the transformed reduced stiffness matrix elements, thickness of each ply, and the location of the middle of each ply with respect to the midplane of the laminate. This is called the parallel axis theorem for the laminate stiffness matrices.

4.13 Find the local stresses at the top of the 60° ply in a [0/60/–60] graphite/epoxy laminate subjected to a bending moment of $M_x = 50$ N-m/m. Use the properties of a unidirectional graphite/epoxy lamina from Table 2.1 and assume the lamina thickness to be 0.125 mm. What is the percentage of the bending moment load taken by each of the three plies?

4.14 Find the forces and moments required in a [0/60/–60] graphite/epoxy laminate to result in bending curvature of $\kappa_x = 0.1$ in.$^{-1}$ and $\kappa_y = 0.1$ in.$^{-1}$. Use the properties of a unidirectional graphite/epoxy lamina from Table 2.2 and assume the lamina thickness to be 0.005 in.

4.15 Find the extensional and flexural engineering elastic moduli of a $[45/-45]_s$ graphite/epoxy laminate. Verify the reciprocal relationships for the Poisson's ratios. Use the properties of a unidirectional graphite/epoxy lamina from Table 2.1.

4.16 Find the residual stresses at the top of the 60° ply in a [0/60/–60] graphite/epoxy laminate subjected to a temperature change of –150°F. Each lamina is 0.005 in. thick; use the properties of a unidirectional graphite/epoxy lamina from Table 2.2.

4.17 For a $[0/45]_s$ glass/epoxy laminate, find the coefficients of the thermal expansion. Use the properties of a unidirectional glass/epoxy lamina from Table 2.1. Assume thickness of each lamina as 0.125 mm. Also, find the change in the volume of the laminate if the cross-sectional area is 100 mm × 50 mm and the temperature change is 100°C.

4.18 Find the coefficients of moisture expansion of a $[0/45]_s$ graphite/epoxy laminate. The properties of a unidirectional graphite/epoxy lamina are given in Table 2.1. Assume thickness of each lamina as 0.125 mm.

4.19 Find the local stresses at the middle of the 30° ply in a [30/45] glass/epoxy laminate that is subjected to the following mechanical and hygrothermal loads: $N_x = 10^8$ lb/in.; $\Delta T = -100$°F; $\Delta C = 5\%$. Use the properties of a unidirectional glass/epoxy lamina given in Table 2.2. The thickness of each lamina is 0.005 m.

4.20 Find the difference between the vertical deflection (through the thickness) at the center and the four corners of a [0/60] graphite/epoxy cuboid laminate. The thickness of each ply is 0.005 in. and the top surface dimensions of the laminate are 20 in. × 10 in. The laminate is subjected to a temperature change of –75°F. Use the properties of a unidirectional graphite/epoxy lamina given in Table 2.2.

References

1. Ashton, J.E., Halphin, J.D., and Petit, P.H., *Primer on Composite Materials: Analysis*, Technomic Publishing Company, West Port, CT, 1969.
2. Soni, S.R., A digital algorithm for composite laminate analysis — Fortran, AFWAL-TR-81-4073, WPAFB report, 1983.
3. Hahn, H.T., Residual stresses in polymer matrix composite laminates, *J. Composite Mater.*, 10, 266, 1976.
4. Zewi, I.G., Daniel, I.M., and Gotro, J.T., Residual stresses and warpage in woven-glass/epoxy laminates, *Exp. Mech.*, 27, 44, 1987.

5

Failure, Analysis, and Design of Laminates

Chapter Objectives

- Understand the significance of stiffness, and hygrothermal and mechanical response of special cases of laminates.
- Establish the failure criteria for laminates based on failure of individual lamina in a laminate.
- Design laminated structures such as plates, thin pressure vessels, and drive shafts subjected to in-plane and hygrothermal loads.
- Introduce other mechanical design issues in laminated composites.

5.1 Introduction

The design of a laminated composite structure, such as a flat floor panel or a pressure vessel, starts with the building block of laminae, in which fiber and matrix are combined in a manufacturing process such as filament winding or prepregs. The material of the fiber and matrix, processing factors such as packing arrangements, and fiber volume fraction determine the stiffness, strength, and hygrothermal response of a single lamina. These properties can be found by using the properties of the individual constituents of the lamina or by experiments, as explained in Chapter 3. Then the laminate can have variations in material systems and in stacking sequence of plies to tailor a composite for a particular application.

In Chapter 4, we developed analysis to find the stresses and strains in a laminate under in-plane and hygrothermal loads. In this chapter, we will first use that analysis and failure theories studied in Chapter 2 to predict failure in a laminate. Then the fundamentals learned in Chapter 4 and the failure analysis discussed in this chapter will be used to design structures using laminated composites.

First, special cases of laminates that are important in the design of laminated structures will be introduced. Then the failure criterion analysis will be shown for a laminate. Eventually, we will be designing laminates mainly on the basis of optimizing for cost, weight, strength, and stiffness. Other mechanical design issues are briefly introduced at the end of the chapter.

5.2 Special Cases of Laminates

Based on angle, material, and thickness of plies, the symmetry or antisymmetry of a laminate may zero out some elements of the three stiffness matrices [*A*], [*B*], and [*D*]. These are important to study because they may result in reducing or zeroing out the coupling of forces and bending moments, normal and shear forces, or bending and twisting moments. This not only simplifies the mechanical analysis of composites, but also gives desired mechanical performance. For example, as already shown in Chapter 4, the analysis of a symmetric laminate is simplified due to the zero coupling matrix [*B*]. Mechanically, symmetric laminates result in no warpage in a flat panel due to temperature changes in processing.

5.2.1 Symmetric Laminates

A laminate is called symmetric if the material, angle, and thickness of plies are the same above and below the midplane. An example of symmetric laminates is $[0/30/60]_s$:

0
30
60
30
0

For symmetric laminates from the definition of [*B*] matrix, it can be proved that [*B*] = 0. Thus, Equation (4.29) can be decoupled to give

$$\begin{bmatrix} N_x \\ N_y \\ N_{xy} \end{bmatrix} = \begin{bmatrix} A_{11} & A_{12} & A_{16} \\ A_{12} & A_{22} & A_{26} \\ A_{16} & A_{26} & A_{66} \end{bmatrix} \begin{bmatrix} \varepsilon_x^0 \\ \varepsilon_y^0 \\ \gamma_{xy}^0 \end{bmatrix} \tag{5.1a}$$

$$\begin{bmatrix} M_x \\ M_y \\ M_{xy} \end{bmatrix} = \begin{bmatrix} D_{11} & D_{12} & D_{16} \\ D_{12} & D_{22} & D_{26} \\ D_{16} & D_{26} & D_{66} \end{bmatrix} \begin{bmatrix} \kappa_x \\ \kappa_y \\ \kappa_{xy} \end{bmatrix}. \tag{5.1b}$$

This shows that the force and moment terms are uncoupled. Thus, if a laminate is subjected only to forces, it will have zero midplane curvatures. Similarly, if it is subjected only to moments, it will have zero midplane strains.

The uncoupling between extension and bending in symmetric laminates makes analyzing such laminates simpler. It also prevents a laminate from twisting due to thermal loads, such as cooling down from processing temperatures and temperature fluctuations during use such as in a space shuttle, etc.

5.2.2 Cross-Ply Laminates

A laminate is called a cross-ply laminate (also called laminates with specially orthotropic layers) if only 0 and 90° plies were used to make a laminate. An example of a cross ply laminate is a $[0/90_2/0/90]$ laminate:

0
90
90
0
90

For cross-ply laminates, $A_{16} = 0$, $A_{26} = 0$, $B_{16} = 0$, $B_{26} = 0$, $D_{16} = 0$, and $D_{26} = 0$; thus, Equation (4.29) can be written as

$$\begin{bmatrix} N_x \\ N_y \\ N_{xy} \\ M_x \\ M_y \\ M_{xy} \end{bmatrix} = \begin{bmatrix} A_{11} & A_{12} & 0 & B_{11} & B_{12} & 0 \\ A_{12} & A_{22} & 0 & B_{12} & B_{22} & 0 \\ 0 & 0 & A_{66} & 0 & 0 & B_{66} \\ B_{11} & B_{12} & 0 & D_{11} & D_{12} & 0 \\ B_{12} & B_{22} & 0 & D_{12} & D_{22} & 0 \\ 0 & 0 & B_{66} & 0 & 0 & D_{66} \end{bmatrix} \begin{bmatrix} \varepsilon_x^0 \\ \varepsilon_y^0 \\ \gamma_{xy}^0 \\ \kappa_x \\ \kappa_y \\ \kappa_{xy} \end{bmatrix}. \tag{5.2}$$

In these cases, uncoupling occurs between the normal and shear forces, as well as between the bending and twisting moments. If a cross-ply laminate is also symmetric, then in addition to the preceding uncoupling, the coupling matrix $[B] = 0$ and no coupling takes place between the force and moment terms.

5.2.3 Angle Ply Laminates

A laminate is called an angle ply laminate if it has plies of the same material and thickness and only oriented at $+\theta$ and $-\theta$ directions. An example of an angle ply laminate is [–40/40/–40/40]:

–40
40
–40
40

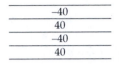

If a laminate has an even number of plies, then $A_{16} = A_{26} = 0$. However, if the number of plies is odd and it consists of alternating $+\theta$ and $-\theta$ plies, then it is symmetric, giving $[B] = 0$, and A_{16}, A_{26}, D_{16}, and D_{26} also become small as the number of layers increases for the same laminate thickness. This behavior is similar to the symmetric cross-ply laminates. However, these angle ply laminates have higher shear stiffness and shear strength properties than cross-ply laminates.

5.2.4 Antisymmetric Laminates

A laminate is called antisymmetric if the material and thickness of the plies are the same above and below the midplane, but the ply orientations at the same distance above and below the midplane are negative of each other. An example of an antisymmetric laminate is:

45
60
–60
–45

From Equation (4.28a) and Equation (4.28c), the coupling terms of the extensional stiffness matrix, $A_{16} = A_{26} = 0$, and the coupling terms of the bending stiffness matrix, $D_{16} = D_{26} = 0$:

$$\begin{bmatrix} N_x \\ N_y \\ N_{xy} \\ M_x \\ M_y \\ M_{xy} \end{bmatrix} = \begin{bmatrix} A_{11} & A_{12} & 0 & B_{11} & B_{12} & B_{16} \\ A_{12} & A_{22} & 0 & B_{12} & B_{22} & B_{26} \\ 0 & 0 & A_{66} & B_{16} & B_{26} & B_{66} \\ B_{11} & B_{22} & B_{16} & D_{11} & D_{12} & 0 \\ B_{12} & B_{22} & B_{26} & D_{12} & D_{22} & 0 \\ B_{16} & B_{26} & B_{66} & 0 & 0 & D_{66} \end{bmatrix} \begin{bmatrix} \varepsilon_x^0 \\ \varepsilon_y^0 \\ \gamma_{xy}^0 \\ \kappa_x \\ \kappa_y \\ \kappa_{xy} \end{bmatrix}. \tag{5.3}$$

5.2.5 Balanced Laminate

A laminate is balanced if layers at angles other than 0 and 90° occur only as plus and minus pairs of $+\theta$ and $-\theta$. The plus and minus pairs do not need to be adjacent to each other, but the thickness and material of the plus and minus pairs need to be the same. Here, the terms $A_{16} = A_{26} = 0$. An example of a balanced laminate is [30/40/–30/30/–30/–40]:

30
40
–30
30
–30
–40

From Equation (4.28a),

$$\begin{bmatrix} N_x \\ N_y \\ N_{xy} \\ M_x \\ M_y \\ M_{xy} \end{bmatrix} = \begin{bmatrix} A_{11} & A_{12} & 0 & B_{11} & B_{12} & B_{16} \\ A_{12} & A_{22} & 0 & B_{12} & B_{22} & B_{26} \\ 0 & 0 & A_{26} & B_{16} & B_{26} & B_{66} \\ B_{11} & B_{12} & B_{16} & D_{11} & D_{12} & D_{16} \\ B_{12} & B_{22} & B_{26} & D_{12} & D_{22} & D_{26} \\ B_{16} & B_{26} & B_{66} & D_{16} & D_{26} & D_{66} \end{bmatrix} \begin{bmatrix} \varepsilon_x^0 \\ \varepsilon_y^0 \\ \gamma_{xy}^0 \\ \kappa_x \\ \kappa_y \\ \kappa_{xy} \end{bmatrix}. \tag{5.4}$$

5.2.6 Quasi-Isotropic Laminates

For a plate of isotropic material with Young's modulus, E, Poisson's ratio, ν, and thickness, h, the three stiffness matrices are

$$[A] = \begin{bmatrix} \dfrac{E}{1-\nu^2} & \dfrac{\nu E}{1-\nu^2} & 0 \\ \dfrac{\nu E}{1-\nu^2} & \dfrac{E}{1-\nu^2} & 0 \\ 0 & 0 & \dfrac{E}{2(1+\nu)} \end{bmatrix} h, \tag{5.5}$$

$$[B] = \begin{bmatrix} 0 & 0 & 0 \\ 0 & 0 & 0 \\ 0 & 0 & 0 \end{bmatrix}, \tag{5.6}$$

$$[D] = \begin{bmatrix} \frac{E}{12(1-\nu^2)} & \frac{\nu E}{12(1-\nu^2)} & 0 \\ \frac{\nu E}{12(1-\nu^2)} & \frac{E}{12(1-\nu^2)} & 0 \\ 0 & 0 & \frac{E}{24(1+\nu)} \end{bmatrix} h^3 . \tag{5.7}$$

A laminate is called quasi-isotropic if its extensional stiffness matrix $[A]$ behaves like that of an isotropic material. This implies not only that $A_{11} = A_{22}$, $A_{16} = A_{26} = 0$, and $A_{66} = \frac{A_{11} - A_{12}}{2}$, but also that these stiffnesses are independent of the angle of rotation of the laminate. The reason for calling such a laminate quasi-isotropic and not isotropic is that the other stiffness matrices, $[B]$ and $[D]$, may not behave like isotropic materials. Examples of quasi-isotropic laminates include [0/±60], $[0/\pm 45/90]_s$, and [0/36/72/–36/–72].

Example 5.1

A [0/±60] graphite/epoxy laminate is quasi-isotropic. Find the three stiffness matrices $[A]$, $[B]$, and $[D]$ and show that

1. $A_{11} = A_{22}$; $A_{16} = A_{26} = 0$; $A_{66} = \frac{A_{11} - A_{12}}{2}$.
2. $[B] \neq 0$, unlike isotropic materials.
3. $[D]$ matrix is unlike isotropic materials.

Use properties of unidirectional graphite/epoxy lamina from Table 2.1. Each lamina has a thickness of 5 mm.

Solution

From Example 2.6, the reduced stiffness matrix $[Q]$ for the 0° graphite/epoxy lamina is

$$[Q] = \begin{bmatrix} 181.8 & 2.897 & 0 \\ 2.897 & 10.35 & 0 \\ 0 & 0 & 7.17 \end{bmatrix} (10^9)\ Pa .$$

From Equation (2.104), the transformed reduced stiffness matrices for the three plies are

$$[\bar{Q}]_0 = \begin{bmatrix} 181.8 & 2.897 & 0 \\ 2.897 & 10.35 & 0 \\ 0 & 0 & 7.17 \end{bmatrix} (10^9)\ Pa\ ,$$

$$[\bar{Q}]_{60} = \begin{bmatrix} 23.65 & 32.46 & 20.05 \\ 32.46 & 109.4 & 54.19 \\ 20.05 & 54.19 & 36.74 \end{bmatrix} (10^9)\ Pa\ ,$$

$$[\bar{Q}]_{-60} = \begin{bmatrix} 23.65 & 32.46 & -20.05 \\ 32.46 & 109.4 & -54.19 \\ -20.05 & -54.19 & 36.74 \end{bmatrix} (10^9)\ Pa.$$

The total thickness of the laminate is $h = (0.005)(3) = 0.015$ m.

The midplane is 0.0075 m from the top and bottom of the laminate. Thus, using Equation (4.20),

$$h_0 = -0.0075 \text{ m}$$

$$h_1 = -0.0025 \text{ m}$$

$$h_2 = 0.0025 \text{ m}$$

$$h_3 = 0.0075 \text{ m}$$

Using Equation (4.28a) to Equation (4.28c), one can now calculate the stiffness matrices $[A]$, $[B]$, and $[D]$, respectively, as shown in Example 4.2:

$$[A] = \begin{bmatrix} 1.146 & 0.3391 & 0 \\ 0.3391 & 1.146 & 0 \\ 0 & 0 & 0.4032 \end{bmatrix} (10^9)\ Pa\text{-}m\ ,$$

$$[B] = \begin{bmatrix} -3.954 & 0.7391 & -0.5013 \\ 0.7391 & 2.476 & -1.355 \\ -0.5013 & -1.355 & 0.7391 \end{bmatrix} (10^6)\ Pa\text{-}m^2\ ,$$

$$[D] = \begin{bmatrix} 28.07 & 5.126 & -2.507 \\ 5.126 & 17.35 & -6.774 \\ -2.507 & -6.774 & 6.328 \end{bmatrix} (10^3)\ Pa\text{-}m^3 .$$

1. From the extensional stiffness matrix $[A]$,

$$A_{11} = A_{22} = 1.146 \times 10^9\ Pa\text{-}m$$

$$A_{16} = A_{26} = 0$$

$$\frac{A_{11} - A_{12}}{2} = \frac{1.146 - 0.3391}{2} \times 10^9$$

$$= 0.4032 \times 10^9\ Pa\text{-}m$$

$$= A_{66}.$$

This behavior is similar to that of an isotropic material. However, a quasi-isotropic laminate should give the same $[A]$ matrix, if a constant angle is added to each of the layers of the laminate. For example, adding 30° to each ply angle of the [0/±60] laminate gives a [30/90/–30] laminate, which has the same $[A]$ matrix as the [0/±60] laminate.

2. Unlike isotropic materials, the coupling stiffness matrix $[B]$ of the [0/±60] laminate is nonzero.
3. In an isotropic material,

$$D_{11} = D_{22} ,$$

$$D_{16} = D_{26} = 0 ,$$

and

$$D_{66} = \frac{D_{11} - D_{12}}{2} .$$

In this example, unlike isotropic materials, $D_{11} \neq D_{22}$ because

$$D_{11} = 28.07 \times 10^3\ Pa\text{-}m^3$$

$$D_{22} = 17.35 \times 10^3 \ Pa\text{-}m^3$$

$$D_{16} \neq 0, D_{26} \neq 0 \text{ as}$$

$$D_{16} = -2.507 \times 10^3 \ Pa\text{-}m^3$$

$$D_{26} = -6.774 \times 10^3 \ Pa\text{-}m^3$$

$$\frac{D_{11} - D_{12}}{2} \neq D_{66}$$

because

$$\frac{D_{11} - D_{12}}{2} = \frac{28.07 \times 10^3 - 5.126 \times 10^3}{2}$$

$$= 11.47 \times 10^3 \ Pa\text{-}m^3$$

$$D_{66} = 6.328 \times 10^3 \ Pa\text{-}m^3.$$

One can make a quasi-isotropic laminate by having a laminate with N ($N \geq 3$) lamina of the same material and thickness, where each lamina is oriented at an angle of $180°/N$ between each other. For example, a three-ply laminate will require the laminae to be oriented at $180°/3 = 60°$ to each other. Thus, [0/60/–60], [30/90/–30], and [45/–75/–15] are all quasi-isotropic laminates. One can make the preceding combinations symmetric or repeated to give quasi-isotropic laminates, such as $[0/\pm60]_s$, $[0/\pm60]_s$, and $[0/\pm60]_{2s}$ laminates. The symmetry of the laminates zeros out the coupling matrix $[B]$ and makes its behavior closer (not same) to that of an isotropic material.

Example 5.2

Show that the extensional stiffness matrix for a general N-ply quasi-isotropic laminate is given by

$$[A] = \begin{bmatrix} U_1 & U_4 & 0 \\ U_4 & U_1 & 0 \\ 0 & 0 & \dfrac{U_1 - U_4}{2} \end{bmatrix} h. \tag{5.8}$$

where U_1 and U_4 are the stiffness invariants given by Equation (2.132) and h is the thickness of the laminate. Also, find the in-plane engineering stiffness constants of the laminate.

Solution

From Equation (2.131a), for a general angle ply with angle θ,

$$\bar{Q}_{11} = U_1 + U_2 \text{ Cos} 2\theta + U_3 \text{ Cos} 4\theta. \tag{5.9}$$

For the kth ply of the quasi-isotropic laminate with an angle θ_k,

$$(\bar{Q}_{11})_k = U_1 + U_2 \text{ Cos} 2\theta_k + U_3 \text{ Cos} 4\theta_k, \tag{5.10}$$

where

$$\theta_1 = \frac{\pi}{N}, \theta_2 = \frac{2\pi}{N}, \ldots, \theta_k = \frac{k\pi}{N}, \ldots, \theta_{N-1} = \frac{(N-1)\pi}{N}, \theta_N = \pi.$$

From Equation (4.28a),

$$A_{11} = \sum_{k=1}^{N} t_k (\bar{Q}_{11})_k \,, \tag{5.11}$$

where t_k = thickness of kth lamina.

Because the thickness of the laminate is h and all laminae are of the same thickness,

$$t_k = \frac{h}{N}, \quad k = 1, 2, \ldots\ldots\ldots\ldots, N, \tag{5.12}$$

and, substituting Equation (5.10) in Equation (5.11),

$$\begin{aligned} A_{11} &= \frac{h}{N}\sum_{k=1}^{N}(U_1 + U_2 \text{ Cos } 2\theta_k + U_3 \text{ Cos } 4\theta_k) \\ &= hU_1 + U_2\frac{h}{N}\sum_{k=1}^{N}\text{Cos } 2\theta_k + U_3\frac{h}{N}\sum_{k=1}^{N}\text{Cos } 4\theta_k. \end{aligned} \tag{5.13}$$

Using the following identity,[1]

$$\sum_{k=1}^{N} \text{Cos } kx \equiv \frac{\text{Sin}\left(N+\frac{1}{2}\right)x}{2\ \text{Sin}\left(\frac{x}{2}\right)} - \frac{1}{2}\,. \tag{5.14}$$

Then,

$$\sum_{k=1}^{N} \text{Cos } 2\theta_k = 0 \ \textit{for}\ N \geq 1 \tag{5.15a}$$

$$\sum_{k=1}^{N} \text{Cos } 4\theta_k = 0 \ \textit{for}\ N \geq 3. \tag{5.15b}$$

Thus,

$$A_{11} = U_1 h\,. \tag{5.16a}$$

Similarly, it can be shown that

$$A_{12} = U_4 h\,, \tag{5.16b}$$

$$A_{22} = U_1 h\,, \tag{5.16c}$$

$$A_{66} = \left(\frac{U_1 - U_4}{2}\right) h\,. \tag{5.16d}$$

Therefore,

$$[A] = \begin{bmatrix} U_1 & U_4 & 0 \\ U_4 & U_1 & 0 \\ 0 & 0 & \frac{U_1 - U_4}{2} \end{bmatrix} h. \tag{5.17}$$

Because Equation (5.15b) is valid only for $N \geq 3$, this proves that one needs at least three plies to make a quasi-isotropic laminate.

For a symmetric quasi-isotropic laminate, the extensional compliance matrix is given by

$$[A^*] = \frac{1}{h}\begin{bmatrix} \frac{U_1}{U_1^2 - U_4^2} & -\frac{U_4}{U_1^2 - U_4^2} & 0 \\ -\frac{U_4}{U_1^2 - U_4^2} & \frac{U_1}{U_1^2 - U_4^2} & 0 \\ 0 & 0 & \frac{2}{U_1 - U_4} \end{bmatrix}. \tag{5.18}$$

From the definitions of engineering constants given in Equations (4.35), (4.37), (4.39), (4.42), and (4.45), and using Equation (5.18), the elastic moduli of the laminate are independent of the angle of the lamina and are given by

$$E_x = E_y = E_{iso} = \frac{1}{A_{11}^* h} = \frac{U_1^2 - U_4^2}{U_1}, \tag{5.19a}$$

$$G_{xy} = G_{iso} = \frac{1}{A_{66}^* h} = \frac{U_1 - U_4}{2}, \tag{5.19b}$$

$$\nu_{xy} = \nu_{yx} = \nu_{iso} = -\frac{A_{12}^*}{A_{22}^*} = \frac{U_4}{U_1}. \tag{5.19c}$$

5.3 Failure Criterion for a Laminate

A laminate will fail under increasing mechanical and thermal loads. The laminate failure, however, may not be catastrophic. It is possible that some layer fails first and that the composite continues to take more loads until all the plies fail. Failed plies may still contribute to the stiffness and strength of the laminate. The degradation of the stiffness and strength properties of each failed lamina depends on the philosophy followed by the user.

- When a ply fails, it may have cracks parallel to the fibers. This ply is still capable of taking load parallel to the fibers. Here, the cracked ply can be replaced by a hypothetical ply that has no transverse stiffness, transverse tensile strength, and shear strength. The longitudinal modulus and strength remain unchanged.

- When a ply fails, fully discount the ply and replace the ply of near zero stiffness and strength. Near zero values avoid singularities in stiffness and compliance matrices.

The procedure for finding the successive loads between first ply failure and last ply failure given next follows the fully discounted method:

1. Given the mechanical loads, apply loads in the same ratio as the applied loads. However, apply the actual temperature change and moisture content.
2. Use laminate analysis to find the midplane strains and curvatures.
3. Find the local stresses and strains in each ply under the assumed load.
4. Use the ply-by-ply stresses and strains in ply failure theories discussed in Section 2.8 to find the strength ratio. Multiplying the strength ratio to the applied load gives the load level of the failure of the first ply. This load is called the *first ply failure* load.
5. Degrade fully the stiffness of damaged ply or plies. Apply the actual load level of previous failure.
6. Go to step 2 to find the strength ratio in the undamaged plies:
   - If the strength ratio is more than one, multiply the strength ratio to the applied load to give the load level of the next ply failure and go to step 2.
   - If the strength ratio is less than one, degrade the stiffness and strength properties of all the damaged plies and go to step 5.
7. Repeat the preceding steps until all the plies in the laminate have failed. The load at which all the plies in the laminate have failed is called the *last ply failure*.

The procedure for partial discounting of fibers is more complicated. The noninteractive maximum stress and maximum strain failure criteria are used to find the mode of failure. Based on the mode of failure, the appropriate elastic moduli and strengths are partially or fully discounted.

Example 5.3

Find the ply-by-ply failure loads for a $[0/\overline{90}]_s$ graphite/epoxy laminate. Assume the thickness of each ply is 5 mm and use properties of unidirectional graphite/epoxy lamina from Table 2.1. The only load applied is a tensile normal load in the *x*-direction — that is, the direction parallel to the fibers in the 0° ply.

Solution

Because the laminate is symmetric and the load applied is a normal load, only the extensional stiffness matrix is required. From Example 4.4, the extensional compliance matrix is

$$[A^*] = \begin{bmatrix} 5.353\times10^{-10} & -2.297\times10^{-11} & 0 \\ -2.297\times10^{-11} & 9.886\times10^{-10} & 0 \\ 0 & 0 & 9.298\times10^{-9} \end{bmatrix} \frac{1}{Pa\text{-}m},$$

which, from Equation (5.1a), gives the midplane strains for symmetric laminates subjected to $N_x = 1$ N/m as

$$\begin{bmatrix} \varepsilon_x^0 \\ \varepsilon_y^0 \\ \gamma_{xy}^0 \end{bmatrix} = \begin{bmatrix} 5.353\times10^{-10} \\ -2.297\times10^{-11} \\ 0 \end{bmatrix}.$$

The midplane curvatures are zero because the laminate is symmetric and no bending and no twisting loads are applied.

The global strains in the top 0° ply at the top surface can be found as follows using Equation (4.16),

$$\begin{bmatrix} \varepsilon_x \\ \varepsilon_y \\ \gamma_{xy} \end{bmatrix} = \begin{bmatrix} 5.353\times10^{-10} \\ -2.297\times10^{-11} \\ 0 \end{bmatrix} + (0.0075)\begin{bmatrix} 0 \\ 0 \\ 0 \end{bmatrix}$$

$$= \begin{bmatrix} 5.353\times10^{-10} \\ -2.297\times10^{-11} \\ 0 \end{bmatrix}.$$

Using Equation (2.103), one can find the global stresses at the top surface of the top 0° ply as

$$\begin{bmatrix} \sigma_x \\ \sigma_y \\ \tau_{xy} \end{bmatrix}_{0°,top} = \begin{bmatrix} 181.8 & 2.897 & 0 \\ 2.897 & 10.35 & 0 \\ 0 & 0 & 7.17 \end{bmatrix} (10^9) \begin{bmatrix} 5.353\times10^{-10} \\ -2.297\times10^{-11} \\ 0 \end{bmatrix}$$

$$= \begin{bmatrix} 9.726 \times 10^1 \\ 1.313 \\ 0 \end{bmatrix} Pa .$$

Using the transformation Equation (2.94), the local stresses at the top surface of the top 0° ply are

$$\begin{bmatrix} \sigma_1 \\ \sigma_2 \\ \tau_{12} \end{bmatrix}_{0°,\ top} = \begin{bmatrix} 1 & 0 & 0 \\ 0 & 1 & 0 \\ 0 & 0 & 1 \end{bmatrix} \begin{bmatrix} 9.726 \times 10^1 \\ 1.313 \times 10^0 \\ 0 \end{bmatrix}$$

$$= \begin{bmatrix} 9.726 \times 10^1 \\ 1.313 \\ 0 \end{bmatrix} Pa$$

All the local stresses and strains in the laminate are summarized in Table 5.1 and Table 5.2.

TABLE 5.1

Local Stresses (Pa) in Example 5.3

Ply no.	Position	σ_1	σ_2	τ_{12}
1 (0°)	Top	9.726×10^1	1.313×10^0	0.0
	Middle	9.726×10^1	1.313×10^0	0.0
	Bottom	9.726×10^1	1.313×10^0	0.0
2 (90°)	Top	-2.626×10^0	5.472×10^0	0.0
	Middle	-2.626×10^0	5.472×10^0	0.0
	Bottom	-2.626×10^0	5.472×10^0	0.0
3 (0°)	Top	9.726×10^1	1.313×10^0	0.0
	Middle	9.726×10^1	1.313×10^0	0.0
	Bottom	9.726×10^1	1.313×10^0	0.0

TABLE 5.2

Local Strains in Example 5.3

Ply no.	Position	ε_1	ε_2	τ_{12}
1 (0°)	Top	5.353×10^{-10}	-2.297×10^{-11}	0.0
	Middle	5.353×10^{-10}	-2.297×10^{-11}	0.0
	Bottom	5.353×10^{-10}	-2.297×10^{-11}	0.0
2 (90°)	Top	-2.297×10^{-11}	5.353×10^{-10}	0.0
	Middle	-2.297×10^{-11}	5.353×10^{-10}	0.0
	Bottom	-2.297×10^{-11}	5.353×10^{-10}	0.0
3 (0°)	Top	5.353×10^{-10}	-2.297×10^{-11}	0.0
	Middle	5.353×10^{-10}	-2.297×10^{-11}	0.0
	Bottom	5.353×10^{-10}	-2.297×10^{-11}	0.0

The Tsai–Wu failure theory applied to the top surface of the top 0° ply is applied as follows. The local stresses are

$\sigma_1 = 9.726 \times 10^1$ Pa

$\sigma_2 = 1.313$ Pa

$\tau_{12} = 0$

Using the parameters H_1, H_2, H_6, H_{11}, H_{22}, H_{66}, and H_{12} from Example 2.19, the Tsai–Wu failure theory Equation (2.152) gives the strength ratio as

$$(0)\,(9.726 \times 10^1)\,SR + (2.093 \times 10^{-8})\,(1.313)\,SR + (0 \times 0) +$$
$$(4.4444 \times 10^{-19})\,(9.726 \times 10^1)^2(SR)^2 + (1.0162 \times 10^{-16})\,(1.313)^2(SR)^2$$
$$+ (2.1626 \times 10^{-16})\,(0)^2 + 2(-3.360 \times 10^{-18})\,(9.726 \times 10^1)\,(1.313)(SR)^2 = 1$$
$$SR = 1.339 \times 10^7.$$

The maximum strain failure theory can also be applied to the top surface of the top 0° ply as follows. The local strains are

$$\begin{bmatrix} \varepsilon_1 \\ \varepsilon_2 \\ \gamma_{12} \end{bmatrix} = \begin{bmatrix} 5.353 \times 10^{-10} \\ -2.297 \times 10^{-11} \\ 0.000 \end{bmatrix}.$$

Then, according to maximum strain failure theory (Equation 2.143), the strength ratio is given by

$$SR = \min\{[(1500 \times 10^6)/(181 \times 10^9)]/(5.353 \times 10^{-10}),$$
$$[(246 \times 10^6)/(10.3 \times 10^9)]/(2.297 \times 10^{-11})\} = 1.548 \times 10^7.$$

The strength ratios for all the plies in the laminate are summarized in Table 5.3 using the maximum strain and Tsai–Wu failure theories. The symbols in

TABLE 5.3

Strength Ratios in Example 5.3

Ply no.	Position	Maximum strain	Tsai–Wu
1 (0°)	Top	1.548×10^7 (1T)	1.339×10^7
	Middle	1.548×10^7 (1T)	1.339×10^7
	Bottom	1.548×10^7 (1T)	1.339×10^7
2 (90°)	Top	7.254×10^6 (2T)	7.277×10^6
	Middle	7.254×10^6 (2T)	7.277×10^6
	Bottom	7.254×10^6 (2T)	7.277×10^6
3 (0°)	Top	1.548×10^7 (1T)	1.339×10^7
	Middle	1.548×10^7 (1T)	1.339×10^7
	Bottom	1.548×10^7 (1T)	1.339×10^7

the parentheses in the maximum strain failure theory column denote the mode of failure and are explained at the bottom of Table 2.3.

From Table 5.3 and using the Tsai–Wu theory, the minimum strength ratio is found for the 90° ply. This strength ratio gives the maximum value of the allowable normal load as

$$N_x = 7.277 \times 10^6 \frac{N}{m}$$

and the maximum value of the allowable normal stress as

$$\frac{N_x}{h} = \frac{7.277 \times 10^6}{0.015},$$

$$= 0.4851 \times 10^9 \ Pa$$

where h = thickness of the laminate.

The normal strain in the x-direction at this load is

$$(\varepsilon_x^0)_{\text{first ply failure}} = (5.353 \times 10^{-10})(7.277 \times 10^6)$$

$$= 3.895 \times 10^{-3}$$

Now, degrading the 90° ply completely involves assuming zero stiffnesses and strengths of the 90° lamina. Complete degradation of a ply does not allow further failure of that ply. For the undamaged plies, the $[0/90]_s$ laminate has two reduced stiffness matrices as

$$[Q] = \begin{bmatrix} 181.8 & 2.897 & 0 \\ 2.897 & 10.35 & 0 \\ 0 & 0 & 7.17 \end{bmatrix} GPa$$

and, for the damaged ply,

$$[Q] = \begin{bmatrix} 0 & 0 & 0 \\ 0 & 0 & 0 \\ 0 & 0 & 0 \end{bmatrix} GPa\ .$$

Using Equation (4.28a), the extensional stiffness matrix

$$A_{ij} = \sum_{k=1}^{3} [\bar{Q}_{ij}]_k (h_k - h_{k-1})$$

$$[A] = \begin{bmatrix} 181.8 & 2.897 & 0 \\ 2.897 & 10.35 & 0 \\ 0 & 0 & 7.17 \end{bmatrix} (10^9)(0.005)$$

$$+ \begin{bmatrix} 0 & 0 & 0 \\ 0 & 0 & 0 \\ 0 & 0 & 0 \end{bmatrix} (10^9)(0.005)$$

$$+ \begin{bmatrix} 181.8 & 2.897 & 0 \\ 2.897 & 10.35 & 0 \\ 0 & 0 & 7.17 \end{bmatrix} (10^9)(0.005)$$

$$[A] = \begin{bmatrix} 181.8 & 2.897 & 0 \\ 2.897 & 10.35 & 0 \\ 0 & 0 & 7.17 \end{bmatrix} (10^7)\ Pa\text{-}m.$$

Inverting the new extensional stiffness matrix $[A]$, the new extensional compliance matrix is

$$[A^*] = \begin{bmatrix} 5.525 \times 10^{-10} & -1.547 \times 10^{-10} & 0 \\ -1.547 \times 10^{-10} & 9.709 \times 10^{-9} & 0 \\ 0 & 0 & 1.395 \times 10^{-8} \end{bmatrix} \frac{1}{Pa\text{-}m},$$

which gives midplane strains subjected to $N_x = 1$ N/m by Equation (5.1a) as

$$\begin{bmatrix} \varepsilon_x^0 \\ \varepsilon_y^0 \\ \gamma_{xy} \end{bmatrix} = \begin{bmatrix} 5.525 \times 10^{-10} & -1.547 \times 10^{-10} & 0 \\ -1.547 \times 10^{-10} & 9.709 \times 10^{-9} & 0 \\ 0 & 0 & 1.395 \times 10^{-8} \end{bmatrix} \begin{bmatrix} 1 \\ 0 \\ 0 \end{bmatrix}$$

TABLE 5.4

Local Stresses after First Ply Failure in Example 5.3

Ply no.	Position	σ_1	σ_2	τ_{12}
1 (0°)	Top	1.0000×10^2	0.0	0.0
	Middle	1.0000×10^2	0.0	0.0
	Bottom	1.0000×10^2	0.0	0.0
2 (90°)	Top	—	—	—
	Middle	—	—	—
	Bottom	—	—	—
3 (0°)	Top	1.0000×10^2	0.0	0.0
	Middle	1.0000×10^2	0.0	0.0
	Bottom	1.0000×10^2	0.0	0.0

TABLE 5.5

Local Strains after First Ply Failure in Example 5.3

Ply no.	Position	ε_1	ε_2	γ_{12}
1 (0°)	Top	5.25×10^{-10}	-1.547×10^{-10}	0.0
	Middle	5.525×10^{-10}	-1.547×10^{-10}	0.0
	Bottom	5.525×10^{-10}	-1.547×10^{-10}	0.0
2 (90°)	Top	—	—	—
	Middle	—	—	—
	Bottom	—	—	—
3 (0°)	Top	5.525×10^{-10}	-1.547×10^{-10}	0.0
	Middle	5.525×10^{-10}	-1.547×10^{-10}	0.0
	Bottom	5.525×10^{-10}	-1.547×10^{-10}	0.0

$$\begin{bmatrix} \varepsilon_x^0 \\ \varepsilon_y^0 \\ \gamma_{xy}^0 \end{bmatrix} = \begin{bmatrix} 5.525 \times 10^{-10} \\ -1.547 \times 10^{-10} \\ 0 \end{bmatrix}.$$

These strains are close to those obtained before the ply failure only because the 90° ply takes a small percentage of the load out of the normal load in the x-direction.

The local stresses in each layer are found using earlier techniques given in this example and are shown in Table 5.4. The strength ratios in each layer are also found using methods given in this example and are shown in Table 5.5.

From Table 5.6 and using Tsai–Wu failure theory, the minimum strength ratio is found in both the 0° plies. This strength ratio gives the maximum value of the normal load as

$$N_x = 1.5 \times 10^7 \frac{N}{m}$$

TABLE 5.6

Strength Ratios after First Ply Failure in Example 5.3

Ply no.	Position	Max strain	Tsai–Wu
1 (0°)	Top	1.5000×10^7 (1T)	1.5000×10^7
	Middle	1.5000×10^7(1T)	1.5000×10^7
	Bottom	1.5000×10^7(1T)	1.5000×10^7
2 (90°)	Top	—	—
	Middle	—	—
	Bottom	—	—
3 (0°)	Top	1.5000×10^7(1T)	1.5000×10^7
	Middle	1.5000×10^7(1T)	1.5000×10^7
	Bottom	1.5000×10^7(1T)	1.5000×10^7

and the maximum value of the allowable normal stress as

$$\frac{N_x}{h} = \frac{1.5 \times 10^7}{0.015},$$

$$= 1.0 \times 10^9 \ Pa$$

where h is the thickness of the laminate.

The normal strain in the x-direction at this load is

$$(\varepsilon_x^o)_{\text{last ply failure}} = (5.525 \times 10^{-10})(1.5 \times 10^7).$$

$$= 8.288 \times 10^{-3}$$

The preceding load is also the last ply failure (LPF) because none of the layers is left undamaged. Plotting the stress vs. strain curve for the laminate until last ply failure shows that the curve will consist of two linear curves, each ending at each ply failure. The slope of the two lines will be the Young's modulus in x direction for the undamaged laminate and for the FPF laminate — that is, using Equation (4.35),

$$E_x = \frac{1}{(0.015)(5.353 \times 10^{-10})},$$

$$= 124.5 \ GPa$$

until first ply failure, and

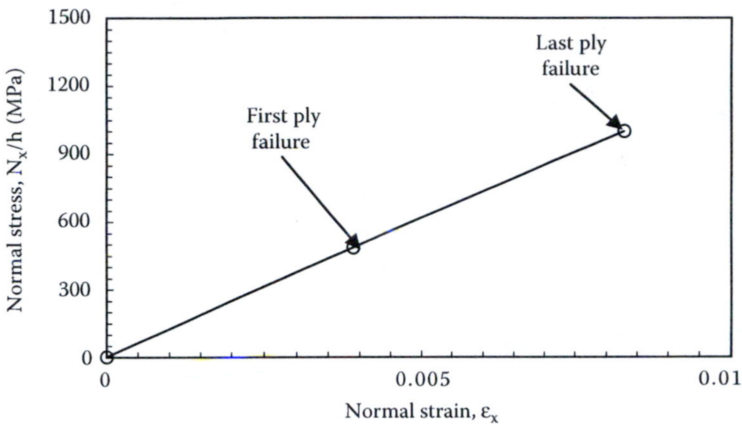

FIGURE 5.1
Stress–strain curve showing ply-by-ply failure of a laminated composite.

$$E_x = \frac{(N_x / h)_{\text{last ply failure}} - (N_x / h)_{\text{first play failure}}}{(\varepsilon_x^o)_{\text{last play failure}} - (\varepsilon_x^o)_{\text{first play failure}}}$$

$$= \frac{0.1 \times 10^{10} - 0.4851 \times 10^9}{8.288 \times 10^{-3} - 3.895 \times 10^{-3}},$$

$$= 117.2 \; GPa$$

after first ply failure and until last ply failure (Figure 5.1).

Example 5.4

Repeat Example 5.3 for the first ply failure and use Tsai–Wu failure theory now with an additional thermal load: a temperature change of –75°C.

Solution

The laminate is symmetric and the load applied is a normal load and a temperature change. Thus, only the extensional stiffness matrix is needed. From Example 5.3,

$$[A^*] = \begin{bmatrix} 5.353 \times 10^{-10} & -2.297 \times 10^{-11} & 0 \\ -2.297 \times 10^{-11} & 9.886 \times 10^{-10} & 0 \\ 0 & 0 & 9.298 \times 10^{-9} \end{bmatrix} \frac{1}{Pa\text{-}m}.$$

Corresponding to a temperature change of –75°C, the mechanical stresses can be found as follows. The fictitious thermal forces given by Equation (4.64) are

$$\begin{bmatrix} N_x^T \\ N_y^T \\ N_{xy}^T \end{bmatrix} = (-75)\begin{bmatrix} 181.8 & 2.897 & 0 \\ 2.897 & 10.35 & 0 \\ 0 & 0 & 7.17 \end{bmatrix}(10^9)\begin{bmatrix} 0.200\times 10^{-7} \\ 0.225\times 10^{-4} \\ 0 \end{bmatrix}[-0.0025-(-0.0075)]$$

$$+ (-75)\begin{bmatrix} 10.35 & 2.897 & 0 \\ 2.897 & 181.8 & 0 \\ 0 & 0 & 7.17 \end{bmatrix}(10^9)\begin{bmatrix} 0.225\times 10^{-4} \\ 0.200\times 10^{-7} \\ 0 \end{bmatrix}[0.0025-(-0.0025)]$$

$$+ (-75)\begin{bmatrix} 181.8 & 2.897 & 0 \\ 2.897 & 10.35 & 0 \\ 0 & 0 & 7.17 \end{bmatrix}(10^9)\begin{bmatrix} 0.200\times 10^{-7} \\ 0.225\times 10^{-4} \\ 0 \end{bmatrix}[0.0075-(-0.0025)]$$

$$= \begin{bmatrix} -1.389\times 10^{5} \\ -2.004\times 10^{5} \\ 0 \end{bmatrix} Pa\text{-}m\ .$$

Because the laminate is symmetric, the fictitious thermal moments are zero. This also then gives only midplane strains in the laminate without any plate curvatures. The midplane strain due to the thermal load is given by

$$\begin{bmatrix} \varepsilon_x^0 \\ \varepsilon_y^0 \\ \gamma_{xy}^0 \end{bmatrix} = \begin{bmatrix} 5.353\times 10^{-10} & -2.297\times 10^{-11} & 0 \\ -2.297\times 10^{-11} & 9.886\times 10^{-10} & 0 \\ 0 & 0 & 9.298\times 10^{-9} \end{bmatrix}\begin{bmatrix} -1.389\times 10^{5} \\ -2.004\times 10^{5} \\ 0 \end{bmatrix}$$

$$= \begin{bmatrix} -0.6977\times 10^{-4} \\ -0.1950\times 10^{-3} \\ 0 \end{bmatrix}.$$

The laminate is symmetric and no bending or torsional moments are applied; therefore, the global strains in the laminate are the same as the midplane strains. The free expansional thermal strains in the top 0° ply are

$$\begin{bmatrix} \alpha_x \\ \alpha_y \\ \alpha_{xy} \end{bmatrix}_{0^\circ} \Delta T$$

$$= \begin{bmatrix} 0.200 \times 10^{-7} \\ 0.225 \times 10^{-4} \\ 0 \end{bmatrix} (-75)$$

$$= \begin{bmatrix} -0.1500 \times 10^{-5} \\ -0.16875 \times 10^{-2} \\ 0 \end{bmatrix}.$$

From Equation (4.70), the global mechanical strain at the top surface of the top 0° ply is

$$\begin{bmatrix} -0.6977 \times 10^{-4} \\ -0.1950 \times 10^{-3} \\ 0 \end{bmatrix} - \begin{bmatrix} -0.1500 \times 10^{-5} \\ -0.16875 \times 10^{-2} \\ 0 \end{bmatrix}$$

$$= \begin{bmatrix} -0.6827 \times 10^{-4} \\ 0.14925 \times 10^{-2} \\ 0 \end{bmatrix}$$

From Equation (2.103), the global mechanical stresses at the top of the top 0° ply are

$$\begin{bmatrix} \sigma_x \\ \sigma_y \\ \tau_{xy} \end{bmatrix} = \begin{bmatrix} 181.8 & 2.897 & 0 \\ 2.897 & 10.35 & 0 \\ 0 & 0 & 7.17 \end{bmatrix} (10^9) \begin{bmatrix} -0.6827 \times 10^{-4} \\ 0.14925 \times 10^{-2} \\ 0 \end{bmatrix}$$

$$= \begin{bmatrix} -8.088 \times 10^{6} \\ 1.524 \times 10^{7} \\ 0 \end{bmatrix} Pa$$

Now, if the mechanical loads were given, the resulting mechanical stresses could then be added to the previous stresses due to the temperature difference.

Then, the failure criteria could be used to find out whether the ply has failed. However, we are asked to find out the mechanical load that could be applied in the presence of the temperature difference. This can be done as follows.

The stress at the top of the 0° ply, per Example 5.3 for a unit load $N_x = 1$ N/m, is

$$\begin{bmatrix} \sigma_x \\ \sigma_y \\ \tau_{xy} \end{bmatrix} = \begin{bmatrix} 9.726 \times 10^1 \\ 1.313 \times 10^0 \\ 0.0 \end{bmatrix} Pa.$$

If the unknown load is N_x, then the overall stress at the top surface of the top 0° ply is

$$\begin{bmatrix} \sigma_x \\ \sigma_y \\ \tau_{xy} \end{bmatrix} = \begin{bmatrix} -8.088 \times 10^6 + 9.726 \times 10^1 N_x \\ 1.524 \times 10^7 + 1.313 \times 10^0 N_x \\ 0 \end{bmatrix} Pa.$$

Now, the failure theories can be applied to find the value of N_x. Using transformation equation (2.94), the local stresses at the top surface of the top 0° ply are

$$\begin{bmatrix} \sigma_1 \\ \sigma_2 \\ \tau_{12} \end{bmatrix} = \begin{bmatrix} -8.088 \times 10^6 + 9.726 \times 10^1 N_x \\ 1.524 \times 10^7 + 1.313 \times 10^0 N_x \\ 0 \end{bmatrix} Pa.$$

Using the parameters H_1, H_2, H_6 H_{11}, H_{22}, H_{66} and H_{12} from Example 2.19, the Tsai–Wu failure criterion (Equation 2.146) is

$$(0)\,[-8.088 \times 10^6 + 9.726 \times 10^1\, N_x] + (2.093 \times 10^{-8})[1.524 \times 10^7 + 1.313 \times 10^0\, N_x] + (0)(0) + 4.4444 \times 10^{-19}\, [-8.088 \times 10^6 + 9.726 \times 10^1\, N_x]^2 + 1.0162 \times 10^{-16}\, [1.524 \times 10^7 + 1.313 \times 10^0\, N_x]^2 + 2.1626 \times 10^{-16}\, [0]^2 + 2[-3.360 \times 10^{-18}]\, [-8.088 \times 10^6 + 9.726 \times 10^1\, N_x]\, [1.524 \times 10^7 + 1.313 \times 10^0\, N_x] < 1.$$

TABLE 5.7

Strength Ratios of Example 5.4

Ply no.	Position	Tsai–Wu
1 (0°)	Top	1.100×10^7
	Middle	1.100×10^7
	Bottom	1.100×10^7
2 (90°)	Top	4.279×10^6
	Middle	4.279×10^6
	Bottom	4.279×10^6
3 (0°)	Top	1.100×10^7
	Middle	1.100×10^7
	Bottom	1.100×10^7

As can be seen, this results in a quadratic polynomial in the left-hand side of the strength criteria — that is,

$$3.521 \times 10^{-15} N_x^2 + 2.096 \times 10^{-8} N_x - 0.6566 = 0 .$$

This gives two roots for which the inequality is satisfied for $N_x < 1.100 \times 10^7$ and $N_x > -1.695 \times 10^7$.

Because the load N_x is tensile, $N_x = 1.100 \times 10^7$ is the valid solution. Similarly, the values of strength ratios for all the plies in the laminate are found and summarized in Table 5.7.

Using the lowest value of strength ratio of 4.279×10^6 gives $N_x = 4.279 \times 10^6$ N/m as the load at which the first ply failure would take place. Compare this with the value of $N_x = 7.277 \times 10^6$ in Example 5.3, in which no temperature change was applied.

5.4 Design of a Laminated Composite

Because we have developed the laminated plate theory for composites subjected to in-plane mechanical loads, temperature, and moisture, the designs in this chapter are also limited to such loads and simple shapes. Factors not covered in this section include stability; out-of plane loads; and fracture, impact, and fatigue resistance; interlaminar strength; damping characteristics; vibration control; and complex shapes. These factors are introduced briefly in Section 5.5.

Design of laminated composites includes constraints on optimizing and constraining factors such as

- Cost
- Mass as related to aerospace and automotive industry to reduce energy cost

- Stiffness (to limit deformations) as related to aircraft skins to avoid buckling
- Thermal and moisture expansion coefficients as related to space antennas to maintain dimensional stability

These factors are similar to those used with designing with monolithic materials; thus, the main issue with designing with composites as opposed to monolithic materials involves understanding the orthotropic nature of composite plies.

The possibility of different fiber-matrix systems combined with the variables such as fiber volume fraction first dictate the properties of a lamina. Then, laminae can be placed at angles and at particular distances from the midplane in the laminate. The material systems and the stacking sequence then determine the stresses and strains in the laminate. The failure of the composite may be based on the first ply failure (FPF) or the last ply failure (LPF). Although one may think that all plies failing at the same time is an ideal laminate, others may argue that differences between the two give time for detection and repair or replacement of the part.

Laminate selection is a computationally intensive and repetitive task due to the many possibilities of fiber-matrix combinations, material systems, and stacking sequence. Computer programs have made these calculations easy and the reader is directed to use the PROMAL[2] program included in this book or any other equivalent program of choice to fully appreciate designing with composites. A more scientific approach to optimization of laminated composites is out of scope of this book, and the reader is referred to Gurdal et al.[3]

Example 5.5

1. An electronic device uses an aluminum plate of cross-section 4 in. × 4 in. to take a pure bending moment of 13,000 lb-in. The factor of safety is 2. Using the properties of aluminum given in Table 3.4, find the thickness of the plate.
2. The designer wants at least to halve the thickness of the plate to make room for additional hardware on the electronic device. The choices include unidirectional laminates of graphite/epoxy, glass/epoxy, or their combination (hybrid laminates). The ply thickness is 0.125 mm (0.0049213 in.). Design a plate with the lowest cost if the manufacturing cost per ply of graphite/epoxy and glass/epoxy is ten and four units, respectively. Use the properties of unidirectional graphite/epoxy and glass/epoxy laminae from Table 2.2.
3. Did your choice of the laminate composite design decrease the mass? If so, by how much?

Solution

1. The maximum normal stress in a plate under bending is given by

$$\sigma = \pm \frac{M\frac{t}{2}}{I}, \tag{5.20}$$

where

M = bending moment (lb-in.)
t = thickness of plate (in.)
I = second moment of area (in.^{4})

For a rectangular cross-section, the second moment of area is

$$I = \frac{bt^3}{12}, \tag{5.21}$$

where b = width of plate (in.).

Using the given factor of safety, $F_s = 2$, and given $b = 4$ in., the thickness of the plate using the maximum stress criterion is

$$t = \sqrt{\frac{6MF_s}{b\sigma_{ult}}}, \tag{5.22}$$

where σ_{ult} = 40.02 Ksi from Table 3.4

$$t = \sqrt{\frac{6(13000)2}{4(40.02)10^3}}$$
$$= 0.9872 \ in.$$

2. Now the designer wants to replace the 0.9872 in. thick aluminum plate by a plate of maximum thickness of 0.4936 in. (half that of aluminum) made of laminated composites. The bending moment per unit width is

$$M_{xx} = \frac{13,000}{4}$$
$$= 3,250 \ lb\text{-}in./in.$$

Using the factor of safety of two, the plate is designed to take a bending moment per unit width of

$$M_{xx} = 3,250 \times 2$$

$$= 6,500 \ lb\text{-}in./in.$$

The simplest choices are to replace the aluminum plate by an all graphite/epoxy laminate or an all glass/epoxy laminate. Using the procedure described in Example 5.3 or using the PROMAL[2] program, the strength ratio for using a single 0° ply for the previous load for glass/epoxy ply is

$$SR = 5.494 \times 10^{-5}.$$

The bending moment per unit width is inversely proportional to the square of the thickness of the plate, so the minimum number of plies required would be

$$N_{Gl/Ep} = \sqrt{\frac{1}{5.494 \times 10^{-5}}}$$

$$= 135 \text{ plies}.$$

This gives the thickness of the all-glass/epoxy laminate as

$$t_{Gl/Ep} = 135 \times 0.0049213 \text{ in.}$$

$$= 0.6643 \text{ in.}$$

The thickness of an all-glass/epoxy laminate is more than 0.4935 in. and is thus not acceptable.

Similarly, for an all graphite/epoxy laminate made of only 0° plies, the minimum number of plies required is

$$N_{Gr/Ep} = 87 \text{ plies}.$$

This gives the thickness of the plate as

$$t_{Gr/Ep} = 87 \times 0.0049213$$

$$= 0.4282 \text{ in.}$$

The thickness of an all-graphite/epoxy laminate is less than 0.4936 in. and is acceptable.

Even if an all-graphite/epoxy laminate is acceptable, because graphite/epoxy is 2.5 times more costly than glass/epoxy, one would suggest the use of a hybrid laminate. The question that arises now concerns the sequence in which the unidirectional plies should be stacked. In a plate under a bending moment, the magnitude of ply stresses is maximum on the top and bottom face. Because the longitudinal tensile and compressive strengths are larger in the graphite/epoxy lamina than in a glass/epoxy lamina, one would put the former as the facing material and the latter in the core.

The maximum number of plies allowed in the hybrid laminate is

$$N_{\max} = \frac{\textit{Maximum Allowable Thickness}}{\textit{Thickness of each ply}}$$

$$= \frac{0.4936}{0.0049213}$$

$$= 100 \text{ plies.}$$

Several combinations of 100-ply symmetric hybrid laminates of the form $[0_n^{Gr}/0_m^{Gl}/0_n^{Gr}]$ are now subjected to the applied bending moment. Minimum strength ratios in each laminate stacking sequence are found. Only if the strength ratios are greater than one — that is, the laminate is safe — is the cost of the stacking sequence determined. A summary of these results is given in Table 5.8.

From Table 5.8, an acceptable hybrid laminate with the lowest cost is case VI, $[0_{16}^{Gr}/0_{68}^{Gl}/0_{16}^{Gr}]$.

TABLE 5.8

Cost of Various Glass/Epoxy–Graphite/Epoxy Hybrid Laminates

	Number of plies			
Case	**Glass/epoxy (m)**	**Graphite/epoxy (2n)**	**Minimum SR**	**Cost**
I	0	87	1.023	870
II	20	80	1.342	880
III	60	40	1.127	640
IV	80	20	0.8032	—
V	70	30	0.9836	—
VI	68	32	1.014	592
VII	66	34	1.043	604

3. The volume of the aluminum plate is

$$V_{Al} = 4 \times 4 \times 0.9871$$

$$= 15.7936 \text{ in.}^3$$

The mass of the aluminum plate is (specific gravity = 2.7 from Table 3.2),

$$M_{Al} = V_{Al}\, \rho_{Al}$$

$$= 15.793 \times [(2.7)\,(3.6127 \times 10^{-2})]$$

$$= 1.540 \text{ lbm.}$$

The volume of the glass/epoxy in the hybrid laminate is

$$V_{Gl/Ep} = 4 \times 4 \times 0.0049213 \times 68$$

$$= 5.354 \text{ in.}^3$$

The volume of graphite/epoxy in the hybrid laminate is

$$V_{Gr/Ep} = 4 \times 4 \times 0.0049213 \times 32$$

$$= 2.520 \text{ in.}^3$$

Using the specific gravities of glass, graphite, and epoxy given in Table 3.1 and Table 3.2 and considering that the density of water is 3.6127×10^{-2} lbm/in.^{3}:

$$\rho_{Gl} = 2.5 \times (3.6127 \times 10^{-2}) = 0.9032 \times 10^{-1} \text{ lbm/in.}^3$$

$$\rho_{Gr} = 1.8 \times (3.6127 \times 10^{-2}) = 0.6503 \times 10^{-1} \text{ lbm/in.}^3$$

$$\rho_{Ep} = 1.2 \times (3.6127 \times 10^{-2}) = 0.4335 \times 10^{-1} \text{ lbm/in.}^3$$

The fiber volume fraction is given in Table 2.1 and, substituting in Equation (3.8), the density of glass/epoxy and graphite/epoxy laminae is

$$\rho_{Gl/Ep} = (0.9032 \times 10^{-1})\,(0.45) + (0.4335 \times 10^{-1})\,(0.55)$$

$$= 0.6449 \times 10^{-1} \text{ lbm/in.}^3$$

$$\rho_{Gr/Ep} = (0.6503 \times 10^{-1})\ (0.70) + (0.4335 \times 10^{-1})\ (0.30)$$

$$= 0.5853 \times 10^{-1}\ \text{lbm/in.}^3$$

The mass of the hybrid laminate then is

$$M_h = (5.354)\ (0.6449 \times 10^{-1}) + (2.520)(0.5853 \times 10^{-1})$$

$$= 0.4928\ \text{lbm.}$$

The percentage savings using the composite laminate over aluminum is

$$= \frac{1.540 - 0.4928}{1.540} \times 100$$

$$= 68\%.$$

This example dictated the use of unidirectional laminates. How will the design change if multiple loads are present? Examples of multiple loads include a leaf spring subjected to bending moment as well as torsion or a thin pressure vessel subjected to an internal pressure to yield a biaxial state of stress. In such cases, one may have a choice not only of material systems and their combination, but also of orientation of plies. Combinations of angle plies can be infinite, so attention may be focused on angle plies of 0°, 90°, 45°, and –45° and their combinations. This reduces the possibilities to a finite number for a limited number of material systems; however, but the number of combinations can still be quite large to handle.

Example 5.6

An electronic device uses an aluminum plate of 1-in. thickness and a top cross–sectional area of 4 in. × 4 in. to take a pure bending moment. The designer wants to replace the aluminum plate with graphite/epoxy unidirectional laminate. The ply thickness of graphite/epoxy is 0.125 mm (0.0049213 in.).

1. Use the properties of aluminum and unidirectional graphite/epoxy as given in Table 3.4 and Table 2.2, respectively, to design a plate of graphite/epoxy with the same bending stiffness in the needed direction of load as that of the aluminum beam.
2. Does the laminate design decrease the mass? If so, by how much?

Solution

1. The bending stiffness, E_b, of the aluminum plate is given by:

$$E_b = EI \tag{5.23}$$

$$= E\left(\frac{1}{12}bh^3\right),$$

where

E = Young's modulus of aluminum
b = width of beam
h = thickness of beam

$$E_b = 10.3 \times 10^6\left(\frac{1}{12}(4)(1)^3\right)$$

$$= 3.433 \times 10^6 \text{ lb–in.}^2$$

To find the thickness of a graphite/epoxy laminate with unidirectional plies and the same flexural rigidity, let us look at the bending stiffness of a laminate of thickness, h:

$$E_b = E_x I$$

$$= E_x \frac{1}{12}bh^3,$$

where E_x = Young's modulus in direction of fibers.
Because $E_x = E_1 = 26.25$ Msi for a 0° ply from Table 2.2,

$$3.433 \times 10^6 = 26.25 \times 10^6\left(\frac{1}{12}\,4h^3\right),$$

giving

$$h = 0.732 \text{ in.}$$

Thus, a 1-in. thick aluminum beam can be replaced with a graphite/epoxy laminate of 0.732 in. thickness. Note that, although the Young's modulus of graphite /epoxy is approximately 2.5 times that of aluminum, the thickness of aluminum plate is approximately only

1.4 times that of the graphite /epoxy of laminate because the bending stiffness of a beam is proportional to the cube of the thickness. Thus, the lightest beam for such bending would be influenced by the cube root of the Young's moduli. From the thickness of 0.732 in. of the laminate and a thickness of 0.0049312 in. of the lamina, the number of 0° graphite/epoxy plies needed is

$$n = \frac{0.732}{0.0049213} = 149 \ .$$

The resulting graphite/epoxy laminate then is $[0_{149}]$.

2. The volume of the aluminum plate V_{Al} is

$$V_{Al} = 4 \times 4 \times 1.0$$

$$= 16 \text{ in.}^3$$

The mass of the aluminum plate is (specific gravity = 2.7 from Table 3.2; density of water is 3.6127×10^{-2} lbm/in.^{3}):

$$M_{Al} = V_{Al}\,\rho_{Al}$$

$$= 16 \times (2.7 \times 3.6127 \times 10^{-2})$$

$$= 1.561 \text{ lbm.}$$

The volume of a $[0_{149}]$ graphite/epoxy laminate is

$$V_{Gr/Ep} = 4 \times 4 \times 0.0049213 \times 149$$

$$= 11.73 \text{ in.}^3$$

The density of a graphite/epoxy from Example 5.5 is

$$\rho_{Gr/Ep} = 0.5853 \times 10^{-1} \frac{lbm}{in^3} \ .$$

The mass of the graphite/epoxy laminate beam is

$$M_{Gr/Ep} = (0.5853 \times 10^{-1})\ (11.73)$$

$$= 0.6866 \text{ lbm.}$$

Therefore, the percentage saving in using graphite/epoxy composite laminate over aluminum is

$$\frac{M_{Al} - M_{Gr/Ep}}{M_{Al}}$$

$$= \frac{1.561 - 0.6866}{1.561} \times 100$$

$$= 56\%.$$

Example 5.7

A 6-ft-long cylindrical pressure vessel (Figure 5.2) with an inner diameter of 35 in. is subjected to an internal gauge pressure of 150 psi. The vessel operates at room temperature and curing residual stresses are neglected. The cost of a graphite/epoxy lamina is 250 units/lbm and cost of a glass/epoxy lamina is 50 units/lbm. The following are other specifications of the design:

1. Only 0°, +45°, –45°, +60°, –60°, and 90° plies can be used.
2. Only symmetric laminates can be used.
3. Only graphite/epoxy and glass/epoxy laminae, as given in Table 2.2, are available, but hybrid laminates made of these two laminae are allowed. The thickness of each lamina is 0.005 in.*

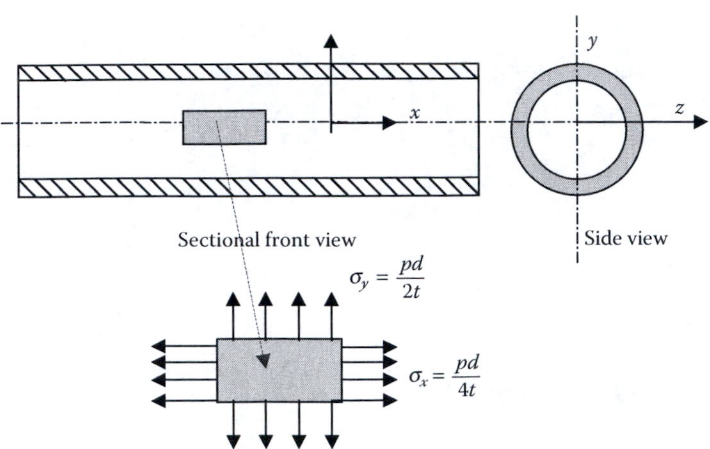

FIGURE 5.2
Fiber composite pressure vessel.

* Note that the thickness of each lamina is given as 0.005 in., and is not 0.125 mm (0.00492 in.), as given in the material database of the PROMAL program. Material properties for two new materials need to be entered in the database.

4. Calculate specific gravities of the laminae using Table 3.3 and Table 3.4 and fiber volume fractions given in Table 2.2.
5. Neglect the end effects and the mass and cost of ends of the pressure vessel in your design.
6. Use Tsai–Wu failure criterion for calculating strength ratios.
7. Use a factor of safety of 1.95.

Design for ply orientation, stacking sequence, number of plies, and ply material and give separate designs (laminate code, including materials) based on each of the following design criteria:

1. Minimum mass
2. Minimum cost
3. Both minimum mass and minimum cost

You may be unable to minimize mass and cost simultaneously — that is, the design of the pressure vessel for the minimum mass may not be same as for the minimum cost. In that case, give equal weight to cost and mass, and use this as your optimization function:

$$F = \frac{A}{B} + \frac{C}{D}, \tag{5.24}$$

where

A = mass of composite laminate
B = mass of composite laminate if design was based only on minimum mass
C = cost of composite laminate
D = cost of composite laminate if design was based only on minimum cost

Solution

LOADING. For thin-walled cylindrical pressure vessels, the circumferential stress or hoop stress σ_y and the longitudinal or axial stress σ_x is given by

$$\sigma_x = \frac{pr}{2t} \tag{5.25a}$$

$$\sigma_y = \frac{pr}{t}, \tag{5.25b}$$

where

p = internal gage pressure, psi
r = radius of cylinder, in.
t = thickness of cylinder, in.

For our case, we have

$$p = 150\ psi$$

$$r = \frac{35}{2} = 17.5\ in$$

giving

$$\sigma_x = \frac{(150)(17.5)}{2t}$$

$$= \frac{1.3125 \times 10^3}{t}$$

$$\sigma_y = \frac{(150)(17.5)}{t}$$

$$= \frac{2.625 \times 10^3}{t}.$$

For the forces per unit length,

$$N_x = \sigma_x t$$

$$= \frac{1.3125 \times 10^3}{t} t \tag{5.26a}$$

$$= 1.3125 \times 10^3 \frac{lb}{in}$$

$$N_y = \sigma_y t$$

$$= \frac{2.625 \times 10^3}{t} t \tag{5.26b}$$

$$= 2.625 \times 10^3 \frac{lb}{in}.$$

MASS OF EACH PLY. The mass of a graphite epoxy ply is

$$m_{Gr/Ep} = V_{Gr/Ep}\rho_{Gr/Ep} \text{ ,}$$

where

$V_{Gr/Ep}$ = volume of a graphite epoxy ply, in.^3
$\rho_{Gr/Ep}$ = density of a graphite/epoxy ply, lbm/in.^3
$V_{Gr/Ep} = \pi L d t_p$

where

L = length of the cylinder, in.
d = diameter of the cylinder, in.
t_p = thickness of graphite/epoxy ply, in.

Because L = 6 ft, d = 35 in., and t_p = 0.005 in,

$$V_{Gr/Ep} = \pi(6 \times 12)(35)(0.005)$$

$$= 39.584 \text{ in.}^3$$

The density of a graphite/epoxy lamina is

$$\rho_{Gr/Ep} = \rho_{Gr}V_f + \rho_{Ep}V_m \text{ .}$$

From Table 2.2, the fiber volume fraction, V_f, of the graphite epoxy is 0.7. Thus,

$$V_f = 0.7$$

The matrix volume fraction, V_m, then is

$$V_m = 1 - V_f$$

$$= 1 - 0.7$$

$$= 0.3$$

The specific gravity of graphite and epoxy is given in Table 3.3 and Table 3.4, respectively, as s_{Gr} = 1.8 and s_{Ep} = 1.2; given that the density of water is 3.6127×10^{-2} lbm/in^3,

$$\rho_{Gr/Ep} = (1.8)(3.6127 \times 10^{-2})(0.7) + (1.2)(3.6127 \times 10^{-2})(0.3)$$

$$= 5.8526 \times 10^{-2} \frac{lbm}{in^3}$$

Therefore, the mass of a graphite/epoxy lamina is

$$m_{Gr/Ep} = V_{Gr/Ep}\rho_{Gr/Ep}$$

$$= (39.584)(5.8526 \times 10^{-2})$$

$$= 2.3167 \; lbm.$$

The mass of a glass/epoxy ply is

$$m_{Gl/Ep} = V_{Gl/Ep}\rho_{Gl/Ep}$$

where

$V_{Gl/Ep}$ = volume of glass/epoxy, in.^{3}
$\rho_{Gl/Ep}$ = density of glass/epoxy, lbm/in.^{3}
$V_{Gl/Ep} = \pi L d t_p$
= 39.584 in.^{3}

The density of a glass/epoxy lamina is

$$\rho_{Gl/Ep} = \rho_{Gl} V_f + \rho_{Ep} V_m \; .$$

From Table 2.2, the fiber volume fraction V_f of the glass/epoxy is 0.45; thus,

$$V_f = 0.45 \; .$$

The matrix volume fraction V_m then is

$$V_m = 1 - V_f$$

$$= 1 - 0.45$$

$$= 0.55.$$

The specific gravity of glass and epoxy is given in Table 3.3 and Table 3.4, respectively, as

$$s_{Gl} = 2.5, s_{Ep} = 1.2$$

and, given that the density of water is 3.6127×10^{-2} lbm/in^3,

$$\rho_{G\ell/Ep} = (2.5)(3.6127\times10^{-2})(0.45) + (1.2)(3.6127\times10^{-2})(0.55)$$

$$= 6.4487\times10^{-2}\,\frac{lbm}{in^3}.$$

Therefore, the mass of glass/epoxy lamina is

$$m_{Gl/Ep} = V_{Gl/Ep}\rho_{Gl/Ep}$$

$$= (39.584)(6.4487\times10^{-2})$$

$$= 2.5526\ lbm.$$

COST OF EACH PLY. The cost of a graphite/epoxy ply is

$$C_{Gr/Ep} = m_{Gr/Ep}c_{Gr/Ep}\,,$$

where

$m_{Gr/Ep}$ = mass of graphite/epoxy ply
$c_{Gr/Ep}$ = unit cost of graphite/epoxy ply

Because

$$m_{Gr/Ep} = 2.3167\ lbm$$

$$c_{Gr/Ep} = 250\,\frac{units}{lbm}\,,$$

the cost of a graphite/epoxy ply is

$$C_{Gr/Ep} = (2.3167)(250)$$

$$= 579.17\ \text{units}.$$

Similarly, the cost of a glass/epoxy ply is

$$C_{Gl/Ep} = m_{Gl/Ep} c_{Gl/Ep} \, .$$

Because

$$m_{Gl/Ep} = 2.5526 \; lbm$$

and

$$c_{Gl/Ep} = 50 \frac{units}{lbm} \, ,$$

the cost of a glass/epoxy ply is

$$C_{Gl/Ep} = (2.5526)(50)$$

$$= 127.63 \text{ units.}$$

1. To find the design for minimum mass, consider a composite laminate made of graphite/epoxy with $[0/90_2]_s$. We simply choose this laminate as $N_y = 2N_x$ and thus choose two 90° plies for every 0° ply. For this laminate, from PROMAL we get the minimum strength ratio as

$$SR = 0.6649.$$

Because the required factor of safety is 1.95, we need

$$\frac{1.95}{0.6649} \times 6$$

$$\cong 18 \text{ plies.}$$

$[0/90_2]_{3s}$ is a possible choice because it gives a strength ratio of 1.995. However, is this laminate with the minimum mass? Choosing some other choices such as laminates with ±60° laminae, a graphite/epoxy $[\pm 60]_{4s}$ laminate gives an SR = 1.192 and that is lower than the required SR of 1.95.

A $[0/90_2]_{3s}$ laminate made of glass/epoxy gives a strength ratio of $SR = 0.5192$ and that is also lower than the needed strength ratio of 1.95. Other combinations tried used more than 18 plies. A summary of possible combinations is shown in Table 5.9.

TABLE 5.9

Mass and Cost of Possible Stacking Sequences for Minimum Mass

Stacking sequence	No. plies	Minimum strength ratio	Mass (lbm)	Cost (units)
$[0/90_2]_{3s}$ (Graphite/epoxy)	18	1.995	41.700	10,425
$[\pm 60]_{4s}$ (Graphite/epoxy)	16	1.192	—	—
$[0/90_2]_{3s}$ (Glass/epoxy)	18	0.5192	—	—
$[\pm 60]_{5s}$ (Graphite/epoxy)	20	1.490	—	—
$[\pm 45_2/\pm 60_3]_s$ (Graphite/epoxy)	20	2.332	46.334	11,583

TABLE 5.10

Mass and Cost of Possible Stacking Sequences for Minimum Cost

Stacking sequence	No. plies	Minimum strength ratio	Mass (lbm)	Cost (units)
$[0/90_2]_{3s}$ (Graphite/epoxy)	18	1.995	41.700	10,425
$[\pm 45_2/\pm 60_3]_s$ (Graphite/epoxy)	20	2.291	46.334	11,583
$[0/90_2]_{12s}$ (Glass/epoxy)	72	2.077	183.79	9,189
$[90/\pm 45]_{10s}$ (Glass/epoxy)	60	1.992	153.16	7,658
$[\pm 60]_{15s}$ (Glass/epoxy)	60	2.033	153.16	7,658

Thus, one can say that the laminate for minimum mass is the first stacking sequence in Table 5.9:

Number of plies: 18

Material of plies: graphite/epoxy

Stacking sequence: $[0/90_2]_{3s}$

Mass of laminate = $(18 \times 2.3167) = 41.700$ lbm

Cost of laminate = $(41.700 \times 250) = 10425$ units

2. To find the design for minimum cost, we found in part (1) that the $[0/90_2]_{3s}$ graphite/epoxy laminate is safe, but the same stacking sequence for glass/epoxy gives a $SR = 0.5192$. Therefore, we may need four times more plies of glass/epoxy to keep it safe to obtain a factor of safety of 1.95. If so, would it be cheaper than the $[0/90_2]_{3s}$ graphite/epoxy laminate? Yes, it would because a glass/epoxy costs 127.63 units per ply as opposed to 579.17 units per ply for graphite/epoxy. Choosing $[0/90_2]_{12s}$ glass/epoxy laminate gives $SR = 2.077$. Are there other combinations that give an $SR > 1.95$ but use less than the 72 plies used in $[0/90_2]_{12s}$? Stacking sequences of 60 plies such as $[90/\pm 45]_{10s}$ and $[\pm 60]_{15s}$ were tried and were acceptable designs. The results from some of the stacking sequences are summarized in Table 5.10.

Therefore, we can say that the laminate for minimum cost is as follows

Number of plies = 60

Material of plies: glass/epoxy

Stacking sequence: $[\pm 60]_{15s}$

Mass of laminate = 60 × 2.5526 = 153.16 lbm

Cost of laminate = 153.16 × 50 = 7658 units

3. Now, how do we find the laminate that minimizes cost and mass? We know that the solutions to part (1) and (2) are different. Thus, we need to look at other combinations. However, before doing so, let us find the minimizing function for parts (1) and (2). The minimizing function is given as

$$F = \frac{A}{B} + \frac{C}{D},$$

where

A = mass of composite laminate
B = mass of composite laminate if design was based only on minimum mass
C = cost of composite laminate
D = cost of composite laminate if design was based only on minimum cost

From part (1), B = 41.700 lbm and, from part (2), D = 7658 units; then, the minimizing function is

$$F = \frac{41.700}{41.700} + \frac{10425}{7658} = 2.361$$

for the $[0/90_2]_{3s}$ graphite/epoxy laminate obtained in part (1).

$$F = \frac{153.16}{41.700} + \frac{7658}{7658} = 4.673$$

for the $[\pm 60]_{15s}$ glass/epoxy laminate obtained in part (2).

Therefore, the question is whether a laminate that has an optimizing function value of less than 2.361 can be found. If not, the answer is the same as the laminate in part (1). Table 5.11 gives the summary of some of the laminates that were tried to find minimum value of F. The third stacking sequence in Table 5.11 is the one in which, in the $[0/90_2]_{3s}$ graphite/epoxy laminate of part (1), six of the graphite/

TABLE 5.11

Optimizing Function Values for Different Stacking Sequences

Stacking sequence	Mass (lbm)	Cost	Minimum strength ratio	F
$[0/90_2]_{3s}$ graphite/epoxy (part a)	41.700	10,425	1.995	2.361
$[\pm 60]_{18s}$ glass/epoxy (part b)	153.16	7,658	2.0768	4.672
$[0_{Gr/Ep}/90_{2\ Gr/Ep}/0_{Gr/Ep}/90_{2\ Gr/Ep}/0_{Gl/Ep}/$ $90_{2\ Gl/Ep}/0_{Gl/Ep}/90_{2\ Gl/Ep}/0_{Gl/Ep}/90_{2Gl/Ep}/0_{Gl/Ep}/$ $90_{2\ Gl/Ep}]_s$	89.063	10,013	2.012	3.443

epoxy plies of $0/90_2$ sublaminate group are substituted with 24 glass/epoxy plies of the $0/90_2$ sublaminate group.

Thus, it seems that $[0/90_2]_{3s}$ graphite/epoxy laminate is the answer to part (3). Although more combinations should have been attempted to come to a definite conclusion, it is left to the reader to try other hybrid combinations using the PROMAL program.

Example 5.8

Drive shafts (Figure 5.3) in cars are generally made of steel. An automobile manufacturer is seriously thinking of changing the material to a composite material. The reasons for changing the material to composite materials are that composites

1. Reduce the weight of the drive shaft and thus reduce energy consumption
2. Are fatigue resistant and thus have a long life
3. Are noncorrosive and thus reduce maintenance costs and increase life of the drive shaft
4. Allow single piece manufacturing and thus reduce manufacturing cost

The design constraints are as follows:

1. Based on the engine overload torque of 140 N-m, the drive shaft needs to withstand a torque of 550 N-m.

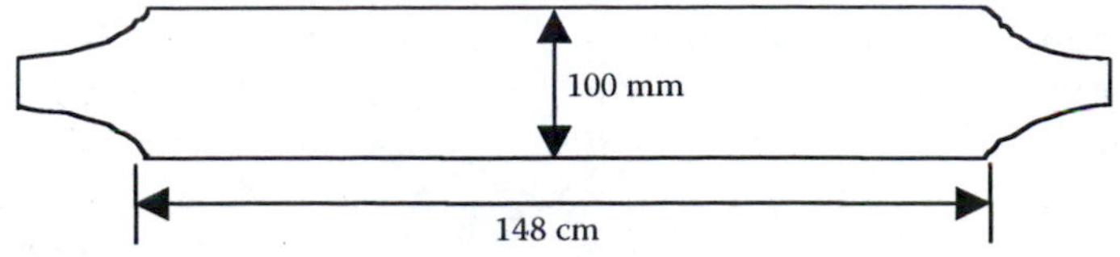

FIGURE 5.3
Fiber composite drive shaft.

2. The shaft needs to withstand torsional buckling.
3. The shaft has a minimum bending natural frequency of at least 80 Hz.
4. Outside radius of drive shaft = 50 mm.
5. Length of drive shaft = 148 cm.
6. Factor of safety = 3.
7. Only 0, +45, –45, +60, –60, and 90° plies can be used.

For steel, use the following properties:

Young's modulus E = 210 GPa,
Poisson's ratio ν = 0.3,
Density of steel ρ = 7800 kg/m^3
Ultimate shear strength τ_{ult} = 80 MPa.

For the composite, use properties of glass/epoxy from Table 2.1 and Table 3.1 and assume that ply thickness is 0.125 mm. Design the drive shaft using

1. Steel
2. Glass/epoxy

Solution

1. STEEL DESIGN.

Torsional strength: The primary load in the drive shaft is torsion. The maximum shear stress, τ_{max}, in the drive shaft is at the outer radius, r_o, and is given as

$$\tau_{max} = \frac{Tr_o}{J}, \tag{5.27}$$

where

T = maximum torque applied in drive shaft (N-m)
r_o = outer radius of shaft (m)
J = polar moment of area (m^4)

Because the ultimate shear strength of steel is 80 MPa and the safety factor used is 3, using Equation (5.27) gives

$$\frac{80\times 10^6}{3} = \frac{(550)(0.050)}{\dfrac{\pi}{2}(0.050^4 - r_i^4)}$$

$$r_i = 0.04863\ m.$$

Therefore, the thickness of the steel shaft is

$$t = r_o - r_i$$

$$= 0.050 - 0.04863$$

$$= 1.368 \text{ mm}.$$

Torsional buckling: This requirement asks that the applied torsion be less than the critical torsional buckling moment. For a thin, hollow cylinder made of isotropic materials, the critical buckling torsion, T_b, is given by[4]

$$T_b = (2\pi r_m^2 t)(0.272)(E)\left(\frac{t}{r_m}\right)^{2/3}, \tag{5.28}$$

where

r_m = mean radius of the shaft (m)
t = wall thickness of the drive shaft (m)
E = Young's modulus (Pa)

Using the thickness t = 1.368 mm calculated in criterion (1) and the mean radius

$$r_m = \frac{r_o + r_i}{2}$$

$$= \frac{0.050 + 0.04863}{2}$$

$$= 0.049315 \ m,$$

$$T_b = 2(0.049315)^2(0.001368)(0.272)(210 \times 10^9)\left(\frac{0.001368}{0.049315}\right)^{3/2}$$

$$= 109442 \ N\text{-}m.$$

The value of critical torsional buckling moment is larger than the applied torque of 550 N-m.

Natural frequency: The lowest natural frequency for a rotating shaft is given by[5]

$$f_n = \frac{\pi}{2}\sqrt{\frac{EI}{mL^4}}, \tag{5.29}$$

where

g = acceleration due to gravity (m/s^2)
E = Young's modulus of elasticity (Pa)
I = second moment of area (m^4)
m = mass per unit length (kg/m)
L = length of drive shaft (m)

Now the second moment of area, I, is

$$I = \frac{\pi}{4}(r_o^4 - r_i^4)$$

$$= \frac{\pi}{4}(0.050^4 - 0.04863^4)$$

$$-5.162 \times 10^{-7}\ m^4.$$

The mass per unit length of the shaft is

$$m = \pi\ (r_o^2 - r_i^2)\ \rho$$

$$= \pi\ (0.050^2 - 0.04863^2)\ (7800)$$

$$= 3.307\ \text{kg/m}.$$

Therefore,

$$f_n = \frac{\pi}{2}\sqrt{\frac{(210 \times 10^9)(5.162 \times 10^{-7})}{(3.307)(1.48)^4}}$$

$$= 129.8\ Hz.$$

This value is greater than the minimum desired natural frequency of 80 Hz.
Thus, the steel design of a hollow shaft of outer radius 50 mm and thickness t = 1.368 mm is an acceptable design.

2. COMPOSITE MATERIALS DESIGN.

Torsional strength: Assuming that the drive shaft is a thin, hollow cylinder, an element in the cylinder can be assumed to be a flat laminate. The only nonzero load on this element is the shear force, N_{xy}. If the average shear stress is $(\tau_{xy})_{average}$, the applied torque then is

$$T = (\text{shear stress})\ (\text{area})\ (\text{moment arm})$$

$$T = (\tau_{xy})_{average} \pi (r_o^2 - r_i^2) r_m \,. \tag{5.30}$$

The shear force per unit width is given by

$$N_{xy} = (\tau_{xy})_{average} t \,.$$

Because

$$t = r_o - r_i$$

$$r_m = \frac{r_o + r_i}{2} ,$$

then

$$N_{xy} = \frac{T}{2\pi r_m^2}$$

$$= \frac{550}{2\pi (0.050)^2} \tag{5.31}$$

$$= 35{,}014 \ N/m.$$

To find approximately how many layers of glass/epoxy may be needed to resist the shear load, choose a four-ply $[\pm 45]_s$ laminate. Inputting a value of N_{xy} = 35,014 N/m into the PROMAL program, the minimum strength ratio obtained using Tsai–Wu theory is 1.261. A strength ratio of at least 3 is needed, so the number of plies is increased proportionately as $\frac{3}{1.261} \times 4 \cong 10$. The next laminate chosen is $[\pm 45_2/45]_s$ laminate. A minimum strength ratio of 3.58 is obtained, so it is an acceptable design based on torsional strength criterion.

Torsional buckling: An orthotropic thin hollow cylinder will buckle torsionally if the applied torque is greater than the critical torsional buckling load given by[4]

$$T_c = (2\pi r_m^2 t)(0.272)(E_x E_y^3)^{1/4} \left(\frac{t}{r_m} \right)^{3/2} . \tag{5.32}$$

From PROMAL, the longitudinal Young's moduli E_x and the transverse Young's moduli E_y of the $[\pm 45_2/45]_s$ glass/epoxy laminate based on properties from Table 2.1 are

$$E_x = 12.51 \text{ GPa}$$

$$E_y = 12.51 \text{ GPa}$$

Because lamina thickness is 0.125 mm, the thickness of the ten-ply $[\pm 45_2/45]_s$ laminate, t, is

$$t = 10 \times 0.125 = 1.25 \text{ mm.}$$

The mean radius, r_m, is

$$r_m = r_o - \frac{t}{2}$$

$$= 50 - \frac{1.25}{2}$$

$$= 49.375\ mm.$$

Therefore,

$$T_c = 2\pi(0.049375)^2(0.00125)(0.272)\times$$

$$[(12.51\times 10^9)(12.51\times 10^9)^3]^{1/4}\left(\frac{0.00125}{0.049375}\right)^{2/3}$$

$$= 262\ N\text{-}m.$$

This is less than the applied torque of 550 N-m. Thus, the $[\pm 45_2/45]_s$ laminate would torsionally buckle. Per the formula, the torsional buckling is proportional to $E_y^{3/4}$ and $E_x^{1/4}$. Because the modulus in the y-direction is more effective in increasing the critical torsional buckling load, it will be necessary to substitute by or add 90° plies.

Natural frequency: Although the $[\pm 45_2/45]_s$ laminate is inadequate, per the torsional buckling criterion, let us still find the minimum natural frequency of the drive shaft, which is given by[5]

$$f_n = \frac{\pi}{2}\sqrt{\frac{E_x I}{mL^4}}\,. \tag{5.33}$$

Now,

$$E_x = 12.51\ GPa$$

$$I = \frac{\pi}{4}(r_0^4 - r_i^4)$$

$$= \frac{\pi}{4}(0.050^4 - 0.04875^4)$$

$$= 4.728 \times 10^{-7}\ m^4.$$

The mass per unit length of the beam is

$$m = \frac{\pi(r_o^2 - r_i^2)L\rho}{L}$$

$$= \pi(r_o^2 - r_i^2)\rho$$

$$= (0.05^2 - 0.04875^2)(1785)$$

$$= 0.6922\frac{kg}{m}.$$

Thus,

$$f_n = \frac{\pi}{2}\sqrt{\frac{(12.51\times 10^9)(4.728\times 10^{-7})}{0.6922\times 1.48^4}}$$

$$= 66.3\ Hz.$$

Because the minimum bending natural frequency is required to be 80 Hz, this requirement is also not met by the $[\pm 45_2/45]_s$ laminate. The minimum natural frequency can be increased by increasing the value of E_x because the natural frequency f_n is proportional to $\sqrt{E_x}$. To achieve this, 0° plies can be added or substituted.

TABLE 5.12

Acceptable and Nonacceptable Designs of Drive Shaft Based on Three Criteria of Torsional Strength, Critical Torsional Buckling Load, and Minimum Natural Frequency

Laminate stacking sequence	No. plies	Minimum strength ratio	Critical torsional buckling load (N-m)	E_x (GPa)	E_y (GPa)	Minimum natural frequency (Hz)	Acceptable design
$[0/\pm45_2/45/90]_s$	14	3.982	797.8	16.44	16.44	***75.6***	No
$[0_2/\pm45_2/90]_s$	14	3.248	828.8	20.16	16.16	83.7	Yes
$[0/\pm45_2/90]_s$	12	3.006	564.1	17.07	17.07	***77.2***	No
$[0/\pm45_2]_s$	10	***2.764***	***291.2***	17.86	12.76	***79.2***	No
$[45/90_3/0_2]_s$	12	4.127	763.5	19.44	24.47	82.4	Yes
Design constraints		*>3*	*>550*			*>80*	

Note: Numbers given in bold italics to show the reason for unacceptable designs.

From the three criteria, we see that ±45° plies increase the torsional strength, 90° plies increase the critical torsional buckling load, and the 0° plies increase the natural frequency of the drive shaft. Therefore, having ±45°, 90°, and 0° plies may be the key to an optimum design.

In Table 5.12, several other combinations have been evaluated to find an acceptable design.

The last stacking sequence $[45/90_3/0_2]_s$ is a 12-ply laminate and meets the three requirements of torsional strength, critical torsional buckling load, and minimum natural frequency.

MASS SAVINGS. The savings in the mass of the drive shaft are calculated as follows:

$$\text{Mass of steel drive shaft} = \pi\,(r_o^2 - r_i^2)\,L\,\rho$$

$$= \pi\,(0.050^2 - 0.04863^2)\,(1.48)\,(7800)$$

$$= 4.894 \text{ kg.}$$

The thickness, t, of the $[45/90_3/0_2]_s$ glass/epoxy shaft is

$$t = 12 \times 0.125$$

$$= 1.5\ mm.$$

The inner radius of the $[45/90_3/0_2]_s$ glass/epoxy shaft then is

$$r_i = r_o - t$$

$$= 0.05 - 0.0015$$

$$= 0.0485 \text{ m.}$$

Mass of $[45/90_3/0_2]_s$ glass/epoxy shaft is

$$= \pi\ (r_o^2 - r_i^2)\ L\ \rho$$

$$= \pi\ (0.05^2 - 0.0485^2)\ (1.48)\ (1758)$$

$$= 1.226 \text{ kg.}$$

Percentage mass saving over steel is

$$= \frac{4.894 - 1.226}{4.894} \times 100$$

$$= 75\%.$$

Would an 11-ply, $[45/90_4/\overline{90}]_s$ glass/epoxy laminate meet all the requirements?

5.5 Other Mechanical Design Issues

5.5.1 Sandwich Composites

One group of laminated composites used extensively is sandwich composites. Sandwich panels consist of thin facings (also called skin) sandwiching a core. The facings are made of high-strength material, such as steel, and composites such as graphite/epoxy; the core is made of thick and lightweight materials such as foam, cardboard, plywood, etc. (Figure 5.4).

The motivation in doing this is twofold. First, if a plate or beam is bent, the maximum stresses occur at the top and bottom surfaces. Therefore, it makes sense to use high-strength materials only at the top and bottom and low- and lightweight strength materials in the middle. The strong and stiff facings also support axial forces. Second, the resistance to bending of a rectangular cross-sectional beam/plate is proportional to the cube of the thickness. Thus, increasing the thickness by adding a core in the middle increases this resistance. Note that the shear forces are maximum in the

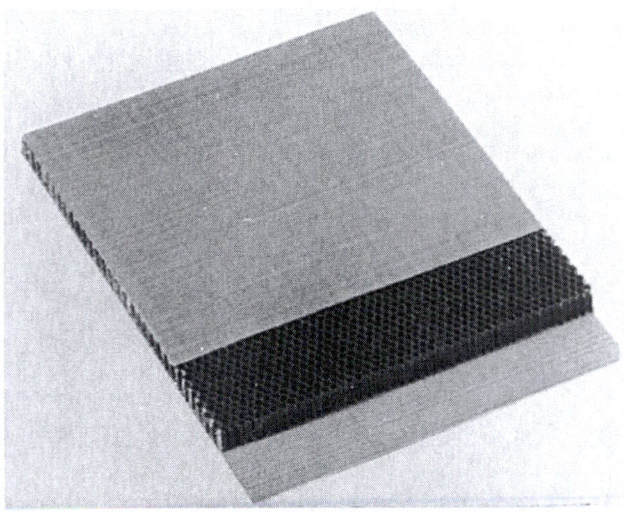

FIGURE 5.4
Fiberglass facings with a Nomex7 honeycomb core. (Picture Courtesy of M.C. Gill Corporation, http://www.mcgillcorp.com).

middle of the sandwich panel, thus requiring the core to support shear. This advantage in weight and bending stiffness makes sandwich panels more attractive than other materials. Sandwich panels are evaluated based on strength, safety, weight, durability, corrosion resistance, dent and puncture resistance, weatherability, and cost.[6]

The most commonly used facing materials are aluminum alloys and fiber-reinforced plastics. Aluminum has high specific modulus, but it corrodes without treatment and is prone to denting. Fiber-reinforced plastics such as graphite/epoxy and glass/epoxy are becoming popular as facing materials because of their high specific modulus and strength and corrosion resistance. Fiber-reinforced plastics may be unidirectional or woven laminae.

The most commonly used core materials are balsa wood, foam, and honeycombs. Balsa wood has high compressive strength (1500 psi), good fatigue life, and high shear strength (200 psi). Foams are low-density polymers such as polyuretherane, phenolic, and polystyrene. Honeycombs are made of plastic, paper, cardboard, etc. The strength and stiffness of honeycomb depend on the material and its cell size and thickness.

Adhesives join the facing and core materials and thus are critical in the overall integrity of the sandwich panel. Adhesives come in forms of film, paste, and liquid. Common examples include vinyl phenolic, modified epoxy, and urethane.

5.5.2 Long-Term Environmental Effects

Section 4.5 has already discussed the effects caused by temperature and moisture, such as residual stresses and strains. What effect do these and other environmental factors such as corrosive atmospheres and temperatures and humidity variations have over the long term on composites? These elements may lessen the adhesion of the fiber-matrix interface, such as between glass and epoxy. Epoxy matrices soften at high temperatures, affecting properties dominated by the matrix, such as transverse and in-plane shear stiffness and strength, and flexural strength. For example, Quinn[7] found that a glass/epoxy composite rod absorbed as much as 0.4% of water over 150 days of immersion. The effect of this moisture absorption on flexural modulus is shown in Figure 5.5.

5.5.3 Interlaminar Stresses

Due to the mismatch of elastic moduli and angle between the layers of a laminated composite, interlaminar stresses are developed between the layers. These stresses, which are normal and shear, can be high enough to cause edge delamination between the layers.[8–10] Delamination eventually limits the life of the laminated structure. Delamination can be further caused due to nonoptimum curing and introduction of foreign bodies in the structure.[11]

In Figure 5.6, theoretical interlaminar shear and normal stresses are plotted as a function of normalized distance — zero at the center line and one at the free edge — from the center line of a $[\pm 45]_s$ graphite/epoxy laminate. The interlaminar stresses given are for the bottom surface of the top ply of the laminate and are found by using equations of elasticity.[9] Away from the edges, these stresses are the same as predicted by the classical lamination theory discussed in Chapter 4. However, near the edges, the normal shear stress τ_{xy} decreases to zero, and the out-of-plane shear stress τ_{xz} becomes infinite (not shown). The classical lamination theory and elasticity results give different results because the former violates equilibrium and boundary conditions at the interface. For example, for a simple state of stress on the $[\pm 45]_s$ laminate, the classical lamination theory predicts nonzero values for the stresses σ_{xx}, σ_{yy}, and τ_{xy} for each ply. This is not true at the edges, where σ_y and τ_{xy} are actually zero because they are free boundaries.

Interlaminar stresses pose a challenge to the designer and there are some ways to counter their effects. Pagano and Pipes[9] found theoretically that keeping the angle, symmetry, and number of plies the same but changing the stacking sequence influences the interlaminar stresses. The key to changing the stacking sequence is to decrease the interlaminar shear stresses without increasing the tensile (if any) interlaminar normal stresses. For example, a laminate stacking sequence of $[\pm 30/90]_s$ produces tensile interlaminar normal stresses under a uniaxial tensile load however, if the sequence is just changed to $[90/\pm 30]_s$, it produces compressive interlaminar normal stresses. This makes the latter sequence less prone to delamination. Other techniques to improve tolerance to delamination include using toughened resin systems[12] and interleaved systems in which a discrete layer of resin with high toughness and strain to failure is added on top of a layer.[13,14]

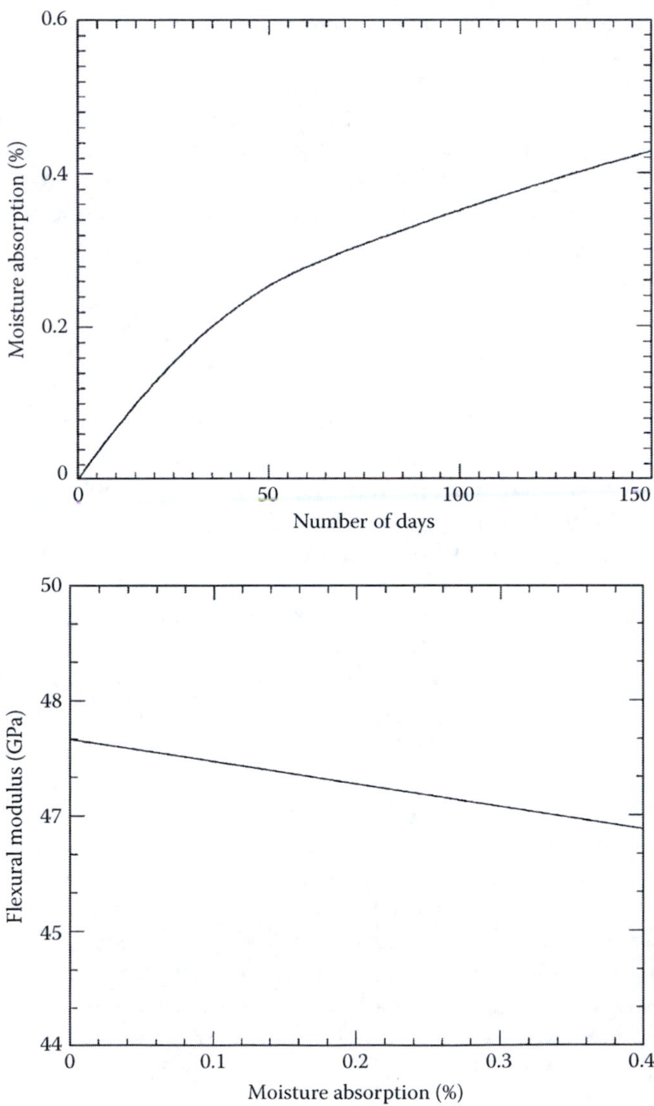

FIGURE 5.5
Moisture absorption as a function of time and its effect on flexural modulus of a glass/polyester composite rod. (Reprinted from Quinn, J.A., in *Design with Advanced Composite Materials*, Phillips, L.N., Ed., 1990, Figure 3.10 (p. 91) and Figure 3.11 (p. 92), Springer–Verlag, Heidelberg.)

5.5.4 Impact Resistance

The resistance to impact of laminated composites is important in applications such as a bullet hitting a military aircraft structure or even the contact of a composite leaf spring in a car to runaway stones on a gravel road. The resistance to impact depends on several factors of the laminate, such as the

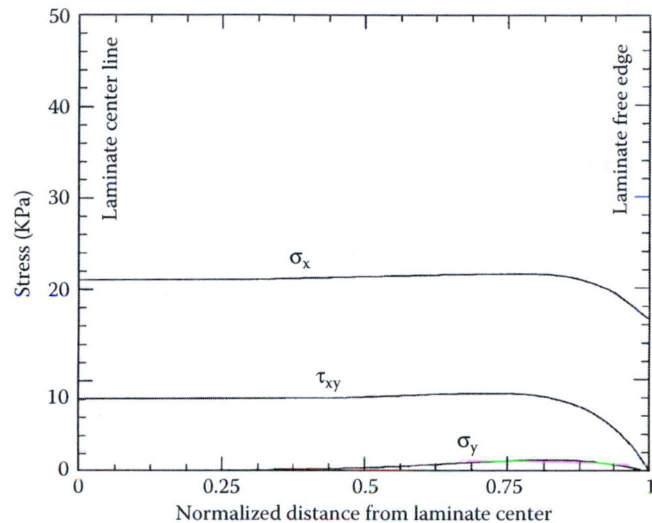

FIGURE 5.6
Normal and shear stresses at the interface of bottom surface of top ply in a four-ply laminate. (Reprinted from Pagano, N.J. and Soni, S.R., in *Interlaminar Response of Composite Materials*, Pagano, N.J., Ed., 1989, p. 9, Elsevier Science, New York, with kind permission from authors.)

material system; interlaminar strengths; stacking sequence; and nature of the impact, such as velocity, mass, and size of the impacting object. Impact reduces strengths of the laminate and also initiates delamination in composites. Delamination becomes more problematic because, many times, visual inspection cannot find it. Solutions for increasing impact resistance and residual impact strength have included toughened epoxies and interleaved laminates. In the former, epoxies are toughened by liquid rubber and, in the latter case, a discrete toughened layer is added to the laminae at selected places.

5.5.5 Fracture Resistance

When a crack develops in an isotropic material, the stresses at the crack tip are infinite. The intensity of these infinite stresses is called the stress intensity factor. If the stress intensity factor is greater than the critical stress intensity factor for that material, the crack is considered to grow catastrophically. Another parameter, called the strain energy release rate, is also used in determining fracture resistance. This is the rate of the energy release as the crack grows. If this rate is greater than the critical strain energy release rate of the material, the crack will grow catastrophically. The strain energy release rate and stress intensity factor are related to each other in isotropic materials.

In composites, the mechanics of fracture is not as simple. First, cracks can grow in the form of fiber breaks, matrix breaks, debonding between fiber and matrix, and debonding between layers. Second, no single critical stress intensity factors and strain energy release rates can determine the fracture mechanics process.

Fiber breaks may occur because of the brittle nature of fibers. Some fibers may break because, statistically, some fibers are weaker than others and thus fail at low strains. The matrix may then break because of high strains caused by the fiber breaks. In ceramic matrix composites, the matrix failure strain is lower than that of the fiber. Therefore, matrix breaks precede fiber breaks. In fact, fiber breaks are seen to occur only close to the ultimate failure of the composite. Also, matrix breaks may keep occurring parallel to the crack length.

When a fiber or matrix breaks, the crack does not grow in a self-similar fashion. It may grow along the interface that blunts the crack and improves the fracture resistance of the composites, or it may grow into the next constituent, resulting in uncontrolled failure. The competition between whether a crack grows along the interface or jumps to the adjoining constituent depends on the material properties of the fiber, matrix, and the interface, as well as the fiber volume fraction.

Fracture mechanics in composites is still an open field because there are several mechanisms of failure and developing uniform criteria for the materials looks quite impossible.

5.5.6 Fatigue Resistance

Structures over time are subjected to repeated cyclic loading, such as the fluctuating loads on an aircraft wing. This cyclic loading weakens the material and gives it a finite life. For example, a composite helicopter blade may have a service life of 10,000 hours.

Fatigue data for composite materials are collected using several different data, such as plotting the peak stress applied during the loading as a function of the number of cycles. The allowable peak stress decreases as the number of cycles to failure is increased. The peak stress is compared to the static strength of the composite structure. If these peak stresses are comparably larger than the allowable ultimate strength of the composite, fatigue does not influence the design of the composite structure. This is the case in graphite/epoxy composites in which the allowable ultimate strength is low due to its low impact resistance.

Other factors that influence the fatigue properties are the laminate stacking sequence, fiber and matrix properties, fiber volume fraction, interfacial bonding, etc. For example, for quasi-isotropic laminates, S–N curves are quite different from those of unidirectional laminates. In this case, the 90° plies develop transverse cracks, which influence the elastic moduli and strength of the laminate. Although the influence is limited because 90° plies do not contribute to the static stiffness and strength in the first place, the stress concentrations caused by these cracks may lead to damage in the 0° plies. Other damage modes include fiber and matrix breaks, interfacial and interlaminar debonding, etc. Laminate stacking sequence influences the onset of edge delamination. For example, Foye and Baker[15] conducted tensile fatigue

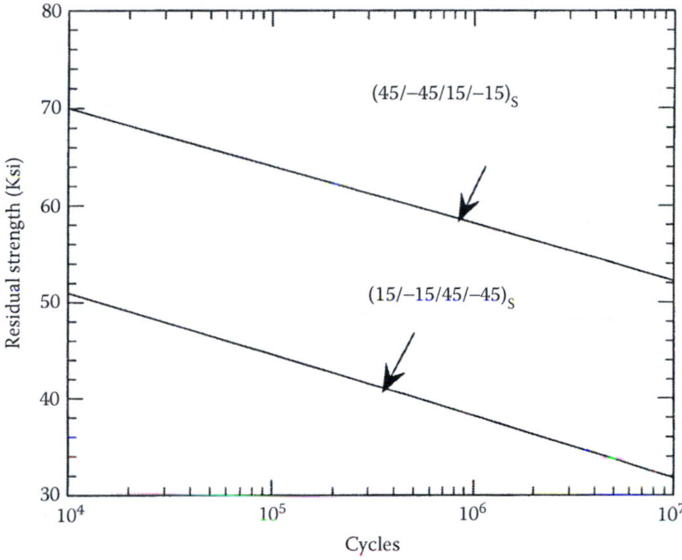

FIGURE 5.7
Comparison of residual strength as a function of number of cycles for two laminates. (Reprinted from Pagano, N.J. and Soni, S.R., in *Interlaminar Response of Composite Materials*, Pagano, N.J., Ed., 1989, p. 12, Elsevier Science, New York, with kind permission from authors.)

testing of boron/epoxy laminates and found the dependence of fatigue life on stacking sequence. A $[\pm 45/\pm 15]_s$ laminate had a higher fatigue life than a $[\pm 15/\pm 45]_s$ laminate (Figure 5.7). Both laminates have the same number and angle of plies, and only the stacking sequence has been changed.

Loading factors such as tension and/or compression, temperature, moisture, and frequency of loading also determine the fatigue behavior of composites. For example, for compressive fatigue loading or tension-compressive fatigue loading, carbon/epoxy composites have very low peak strains because compression can cause layer buckling, etc. In such cases, the dominance of fiber effects is not present, but matrix, fiber-matrix interfaces, and the layers play a more important role.

Nonmechanical issues are also important in design of composite structures. These include fire resistance, smoke emission, lightning strikes, electrical and thermal conductivity, recycling potential, electromagnetic interference, etc.

5.6 Summary

In this chapter, we introduced the special case of laminates and their effect on the stiffness matrices, and response to external loads. We established failure criteria for laminates using the ply-by-ply failure theory. Examples of designing laminated structures such as plates, thin pressure vessels, and drive shafts were given. Other mechanical design issues such as environmental effects, interlaminar stresses, impact resistance, fracture resistance, and fatigue resistance were discussed.

Key Terms

Special laminates
Cross-ply laminates
Angle ply laminates
Antisymmetric laminates
Balanced laminates
Quasi-isotropic laminates
Failures of laminates
Design of laminates
Sandwich composites
Environmental effects
Interlaminar stresses
Impact resistance
Fracture resistance
Fatigue resistance

Exercise Set

5.1 Classify the following laminates:

[–30/45/–45/–30]

[–30/30/–30/30]

[30/–30/30]

[45/30/–30/–45]

[0/90/0/90/0/90/90]

[0/90/90/90/90/0]

[0/18/36/54/72/90/–18/–36/–54/–72]

5.2 Write an example of laminate code for the following:

Symmetric laminate

Antisymmetric laminate

Symmetric cross-ply laminate

Symmetric angle-ply laminate

Balanced angle-ply laminate

5.3 Give an example of a laminate with zero coupling stiffness matrix [B].

5.4 Is a nonzero [B] matrix attributed to the orthotropy of layers?

5.5 Is a nonzero [B] matrix attributed to the unsymmetrical stacking of laminae in a laminate?

5.6 Show numerically that a [0/90] laminate is not a quasi-isotropic laminate. Use the properties of unidirectional glass/epoxy lamina from Table 2.2.

5.7 Does a symmetric quasi-isotropic laminate have [A], [B], and [D] stiffness matrices like that of an isotropic material?

5.8 Are [0/60/–60] and [60/–60/60] quasi-isotropic laminates?

5.9 Are midplane strains and/or midplane curvatures always zero for symmetric laminates?

5.10 Find (1) the extensional stiffness matrix and (2) the extensional elastic moduli of the following graphite/epoxy laminate: $[0/18/36/54/72/90/-18/-36/-54/-72]_s$. Use properties of unidirectional graphite/epoxy lamina from Table 2.1.

5.11 Show that $A_{12} = U_4 h$ for a quasi-isotropic laminate.

5.12 A $[0/90]_s$ laminate made of glass/epoxy is subjected to an axial load N_x. Use properties of unidirectional glass/epoxy lamina from Table 2.2 and assume that each layer is 0.005 in. thick.

1. Use the maximum stress failure theory to find the first and last ply failure of the laminate.
2. Draw the stress–strain curve for the laminate till the last ply failure.

5.13 Using Tsai–Wu theory, find the ply-by-ply failure of a $[45/-45]_s$ graphite/epoxy laminate under a pure bending moment, M_x. Use properties of unidirectional graphite/epoxy lamina from Table 2.1 and assume each layer is 0.125 mm thick.

5.14 Repeat the preceding exercise in the presence of a temperature change of $\Delta T = -150°F$ and a moisture content of $\Delta C = 0.4\%$.

5.15 Develop a comparison table to show the elastic moduli E_x, E_y, ν_{xy}, and G_{xy} and the tensile strengths in x and y directions, shear strength in the x–y plane of the two laminates $[0/90]_s$ and $[45/-45]_s$ glass/epoxy laminate. Use properties of unidirectional glass/epoxy lamina from Table 2.2 and assume failure based on first ply failure (FPF).

5.16 Find the angle in $[\pm\theta]_{ns}$ graphite/epoxy sublaminate for maximum value of each of the elastic moduli:

1. E_x

2. E_y
3. G_{xy}

Use properties of unidirectional graphite/epoxy lamina from Table 2.1.

5.17 The bending stiffness of a laminate does not decrease substantially by replacing some of the plies at the midplane:

1. Find the percentage decrease in the longitudinal bending modulus of a $[0]_8$ glass/epoxy laminate if four of the plies closest to the midplane are replaced by a core of negligible stiffness.
2. What is the percentage decrease in the longitudinal bending modulus of a $[0/90/-45/45]_s$ glass/epoxy laminate if four of the plies closest to the midplane are replaced by a core of negligible stiffness?

Use properties of unidirectional glass/epoxy lamina from Table 2.1.

5.18 A designer uses a $[0]_8$ glass/epoxy laminate to manufacture a rotating blade. The in-plane longitudinal modulus is adequate, but the in-plane shear modulus is not. A suggestion is to replace the $[0]_8$ glass/epoxy laminate by a $[\pm 45]_{2s}$ graphite/epoxy laminate. Use the properties of unidirectional glass/epoxy lamina and unidirectional graphite/epoxy lamina from Table 2.2 to find:

1. Whether the longitudinal modulus increases or decreases and by how much
2. Percentage increase or decrease in the in-plane shear modulus with the replacement

5.19 Design a symmetric graphite/epoxy cross-ply sublaminate such that the thermal expansion coefficient in the x-direction is zero. Use the properties of unidirectional graphite/epoxy laminate from Table 2.1; however, assume that the longitudinal coefficient of thermal expansion is -0.3×10^{-6} m/m/°C.

5.20 1. Find the coefficient of thermal expansion of a symmetric quasi-isotropic graphite/epoxy laminate.

2. If you were able to change the longitudinal Young's modulus of the unidirectional graphite/epoxy lamina without affecting the value of other properties, what value would you choose to get zero thermal expansion coefficient for the quasi-isotropic laminate?

Use the properties of unidirectional graphite/epoxy lamina given in Table 2.1, *except* choose the longitudinal coefficient of thermal expansion as -0.3×10^{-6} m/m/°C.

5.21 Certain laminated structures, such as thin walled hollow drive shafts, are designed for maximum shear stiffness. Find the angle, θ, for a symmetric $[\pm\theta]_{ns}$ graphite/epoxy laminate such that the in-plane shear stiffness is a maximum. Use the properties of unidirectional graphite/epoxy lamina from Table 2.2.

5.22 A thin-walled pressure vessel is manufactured by a filament winding method using glass/epoxy prepregs. Find the optimum angles, θ, if the pressure vessel is made of $[\pm\theta]_{ns}$ sublaminate with

1. Spherical construction for maximum strength
2. Cylindrical construction for maximum strength
3. Cylindrical construction for no change in the internal diameter

Apply Tsai–Wu failure theory. Use properties of unidirectional glass/epoxy lamina from Table 2.2.

5.23 A cylindrical pressure vessel with flat ends of length 6 ft and inner diameter of 35 in. is subjected to an internal gauge pressure of 150 psi. Neglect the end effects and the mass of ends of the pressure vessel in your design. Take the factor of safety as 1.95:

1. Design the radial thickness of the pressure vessel using steel. For steel, assume that the Young's modulus is 30 Msi, Poisson's ratio is 0.3, specific gravity of steel is 7.8, and the ultimate normal tensile and compressive strength is 36 ksi.
2. Find the axial elongation of the steel pressure vessel designed in part (1), assuming plane stress conditions.
3. Find whether graphite/epoxy would be a better material to use for minimizing mass if, in addition to resisting the applied pressure, the axial elongation of the pressure vessel does not exceed that of the steel pressure vessel. The vessel operates at room temperature and curing residual stresses are neglected for simplification. The following are other specifications of the design:

   Only 0°, +45°, –45°, +60°, –60°, and 90° plies can be used.

   The thickness of each lamina is 0.005 in.

   Use specific gravities of the laminae from Example 5.6.

   Use Tsai–Wu failure criterion for calculating strength ratios.

5.24 Revisit the design problem of the drive shaft in Example 5.8. Use graphite/epoxy laminate with ply properties given in Table 2.1 to design the drive shaft.

1. If minimizing mass is still an issue, would a graphite/epoxy laminate be a better choice than glass/epoxy?
2. If cost is the only issue, is glass/epoxy laminate, steel, or graphite/epoxy the best choice? Assume total manufacturing cost of graphite/epoxy is five times that of glass/epoxy on a per-unit-mass basis and that the glass/epoxy and steel cost the same on a per-unit-mass basis.

References

1. Gradshetyn, I.S. and Ryzhik, I.M., *Table of Integrals, Series, and Products*, Academic Press, New York, 1980.
2. PROMAL for Windows software, Mechanical Engineering Department, University of South Florida, Tampa, 1996.
3. Gurdal, Z., Haftka, R.T., and Hajela, P., *Design and Optimization of Laminated Composite Materials*, John Wiley & Sons, New York, 1999.
4. James, M.L., Smith, G.M., Wolford, J.C., and Whaley, P.W., *Vibration of Mechanical and Structural Systems*, Harper and Row, New York, 1989.
5. Column Research Committee of Japan, Eds., *Handbook of Structural Stability*, Tokyo, Corona Publishing, 1971.
6. Sandwich panel review — Part I–IV, *M.C. Gill Doorway*, 28, 1991.
7. Quinn, J.A., Properties of thermoset polymer composites and design of pultrusions, in *Design with Advanced Composite Materials*, Philips, L.N., Ed., Springer–Verlag, New York, 1989, Chap. 3.
8. Pipes, R.B. and Pagano, N.J., Interlaminar stresses in composite laminates under uniform axial extension, *J. Composite Mater.*, 4, 538, 1970.
9. Pagano, N.J. and Pipes, R.B., The influence of stacking sequence on laminate strengths, *Int. J. Mech. Sci.*, 15, 679, 1973.
10. Wang, S.S., Edge delamination in angle-ply composite laminates, *AIAA J.*, 21, 256, 1984.
11. Sela, N. and Ishai, O., Interlaminar fracture toughness and toughening of laminated composite materials: a review, *Composites*, 20, 423, 1989.
12. Williams, J.G., O'Brien, T.K., and Chapman III, A.J., Comparison of toughened composite laminates using NASA standard damage tolerance tests, NASA CP 2321, ACEE Composite Structure Technology Conference, Seattle, WA, August 1984.
13. Chen, S.F. and Jeng, B.Z., Fracture behavior of interleaved fiber–resin composites, *Composites Sci. Technol.*, 41, 77, 1991.
14. Kaw, A.K. and Goree, J.G., Effect of Interleaves on fracture of laminated composites — Part II, *ASME J. Appl. Mech.*, 57, 175, 1990.
15. Foye, R.L. and Baker, D.J., Design of orthotropic laminates, 11th Annual AIAA Conference on Structures, Structural Dynamics and Materials, Denver, CO, April, 1970.

6

Bending of Beams

Chapter Objectives

- Develop formulas to find the deflection and stresses in a beam made of composite materials.
- Develop formulas for symmetric beams that are narrow or wide.
- Develop formulas for nonsymmetric beams that are narrow or wide.

6.1 Introduction

To study mechanics of beams made of laminated composite materials, we need to review the beam analysis of isotropic materials. Several concepts applied to beams made of isotropic materials will help in understanding beams made of composite materials. We are limiting our study to beams with transverse loading or applied moments.

The bending stress in an isotropic beam (Figure 6.1 and Figure 6.2) under an applied bending moment, M, is given by[1,2]

$$\sigma = \frac{Mz}{I}, \tag{6.1}$$

where

z = distance from the centroid
I = second moment of area

The bending deflections, w, are given by solving the differential equation

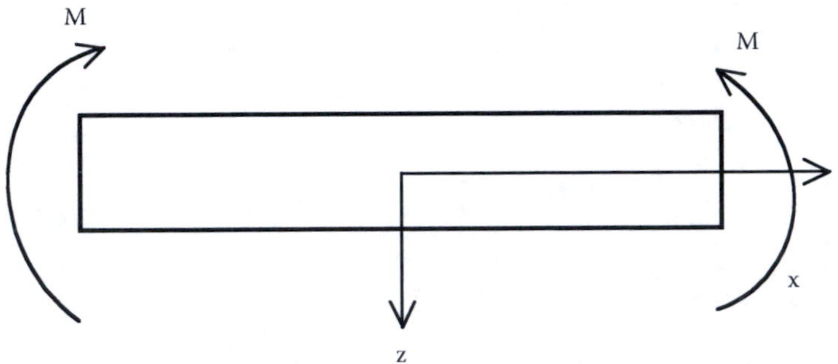

FIGURE 6.1
Bending of a beam.

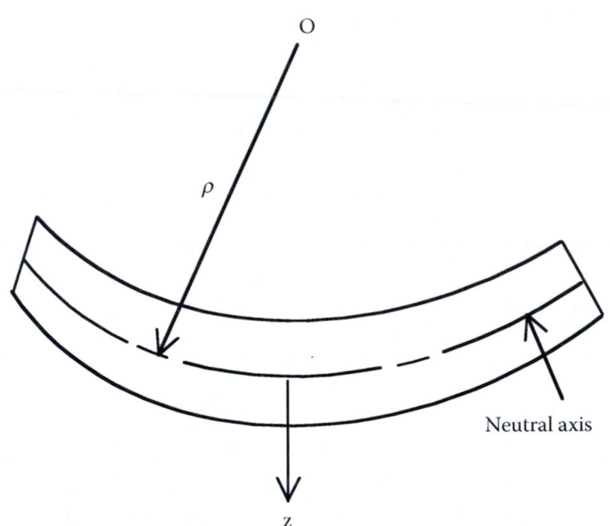

FIGURE 6.2
Curvature of a bended beam.

$$EI\frac{d^2w}{dx^2} = -M , \tag{6.2}$$

where E = Young's modulus of the beam material.

The term of $\frac{d^2w}{dx^2}$ is defined as the curvature

$$\kappa_x = -\frac{\partial^2 w}{\partial x^2} , \tag{6.3}$$

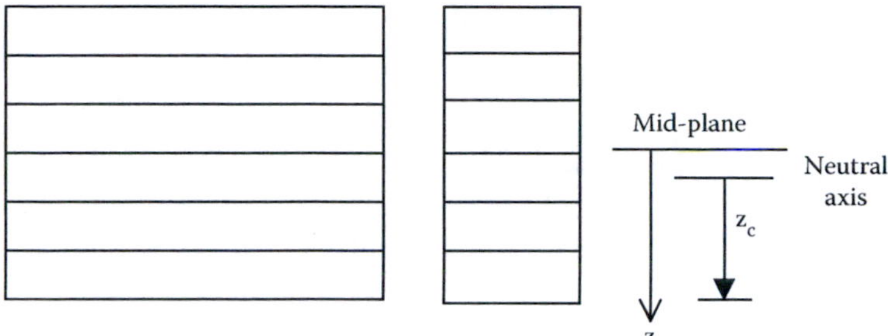

FIGURE 6.3
Laminated beam showing the midplane and the neutral axis.

giving

$$EI\kappa_x = M \, . \tag{6.4}$$

The formula for the bending stress is only valid for an isotropic material because it assumes that the elastic moduli is uniform in the beam. In the case of laminated materials, elastic moduli vary from layer to layer.

6.2 Symmetric Beams

To keep the introduction simple, we will discuss beams that are symmetric and have a rectangular cross-section[3] (Figure 6.3). Because the beam is symmetric, the loads and moments are decoupled in Equation (4.29) to give

$$\begin{bmatrix} M_x \\ M_y \\ M_{xy} \end{bmatrix} = [D] \begin{bmatrix} \kappa_x \\ \kappa_y \\ \kappa_{xy} \end{bmatrix} \tag{6.5}$$

or

$$\begin{bmatrix} \kappa_x \\ \kappa_y \\ \kappa_{xy} \end{bmatrix} = [D]^{-1} \begin{bmatrix} M_x \\ M_y \\ M_{xy} \end{bmatrix} . \tag{6.6}$$

Now, if bending is only taking place in the x-direction, then

$$M_y = 0 \text{ , } M_{xy} = 0$$

$$\begin{bmatrix} \kappa_x \\ \kappa_y \\ \kappa_{xy} \end{bmatrix} = [D]^{-1} \begin{bmatrix} M_x \\ 0 \\ 0 \end{bmatrix}, \tag{6.7}$$

that is,

$$\kappa_x = D_{11}^* M_x \tag{6.8a}$$

$$\kappa_y = D_{12}^* M_x \tag{6.8b}$$

$$\kappa_{xy} = D_{16}^* M_x \text{ ,} \tag{6.8c}$$

where D_{ij}^* are the elements of the $[D]^{-1}$ matrix as given in Equation (4.28c).
Because defining midplane curvatures (Equation 4.15),

$$\kappa_x = -\frac{\partial^2 w_0}{\partial x^2} \text{ ,}$$

$$\kappa_y = -\frac{\partial^2 w_0}{\partial y^2} \text{ ,} \tag{6.9}$$

$$\kappa_{xy} = -2\frac{\partial^2 w_0}{\partial x \partial y} \text{ ,}$$

the midplane deflection w_0 is not independent of y. However, if we have a narrow beam — that is, the length to width ratio (L/b) is sufficiently high, we can assume that $w_0 = w_0(x)$ only.

$$\kappa_x = -\frac{d^2 w_0}{dx^2} = D_{11}^* M_x \text{ .} \tag{6.10}$$

Writing in the form similar to Equation (6.2) for isotropic beams,

$$\frac{d^2w_0}{dx^2} = -\frac{M_x b}{E_x I}, \tag{6.11}$$

where

b = width of beam
E_x = effective bending modulus of beam
I = second moment of area with respect to the x–y-plane

From Equation (6.8a) and (6.11), we get

$$E_x = \frac{12}{h^3 D_{11}^*}. \tag{6.12}$$

Also,

$$I = \frac{bh^3}{12} \tag{6.13}$$

$$M = M_x b. \tag{6.14}$$

To find the strains, we have, from Equation (4.16),

$$\epsilon_x = z\kappa_x \tag{6.15a}$$

$$\epsilon_y = z\kappa_y \tag{6.15b}$$

$$\gamma_{xy} = z\kappa_{xy}. \tag{6.15c}$$

These global strains can be transformed to the local strains in each ply using Equation (2.95):

$$\begin{bmatrix} \epsilon_1 \\ \epsilon_2 \\ \gamma_{12} \end{bmatrix}_k = [R][T][R]^{-1} \begin{bmatrix} \epsilon_x \\ \epsilon_y \\ \gamma_{xy} \end{bmatrix}_k. \tag{6.16}$$

The local stresses in each ply are obtained using Equation (2.73) as

$$\begin{bmatrix} \sigma_1 \\ \sigma_2 \\ \tau_{12} \end{bmatrix}_k = \left[Q \right] \begin{bmatrix} \epsilon_1 \\ \epsilon_2 \\ \gamma_{12} \end{bmatrix}_k . \tag{6.17}$$

The global stresses in each ply are then obtained using Equation (2.89) as

$$\begin{bmatrix} \sigma_x \\ \sigma_y \\ \tau_{xy} \end{bmatrix}_k = \left[T \right]^{-1} \begin{bmatrix} \sigma_1 \\ \sigma_2 \\ \tau_{12} \end{bmatrix}_k . \tag{6.18}$$

Example 6.1

A simply supported laminated composite beam of length 0.1 m and width 5 mm (Figure 6.4) made of graphite/epoxy has the following layup of $[0/90/-30/30]_s$. A uniform load of 200 N/m is applied on the beam. What is the maximum deflection of the beam? Find the local stresses at the top of the third ply (–30°) from the top. Assume that each ply is 0.125 mm thick and the properties of unidirectional graphite/epoxy are as given in Table 2.1.

Solution

The shear and bending moment diagrams for the beam are given in Figure 6.5. The bending moment is maximum at the center of the beam and is given by

$$M = \frac{qL^2}{8}, \tag{6.19}$$

where

q = load intensity (N/m)
L = length of the beam (m)

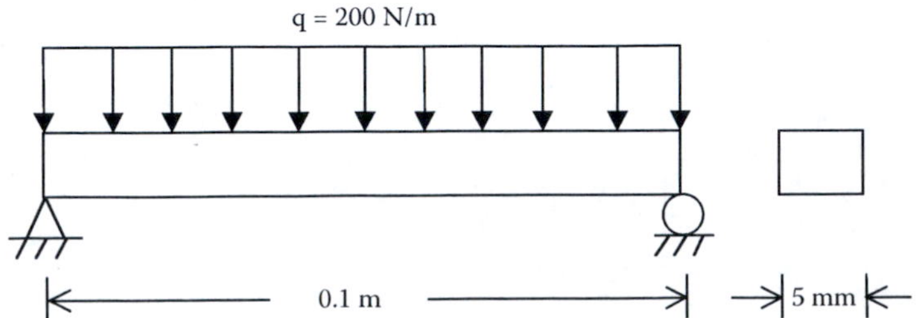

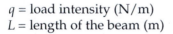

FIGURE 6.4
Uniformly loaded simply supported beam.

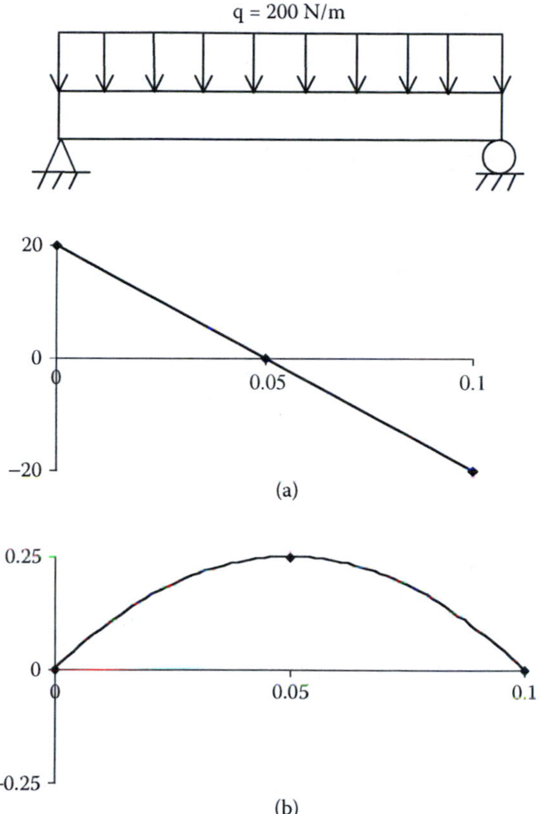

FIGURE 6.5
Shear (a) and bending moment (b) diagrams of a simply supported beam.

The maximum bending moment then is

$$M = \frac{200 \times 0.1^2}{8}$$

$$= 0.25 \text{ N-m.}$$

Without showing the calculations because they are shown in detail in Chapter 4 (see Example 4.2), we get

$$[D] = \begin{bmatrix} 1.015 \times 10^1 & 5.494 \times 10^{-1} & -4.234 \times 10^{-1} \\ 5.494 \times 10^{-1} & 5.243 \times 10^0 & -1.567 \times 10^{-1} \\ -4.234 \times 10^{-1} & -1.567 \times 10^{-1} & 9.055 \times 10^{-1} \end{bmatrix} Pa\text{-}m^3$$

$$[D]^{-1} = \begin{bmatrix} 1.009\times10^{1} & -9.209\times10^{-3} & 4.557\times10^{-2} \\ -9.209\times10^{-3} & 1.926\times10^{-1} & 2.901\times10^{-2} \\ 4.557\times10^{-2} & 2.901\times10^{-2} & 1.131\times10^{0} \end{bmatrix} \frac{1}{Pa\text{-}m^3}.$$

To find the maximum deflection of the beam, δ, we use the isotropic beam formula:

$$\delta = \frac{5qL^4}{384E_xI}. \tag{6.20}$$

Now, in Equation (6.12),

$$h = (8)(0.125\times10^{-3})$$

$$= 0.001m$$

$$D_{11}^* = 1.009\times10^{-1}\frac{1}{Pa-m^3}.$$

Thus,

$$E_x = \frac{12}{h^3D_{11}^*}$$

$$= \frac{12}{(0.001)^3(1.009\times10^{-1})}$$

$$= 1.189\times10^{11}\,Pa$$

From Equation (6.13),

$$I = \frac{bh^3}{12}$$

$$= \frac{(5\times10^{-3})(0.001)^3}{12}$$

$$= 4.167\times10^{-13}\,m^4.$$

Therefore, from Equation (6.20),

$$\delta = \frac{(5)(200)(0.1)^4}{(384)(1.189\times10^{11})(4.167\times10^{-13})}$$

$$= 5.256\times10^{-3}\,m$$

$$= 5.256\ mm\ .$$

The maximum curvature is at the middle of the beam and is given by

$$\begin{bmatrix}\kappa_x\\ \kappa_y\\ \kappa_{xy}\end{bmatrix} = \begin{bmatrix}D_{11}^*\\ D_{12}^*\\ D_{16}^*\end{bmatrix}\frac{qL^2}{8b}$$

$$= \begin{bmatrix}1.009\times10^{-1}\\ -9.209\times10^{-3}\\ 4.557\times10^{-2}\end{bmatrix}\frac{200\times0.1^2}{8\times0.005}$$

$$= \begin{bmatrix}1.009\times10^{-1}\\ -9.209\times10^{-3}\\ 4.557\times10^{-2}\end{bmatrix}50$$

$$= \begin{bmatrix}5.045\\ -0.4605\\ 2.279\end{bmatrix}\frac{1}{m}\ .$$

The global strains (Equation 6.15) at the top of the third ply (–30°) are

$$\begin{bmatrix}\varepsilon_x\\ \varepsilon_y\\ \gamma_{xy}\end{bmatrix} = z\begin{bmatrix}\kappa_x\\ \kappa_y\\ \kappa_{xy}\end{bmatrix}$$

$$= (-0.00025)\begin{bmatrix}5.045\\ -0.4605\\ 2.279\end{bmatrix}$$

$$= \begin{bmatrix} -1.261 \times 10^{-3} \\ 1.151 \times 10^{-4} \\ -5.696 \times 10^{-4} \end{bmatrix} \frac{m}{m} .$$

The global stresses (Equation 6.18) at the top of the third ply (–30°) then are

$$\begin{bmatrix} \sigma_x \\ \sigma_y \\ \tau_{xy} \end{bmatrix} = \left[\bar{Q}\right] \begin{bmatrix} \varepsilon_x \\ \varepsilon_y \\ \gamma_{xy} \end{bmatrix}$$

$$= \begin{bmatrix} 1.094 \times 10^{11} & 3.246 \times 10^{10} & -5.419 \times 10^{10} \\ 3.246 \times 10^{10} & 2.365 \times 10^{10} & -2.005 \times 10^{10} \\ -5.419 \times 10^{10} & -2.005 \times 10^{10} & 3.674 \times 10^{10} \end{bmatrix} \begin{bmatrix} -1.261 \times 10^{-3} \\ 1.151 \times 10^{-4} \\ -5.698 \times 10^{-4} \end{bmatrix}$$

$$= \begin{bmatrix} -1.034 \times 10^{8} \\ -2.680 \times 10^{7} \\ 4.511 \times 10^{7} \end{bmatrix} Pa .$$

Example 6.2

In Example 6.1, the width-to-height ratio in the cross-section of the beam is $b/h = 5/1 = 5$. This may be considered as a narrow-beam cross-section. If the b/h ratio were large, the cross-section may be considered to be wide beam. What are the results of Example 6.1 if one considers the beam to be a wide beam?

Solution

In the case of wide beams, we consider

$$\kappa_y = \kappa_{xy} = 0 .$$

Then, from Equation (6.5),

$$\begin{bmatrix} M_x \\ M_y \\ M_{xy} \end{bmatrix} = \begin{bmatrix} D_{11} & D_{12} & D_{16} \\ D_{12} & D_{22} & D_{26} \\ D_{16} & D_{26} & D_{66} \end{bmatrix} \begin{bmatrix} \kappa_x \\ 0 \\ 0 \end{bmatrix} ,$$

giving

$$M_x = D_{11}\kappa_x \tag{6.21}$$

$$M = M_x b = D_{11}\kappa_x b\ . \tag{6.22}$$

Thus, from Equation (6.9a), Equation (6.11), and Equation (6.21),

$$E_x = \frac{12D_{11}}{h^3}$$

$$= \frac{(12)(1.015\times10^1)}{(0.001)^3}$$

$$= 1.218\times10^{11}\,Pa$$

and, from Equation (6.20),

$$\delta = \frac{(5)(200)(0.1)^4}{(384)(1.218\times10^{11})(4.167\times10^{-13})}$$

$$= 5.131\times10^{-3}\,m$$

$$= 5.131\ mm.$$

The relative difference in the value of deflection between the assumption of a wide and narrow beam is

$$|\epsilon_a| = \left|\frac{\delta_{narrow} - \delta_{wide}}{\delta_{narrow}}\right|\times100$$

$$= \left|\frac{5.256 - 5.131}{5.256}\right|\times100$$

$$= 2.357\%\ .$$

Because there is only 2.357% difference in the maximum deflection, does this mean that the assumption of wide beams influences the stresses only by a similar amount?

From Equation (6.21),

$$\kappa_x = \frac{M_x}{D_{11}}$$

$$= \frac{50}{1.015 \times 10^1}$$

$$= 4.926 \frac{1}{m}.$$

Because $\kappa_y = 0$, $\kappa_{xy} = 0$,

$$\begin{bmatrix} \kappa_x \\ \kappa_y \\ \kappa_{xy} \end{bmatrix} = \begin{bmatrix} 4.926 \\ 0 \\ 0 \end{bmatrix} \frac{1}{m}.$$

The global strains (Equation 6.15) at the top of the third ply (–30°) are

$$\begin{bmatrix} \epsilon_x \\ \epsilon_y \\ \gamma_{xy} \end{bmatrix} = z \begin{bmatrix} \kappa_x \\ \kappa_y \\ \kappa_{xy} \end{bmatrix}$$

$$= (-0.00025) \begin{bmatrix} 4.926 \\ 0 \\ 0 \end{bmatrix}$$

$$= \begin{bmatrix} -1.232 \times 10^{-3} \\ 0 \\ 0 \end{bmatrix} \frac{m}{m}.$$

The global stresses (Equation 6.18) at the top of the third ply (–30°) then are

$$\begin{bmatrix}\sigma_x\\ \sigma_y\\ \tau_{xy}\end{bmatrix}=\left[\bar{Q}\right]\begin{bmatrix}\epsilon_x\\ \epsilon_y\\ \gamma_{xy}\end{bmatrix}$$

$$=\begin{bmatrix}1.094\times10^{11} & 3.246\times10^{10} & -5.419\times10^{10}\\ 3.246\times10^{10} & 2.365\times10^{10} & -2.005\times10^{10}\\ -5.419\times10^{10} & -2.005\times10^{10} & 3.674\times10^{10}\end{bmatrix}\begin{bmatrix}-1.232\times10^{-3}\\ 0\\ 0\end{bmatrix}$$

$$=\begin{bmatrix}-1.348\times10^{8}\\ -3.999\times10^{7}\\ 6.676\times10^{7}\end{bmatrix}Pa\ .$$

The relative differences in the stresses obtained using wide and narrow beam assumptions are

$$\left|\epsilon_a\right|_{\sigma_x}=\left|\frac{\sigma_{x|narrow}-\sigma_{x|wide}}{\sigma_{x|narrow}}\right|\times100$$

$$=\left|\frac{-1.034\times10^{8}-\left(-1.348\times10^{8}\right)}{-1.034\times10^{8}}\right|$$

$$=30.37\%$$

$$\left|\epsilon_a\right|_{\sigma_y}=\left|\frac{\sigma_{y|narrow}-\sigma_{y|wide}}{\sigma_{y|narrow}}\right|\times100$$

$$=\left|\frac{-2.680\times10^{7}-\left(-3.999\times10^{7}\right)}{2.680\times10^{7}}\right|\times100$$

$$=49.22\%\ .$$

$$\left|\varepsilon_a\right|_{\tau_{xy}} = \left|\frac{\tau_{xy|narrow} - \tau_{xy|wide}}{\tau_{xy|narrow}}\right| \times 100$$

$$= \left|\frac{4.511\times 10^7 - 6.676\times 10^7}{4.511\times 10^7}\right| \times 100$$

$$= 48.00\% \,.$$

6.3 Nonsymmetric Beams

In the case of nonsymmetric beams, the loads and moment are not decoupled. The relationship given by Equation (4.29) is

$$\begin{bmatrix} N \\ \hline M \end{bmatrix} = \left[\begin{array}{c|c} A & B \\ \hline B & D \end{array}\right] \begin{bmatrix} \varepsilon_0 \\ \hline \kappa \end{bmatrix}$$

or

$$\begin{bmatrix} \varepsilon_0 \\ \hline \kappa \end{bmatrix} = \left[\begin{array}{c|c} A & B \\ \hline B & D \end{array}\right]^{-1} \begin{bmatrix} N \\ \hline M \end{bmatrix}.$$

Assuming that the preceding 6×6 inverse matrix is denoted by $[J]$ — that is,

$$\left[\begin{array}{c|c} A & B \\ \hline B & D \end{array}\right]^{-1} = [J], \tag{6.23}$$

then

$$\begin{bmatrix} \varepsilon_x^0 \\ \varepsilon_y^0 \\ \gamma_{xy}^0 \\ \kappa_x \\ \kappa_y \\ \kappa_{xy} \end{bmatrix} = \begin{bmatrix} J_{11} & J_{12} & J_{13} & J_{14} & J_{15} & J_{16} \\ J_{21} & J_{22} & J_{23} & J_{24} & J_{25} & J_{26} \\ J_{31} & J_{32} & J_{33} & J_{34} & J_{35} & J_{36} \\ J_{41} & J_{42} & J_{43} & J_{44} & J_{45} & J_{46} \\ J_{51} & J_{52} & J_{53} & J_{54} & J_{55} & J_{56} \\ J_{61} & J_{62} & J_{63} & J_{64} & J_{65} & J_{66} \end{bmatrix} \begin{bmatrix} N_x \\ N_y \\ N_{xy} \\ M_x \\ M_y \\ M_{xy} \end{bmatrix}. \tag{6.24}$$

If bending is taking place only in the x-direction, then M_x is the only nonzero component, giving

$$\varepsilon_x^0 = J_{14} M_x$$

$$\varepsilon_y^0 = J_{24} M_x$$

$$\gamma_{xy}^0 = J_{34} M_x$$

$$\kappa_x = J_{44} M_x$$

$$\kappa_y = J_{54} M_y$$

$$\kappa_{xy} = J_{64} M_{xy} \; . \tag{6.25}$$

The strain distribution in the beam, then, from Equation (4.16) is

$$\varepsilon_x = \varepsilon_x^0 + z\kappa_x \tag{6.26a}$$

$$\varepsilon_y = \varepsilon_y^0 + z\kappa_y \tag{6.26b}$$

$$\gamma_{xy} = \gamma_{xy}^0 + z\kappa_{xy} \; . \tag{6.26c}$$

Because the beam is unsymmetric, the neutral axis does not coincide with the midplane. The location of the neutral axis, z_n, is where $\varepsilon_x = 0$. From Equation (6.26a),

$$0 = \varepsilon_x^0 + z_n \kappa_x$$

$$= J_{14} M_x + z_n J_{44} M_x \; ,$$

giving

$$z_n = -\frac{J_{14}}{J_{44}} \; . \tag{6.27}$$

Because, from Equation (4.15),

$$\kappa_x = -\frac{\partial^2 w_0}{\partial x^2}$$

$$\kappa_y = -\frac{\partial^2 w_0}{\partial y^2}$$

$$\kappa_{xy} = -2\frac{\partial^2 w_0}{\partial x \partial y},$$

the deflection w_0 is not independent of y. However, if we have a narrow beam — that is, the length-to-width ratio (L/b) is sufficiently high, we can assume that $w_0 = w_0(x)$ only.

$$\kappa_x = \frac{d^2 w_0}{dx^2} = -J_{44} M_x, \tag{6.28}$$

writing in the form

$$\frac{d^2 w_0}{dx^2} = -\frac{M_x b}{E_x I}, \tag{6.29}$$

where

b = width of beam
E_x = effective bending modulus of beam
I = second moment of area with respect to the x–y-plane

From Equation (6.28) and Equation (6.29), we get

$$E_x = \frac{12}{h^3 J_{44}}. \tag{6.30}$$

Also,

$$I = \frac{bh^3}{12}$$

$$M = M_x b.$$

To find the strains, we have, from Equation (4.16),

$$\varepsilon_x = \varepsilon_x^o + z\kappa_x \tag{6.31a}$$

$$\varepsilon_y = \varepsilon_y^o + z\kappa_y \tag{6.31b}$$

$$\gamma_{xy} = \gamma_{xy}^0 + z\kappa_{xy} \, . \tag{6.31c}$$

These global strains can be transformed to the local strains in each ply using Equation (2.95):

$$\begin{bmatrix} \varepsilon_1 \\ \varepsilon_2 \\ \gamma_{12} \end{bmatrix}_k = [R][T][R]^{-1} \begin{bmatrix} \varepsilon_x \\ \varepsilon_y \\ \gamma_{xy} \end{bmatrix}_k . \tag{6.32}$$

The local stresses in each ply are obtained using Equation (2.73) as

$$\begin{bmatrix} \sigma_1 \\ \sigma_2 \\ \tau_{12} \end{bmatrix}_k = [Q] \begin{bmatrix} \varepsilon_1 \\ \varepsilon_2 \\ \gamma_{12} \end{bmatrix}_k . \tag{6.33}$$

The global stresses in each ply are then obtained using Equation (2.89) as

$$\begin{bmatrix} \sigma_x \\ \sigma_y \\ \tau_{xy} \end{bmatrix}_k = [T]^{-1} \begin{bmatrix} \sigma_1 \\ \sigma_2 \\ \tau_{12} \end{bmatrix}_k . \tag{6.34}$$

Example 6.3

A simply supported laminated composite beam (Figure 6.4) of length 0.1 m and width 5 mm made of graphite/epoxy has the following layup: $[0/90/-30/30]_2$. A uniform load of 200 N/m is applied on the beam. What is the maximum deflection of the beam? Find the local stresses at the top of the third ply (–30°) from the top. Assume that each ply is 0.125 mm thick and the properties of unidirectional graphite/epoxy are as given in Table 2.1.

Solution

The stiffness matrix found by using Equation (4.28) and Equation (4.29) is

$$\begin{bmatrix} 1.027\times10^{8} & 1.768\times10^{7} & 3.497\times10^{-10} & -1.848\times10^{3} & 1.848\times10^{3} & 1.694\times10^{3} \\ 1.768\times10^{7} & 5.986\times10^{7} & 2.608\times10^{-9} & 1.848\times10^{3} & -1.848\times10^{3} & 6.267\times10^{2} \\ 3.497\times10^{-10} & 2.608\times10^{-9} & 2.195\times10^{7} & 1.694\times10^{3} & 6.267\times10^{2} & 1.848\times10^{3} \\ -1.848\times10^{3} & 1.848\times10^{3} & 1.694\times10^{3} & 9.231 & 1.473 & 4.234\times10^{-1} \\ 1.848\times10^{3} & -1.848\times10^{3} & 6.267\times10^{2} & 1.473 & 4.319 & 1.567\times10^{-1} \\ 1.694\times10^{3} & 6.267\times10^{2} & 1.848\times10^{3} & 4.234\times10^{-1} & 1.567\times10^{-1} & 1.829 \end{bmatrix}.$$

The inverse of the matrix is

$$\begin{bmatrix} 1.068\times10^{-8} & -3.409\times10^{-9} & 7.009\times10^{-10} & 4.298\times10^{-6} & -7.241\times10^{-6} & -9.809\times10^{-6} \\ -3.409\times10^{-9} & 1.829\times10^{-8} & 4.042\times10^{-10} & -6.097\times10^{-6} & 1.142\times10^{-5} & -3.083\times10^{-6} \\ 7.009\times10^{-10} & 4.042\times10^{-10} & 5.035\times10^{-8} & -6.339\times10^{-6} & -3.460\times10^{-6} & -4.989\times10^{-5} \\ 4.298\times10^{-6} & -6.097\times10^{-6} & -6.339\times10^{-6} & 1.194\times10^{-1} & -4.335\times10^{-2} & -1.940\times10^{-2} \\ -7.241\times10^{-6} & 1.142\times10^{-5} & -3.460\times10^{-6} & -4.355\times10^{-2} & 2.551\times10^{-1} & -5.480\times10^{-3} \\ -9.809\times10^{-6} & -3.083\times10^{-6} & -4.989\times10^{-5} & -1.940\times10^{-2} & -5.480\times10^{-3} & 6.123\times10^{-1} \end{bmatrix}$$

$$h = 8\times\left(0.125\times10^{-3}\right)$$

$$= 0.001\ m$$

$$J_{44} = 1.194\times10^{-1}\frac{1}{Pa\text{-}m^{3}}.$$

Now, in Equation (6.30),

$$E_x = \frac{12}{h^3 J_{44}}$$

$$= \frac{12}{(0.001)^3\left(1.194\times10^{-1}\right)}$$

$$= 1.005\times10^{11}\ Pa.$$

From Equation (6.13),

$$I = \frac{bh^3}{12}$$

$$= \frac{\left(5 \times 10^{-3}\right)\left(0.001\right)^3}{12}$$

$$= 4.167 \times 10^{-13} m^4 .$$

Thus, from Equation (6.20),

$$\delta = \frac{(5)(200)(0.1)^4}{(384)\left(1.005 \times 10^{11}\right)\left(4.167 \times 10^{-13}\right)}$$

$$= 6.219 \times 10^{-3}\ m$$

$$= 6.219\ mm.$$

The maximum bending moment occurs at the middle of the beam and is given by

$$M_{\max} = \frac{qL^2}{8}$$

$$= \frac{200 \times 0.1^2}{8}$$

$$= 0.25\ N\text{-}m$$

$$M_{x|max} = \frac{M_{max}}{b}$$

$$= \frac{0.25}{0.005}$$

$$= 50 \frac{N\text{-}m}{m}.$$

Calculating the midplane strains and curvature from Equation (6.24) gives

$$\begin{bmatrix} \varepsilon_x^0 \\ \varepsilon_y^0 \\ \gamma_{xy}^0 \\ \kappa_x \\ \kappa_y \\ \kappa_{xy} \end{bmatrix} =$$

$$\begin{bmatrix} 1.068\times10^{-8} & -3.409\times10^{-9} & 7.009\times10^{-10} & 4.298\times10^{-6} & -7.241\times10^{-6} & -9.809\times10^{-6} \\ -3.409\times10^{-9} & 1.829\times10^{-8} & 4.042\times10^{-10} & -6.097\times10^{-6} & 1.142\times10^{-5} & -3.083\times10^{-6} \\ 7.009\times10^{-10} & 4.042\times10^{-10} & 5.035\times10^{-8} & -6.339\times10^{-6} & -3.460\times10^{-6} & -4.989\times10^{-5} \\ 4.298\times10^{-6} & -6.097\times10^{-6} & -6.339\times10^{-6} & 1.194\times10^{-1} & -4.335\times10^{-2} & -1.940\times10^{-2} \\ -7.241\times10^{-6} & 1.142\times10^{-5} & -3.460\times10^{-6} & -4.355\times10^{-2} & 2.551\times10^{-1} & -5.480\times10^{-3} \\ -9.809\times10^{-6} & -3.083\times10^{-6} & -4.989\times10^{-5} & -1.940\times10^{-2} & -5.480\times10^{-3} & 6.123\times10^{-1} \end{bmatrix} \begin{bmatrix} 0 \\ 0 \\ 0 \\ 50 \\ 0 \\ 0 \end{bmatrix},$$

giving

$$\begin{bmatrix} \varepsilon_x^0 \\ \varepsilon_y^0 \\ \gamma_{xy}^0 \end{bmatrix} = \begin{bmatrix} 2.149\times10^{-4} \\ -3.048\times10^{-4} \\ -3.169\times10^{-4} \end{bmatrix}$$

$$\begin{bmatrix} \kappa_x \\ \kappa_y \\ \kappa_{xy} \end{bmatrix} = \begin{bmatrix} 5.970 \\ -2.178 \\ -9.700\times10^{-1} \end{bmatrix}.$$

The global strains (Equation 6.31) at the top of the third ply (–30°) are

$$\begin{bmatrix} \varepsilon_x \\ \varepsilon_y \\ \gamma_{xy} \end{bmatrix} = \begin{bmatrix} \varepsilon_x^0 \\ \varepsilon_y^0 \\ \gamma_{xy}^0 \end{bmatrix} + z \begin{bmatrix} \kappa_x \\ \kappa_y \\ \kappa_{xy} \end{bmatrix}$$

$$= \begin{bmatrix} 2.149\times10^{-4} \\ -3.048\times10^{-4} \\ -3.169\times10^{-4} \end{bmatrix} + (-0.00025) \begin{bmatrix} 5.970 \\ -2.178 \\ -9.700\times10^{-1} \end{bmatrix}$$

$$= \begin{bmatrix} -1.278\times10^{-3} \\ 2.397\times10^{-4} \\ -7.431\times10^{-5} \end{bmatrix} \frac{m}{m}.$$

The global stresses (Equation 6.34) at the top of the third ply (–30°) are

$$\begin{bmatrix} \sigma_x \\ \sigma_y \\ \tau_{xy} \end{bmatrix} = \left[\bar{Q}\right] \begin{bmatrix} \varepsilon_x \\ \varepsilon_y \\ \gamma_{xy} \end{bmatrix}$$

$$= \begin{bmatrix} 1.094\times10^{11} & 3.246\times10^{10} & -5.419\times10^{10} \\ 3.246\times10^{10} & 2.365\times10^{10} & -2.005\times10^{10} \\ -5.419\times10^{10} & -2.005\times10^{10} & 3.674\times10^{10} \end{bmatrix} \begin{bmatrix} -1.278\times10^{-3} \\ 2.397\times10^{-4} \\ 7.431\times10^{-5} \end{bmatrix}$$

$$= \begin{bmatrix} -1.280\times10^{8} \\ -3.431\times10^{7} \\ 6.170\times10^{7} \end{bmatrix} Pa\ .$$

Example 6.4

In Example 6.3, the width-to-height ratio in the cross-section of the beam is $b/h = 5/1 = 5$. This may be considered as a narrow-beam cross-section. If the b/h ratio were large, the cross-section may be considered to be wide beam. What are the results of Example 6.3 if one considers the beam to be a wide beam?

Solution

In the case of the wide beams, we consider

$$\kappa_y = 0\ .$$

Then, from Equation (6.24),

$$\begin{bmatrix} \varepsilon_x^0 \\ \varepsilon_y^0 \\ \gamma_{xy}^0 \\ \kappa_x \\ 0 \\ \kappa_{xy} \end{bmatrix} = \begin{bmatrix} J_{11} & J_{12} & J_{13} & J_{14} & J_{15} & J_{16} \\ J_{21} & J_{22} & J_{23} & J_{24} & J_{25} & J_{26} \\ J_{31} & J_{32} & J_{33} & J_{34} & J_{35} & J_{36} \\ J_{41} & J_{42} & J_{43} & J_{44} & J_{45} & J_{46} \\ J_{51} & J_{52} & J_{53} & J_{54} & J_{55} & J_{56} \\ J_{61} & J_{62} & J_{63} & J_{64} & J_{65} & J_{66} \end{bmatrix} \begin{bmatrix} 0 \\ 0 \\ 0 \\ M_x \\ M_y \\ 0 \end{bmatrix}, \tag{6.35}$$

we get

$$\varepsilon_x^0 = J_{14}M_x + J_{15}M_y \tag{6.36a}$$

$$\varepsilon_y^0 = J_{24}M_x + J_{25}M_y \tag{6.36b}$$

$$\gamma_{xy}^0 = J_{34}M_x + J_{35}M_y \tag{6.36c}$$

$$\kappa_x = J_{44}M_x + J_{45}M_y \tag{6.36d}$$

$$0 = J_{54}M_x + J_{55}M_y \tag{6.36e}$$

$$\kappa_{xy} = J_{64}M_x + J_{65}M_y\ . \tag{6.36f}$$

To find the neutral axis, $\varepsilon_x = 0$, we use Equation (6.36a) and Equation (6.36e) to give

$$z_n = -\frac{J_{14}J_{55} - J_{15}J_{54}}{J_{44}J_{55} - J_{45}J_{54}} \tag{6.37}$$

$$M_{beam} = bM_x = b\frac{J_{55}}{J_{44}J_{55} - J_{54}J_{45}}\kappa_x\ . \tag{6.38}$$

From Equation (6.9a), Equation (6.11), and Equation (6.38),

$$E_x = \frac{12}{h^3}\frac{J_{55}}{\left(J_{44}J_{55} - J_{45}J_{54}\right)}$$

$$= \frac{12}{(0.001)^3}\frac{2.551\times10^{-1}}{\left(1.194\times10^{-1}\right)\left(2.551\times10^{-1}\right)-\left(-4.355\times10^{-2}\right)\left(-4.355\times10^{-2}\right)}$$

$$= 1.071\times10^{11}\,Pa.$$

Thus, from Equation (6.20), we get

$$\delta = \frac{(5)(200)(0.1)^4}{(384)\left(1.072\times10^{11}\right)\left(4.167\times10^{-13}\right)}$$

$$= 5.830 \times 10^{-3} m$$

$$= 5.830\ mm.$$

From Example 6.3, the maximum bendings' moment per unit width is

$$M_{x|\max} = 50 \frac{N\text{-}m}{m}.$$

From Equation (6.36e),

$$M_y \Big|_{\max} = -\frac{J_{54}}{J_{55}} M_x$$

$$= -\frac{-4.355 \times 10^{-2}}{2.551 \times 10^{-1}} (50)$$

$$= 8.497 \frac{N\text{-}m}{m}.$$

From Equation (6.35),

$$\begin{bmatrix} \epsilon_x^0 \\ \epsilon_y^0 \\ \gamma_{xy}^0 \\ \kappa_x \\ 0 \\ \kappa_y \end{bmatrix} =$$

$$\begin{bmatrix} 1.068 \times 10^{-8} & -3.409 \times 10^{-9} & 7.009 \times 10^{-10} & 4.298 \times 10^{-6} & -7.241 \times 10^{-6} & -9.809 \times 10^{-6} \\ -3.409 \times 10^{-9} & 1.829 \times 10^{-8} & 4.042 \times 10^{-10} & -6.097 \times 10^{-6} & 1.142 \times 10^{-5} & -3.083 \times 10^{-6} \\ 7.009 \times 10^{-10} & 4.042 \times 10^{-10} & 5.035 \times 10^{-8} & -6.339 \times 10^{-6} & -3.460 \times 10^{-6} & -4.989 \times 10^{-5} \\ 4.298 \times 10^{-6} & -6.097 \times 10^{-6} & -6.339 \times 10^{-6} & 1.194 \times 10^{-1} & -4.335 \times 10^{-2} & -1.940 \times 10^{-2} \\ -7.241 \times 10^{-6} & 1.142 \times 10^{-5} & -3.460 \times 10^{-6} & -4.355 \times 10^{-2} & 2.551 \times 10^{-1} & -5.480 \times 10^{-3} \\ -9.809 \times 10^{-6} & -3.083 \times 10^{-6} & -4.989 \times 10^{-5} & -1.940 \times 10^{-2} & -5.480 \times 10^{-3} & 6.123 \times 10^{-1} \end{bmatrix} \begin{bmatrix} 0 \\ 0 \\ 0 \\ 50 \\ 8.497 \\ 0 \end{bmatrix}.$$

$$= \begin{bmatrix} 1.534 \times 10^{-4} \\ -2.078 \times 10^{-4} \\ -3.463 \times 10^{-4} \\ 5.602 \\ 0 \\ -1.017 \end{bmatrix}$$

The global strains (Equation 6.15) at the top of the third ply (–30°) are

$$\begin{bmatrix} \varepsilon_x \\ \varepsilon_y \\ \gamma_{xy} \end{bmatrix} = \begin{bmatrix} \varepsilon_x^0 \\ \varepsilon_y^0 \\ \gamma_{xy}^0 \end{bmatrix} + z \begin{bmatrix} \kappa_x \\ \kappa_y \\ \kappa_{xy} \end{bmatrix}$$

$$= \begin{bmatrix} 1.534\times10^{-4} \\ -2.078\times10^{-4} \\ -3.463\times10^{-4} \end{bmatrix} + (-0.00025) \begin{bmatrix} 5.602 \\ 0 \\ -1.017 \end{bmatrix}$$

$$= \begin{bmatrix} -1.247\times10^{-3} \\ -2.078\times10^{-4} \\ -9.221\times10^{-5} \end{bmatrix} \frac{m}{m}.$$

The global stresses (Equation 6.18) at the top of the third ply (–30°) are

$$\begin{bmatrix} \sigma_x \\ \sigma_y \\ \tau_{xy} \end{bmatrix} = \left[\bar{Q}\right] \begin{bmatrix} \varepsilon_x \\ \varepsilon_y \\ \gamma_{xy} \end{bmatrix}$$

$$= \begin{bmatrix} 1.094\times10^{11} & 3.246\times10^{10} & -5.419\times10^{10} \\ 3.246\times10^{10} & 2.365\times10^{10} & -2.005\times10^{10} \\ -5.419\times10^{10} & -2.005\times10^{10} & 3.674\times10^{10} \end{bmatrix} \begin{bmatrix} -1.247\times10^{-3} \\ -2.078\times10^{-4} \\ -9.221\times10^{-3} \end{bmatrix}$$

$$= \begin{bmatrix} -1.382\times10^{8} \\ -4.354\times10^{7} \\ 6.833\times10^{7} \end{bmatrix}.$$

The relative differences $|\varepsilon_a|$ in the stresses obtained using wide and narrow beam assumptions are

$$\left|\varepsilon_a\right|_{\sigma_x} = \left|\frac{\sigma_{x|narrow} - \sigma_{x|wide}}{\sigma_{x|narrow}}\right| \times 100$$

$$= \left| \frac{-1.280 \times 10^8 - \left(-1.382 \times 10^8\right)}{-1.280 \times 10^8} \right| \times 100$$

$$= 7.97\%$$

$$\left| \varepsilon_a \right|_{\sigma_x} = \left| \frac{\sigma_{y|narrow} - \sigma_{y|wide}}{\sigma_{y|narrow}} \right| \times 100$$

$$= \left| \frac{-3.431 \times 10^7 - \left(-4.354 \times 10^7\right)}{-3.431 \times 10^7} \right| \times 100$$

$$= 26.90\%$$

$$\left| \varepsilon_a \right|_{\tau_{xy}} = \left| \frac{\tau_{xy|narrow} - \tau_{xy|wide}}{\tau_{xy|narrow}} \right| \times 100$$

$$= \left| \frac{6.170 \times 10^7 - 6.836 \times 10^7}{6.170 \times 10^7} \right| \times 100$$

$$= 10.79\%.$$

6.4 Summary

In this chapter, we reviewed the bending of isotropic beams and then extended the knowledge to study stresses and deflection in laminated composite beams. The beams could be symmetric or unsymmetric, and wide or narrow cross-sectioned. Differences in the deflection and stress are calculated between the results of a wide and a narrow beam.

Key Terms

Bending stress
Second moment of area

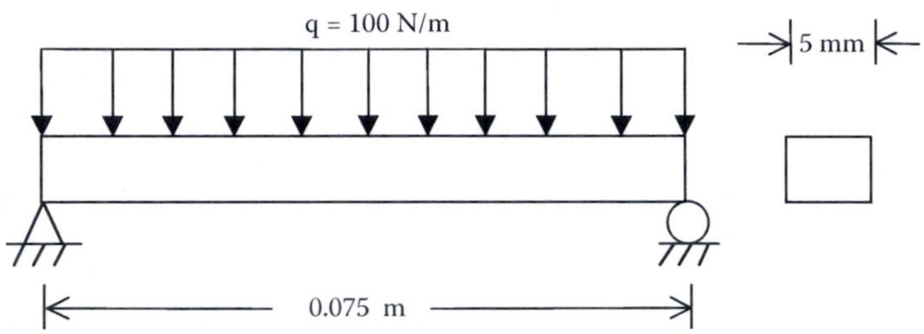

FIGURE 6.6
Uniformly loaded simply supported beam.

Symmetric beams
Wide beams
Narrow beams
Unsymmetric beams

Exercise Set

6.1 A simply supported laminated composite beam (Figure 6.6) made of glass/epoxy is 75 mm long and has the layup of $[\pm 30]_{2s}$. A uniform load is applied on the beam that is 5 mm in width. Assume each ply is 0.125 mm thick and the properties of glass/epoxy are from Table 2.1.

1. What is the maximum deflection of the beam?
2. Find the local stresses at the top of the laminate.

6.2 A simply supported laminated composite beam (Figure 6.6) made of glass/epoxy is 75 mm long and has the layup of $[\pm 30]_4$. A uniform load is applied on the beam that is 5 mm in width. Assume each ply is 0.125 mm thick and the properties of glass/epoxy are from Table 2.1.

1. What is the maximum deflection of the beam?
2. Find the local stresses at the top of the laminate.

6.3 Calculate the bending stiffness of a narrow beam cross-ply laminate $[0/90]_{2s}$. Now compare it by using the average modulus of the laminate. Assume that each ply is 0.125 mm thick and the properties of glass/epoxy are from Table 2.1.

References

1. Buchanan, G.R., *Mechanics of Materials*, HRW Inc., New York, 1988.
2. Ugural, A.C. and Fenster, S.K., *Advanced Strength and Applied Elasticity*, 3rd ed. Prentice Hall, Englewood Cliffs, NJ, 1995.
3. Swanson, S.R., *Introduction to Design and Analysis with Advanced Composite Materials,* Prentice Hall, Englewood Cliffs, NJ, 1997.

Index

C

D

E

F

G

H

I

J

K

L

M

N

O

P

Q

R

S

T

U

V

W

X

Y

Z